Praise for the previous edition:

'This book is a brilliant contribution to a comprehensive, holistic approach to human variation. Sobo examines the interaction of biology and culture in the context of a broad systems perspective on human communities. With examples ranging from the evolution of our species to how we shape our bodies in response to cultural ideals, she offers a refreshing view of us as biological entities suspended in sociocultural systems – inseparable dimensions of human existence'.

Bill Dressler, *University of Alabama*

'Being and becoming human is biocultural, and very few books drive that point home as well as this one. With a friendly and engaging narrative, Sobo guides the reader through a meaningful, and anthropological, engagement with humanity. This book should be used as a core text for both introductory anthropology and biocultural courses'.

Agustín Fuentes, *University of Notre Dame*

'I am delighted to see an introductory text that finally takes seriously the biocultural idea that we are creatures of both biology and culture, moving beyond the tired old nature-nurture dichotomy. Sobo deftly weaves together concepts and examples that students can relate to. She skilfully tackles critical contemporary topics such as globalization, sustainability, inequality, health, and social change'.

Wenda Trevathan, *New Mexico State University*

'Sobo's remarkable textbook is also a benchmark for the field. The material is deftly presented, immensely accessible, innovative in scope, and poised to advance biocultural approaches in deeply compelling and generative new directions'.

Justin Nolan, *University of Arkansas*

'Sobo's approach to the study of human diversity is a gracefully written and creative application of the unifying notion of epigenetic complexity. This introductory text is a readable and gripping account of how observable traits of human behaviour (phenotypes) are the product of interactions of numerous sorts, between genes and their own biochemical products, and between biochemical systems, environmental signals, historical realities and mental experiences. Highly recommended!'

Richard A. Shweder, *University of Chicago*

DYNAMICS OF HUMAN BIOCULTURAL DIVERSITY

This lively text by leading medical anthropologist Elisa J. Sobo offers a unique, holistic approach to human diversity, rising to the challenge of truly integrating biology and culture. The book's inviting writing style and fascinating examples make important ideas accessible to students. Readers learn to conceptualize human biology and culture concurrently—as an adaptive biocultural capacity that has helped to produce the rich range of human diversity seen today. With clearly structured topics, an extensive glossary, focused case studies, and suggestions for further reading, this text makes a complex, interdisciplinary topic a joy to teach.

Elisa J. Sobo is Professor and Chair of Anthropology at San Diego State University, California, USA. A sociocultural anthropologist specializing in health, illness, and medicine, she is Past President of the Society for Medical Anthropology, and currently Section Convener for the American Anthropological Association.

DYNAMICS OF HUMAN BIOCULTURAL DIVERSITY

A Unified Approach

Second Edition

Elisa J. Sobo

NEW YORK AND LONDON

Second edition published 2020
by Routledge
52 Vanderbilt Avenue, New York, NY 10017

and by Routledge
2 Park Square, Milton Park, Abingdon, Oxon, OX14 4RN

Routledge is an imprint of the Taylor & Francis Group, an informa business

First edition published by Left Coast Press 2013

Library of Congress Cataloging-in-Publication Data
A catalog record for this title has been requested

ISBN: 978-1-138-58970-4 (hbk)
ISBN: 978-1-138-58971-1 (pbk)
ISBN: 978-1-138-58972-8 (ebk)

Typeset in Bembo
by Deanta Global Publishing Services, Chennai, India

E-resources are located at http://www.routledge.com/9781138589711

CONTENTS

FIGURES

TABLES

SUPPLEMENTAL CASE MATERIALS

AUTHOR

Elisa J. Sobo is Professor and Chair of Anthropology at San Diego State University, USA. Presently Section Convener for the American Anthropological Association, and a past president of the Society for Medical Anthropology, Dr. Sobo also has served on the medical committee of the Royal Anthropological Institute in the United Kingdom, and as co-chair of the Committee on Public Policy of the American Anthropological Association. A current member of the editorial boards of *Anthropology & Medicine, Medical Anthropology*, and *Medical Anthropology Quarterly*, Dr. Sobo has written numerous peer-reviewed journal articles and has authored, co-authored, and co-edited eleven previous books on various topics relevant to our biocultural experience. Her most recent research examines cultural models of child development as applied in classroom teaching; vaccination acceptance, refusal, and selectivity; cannabis use for children with epilepsy; and health-related conspiracy theories.

PREFACE TO THE SECOND EDITION

Through this book, I aim to bring us face to face with our species in a way that promotes an intellectual understanding of human biocultural diversity while enriching our sense of connection to every other past and present member of the human race. I aim to provide the kinds of opportunities for learning that strengthen our commitment to future generations as well, and that enhance our understanding of the various local, regional, and global systems that make humanity possible.

In the decade that has passed since this book's first incarnation, our understanding of human biocultural diversity has grown richer. Topics at the cutting edge when introduced in the lectures that made up the first edition (e.g., epigenetics, gender fluidity, the microbiome) now appear commonly in our daily news media. Although I take pleasure in knowing that my earlier students had a head start on interpreting recent developments in these arenas of and for biocultural diversity, the depth of coverage originally afforded them seems superficial in hindsight. Revising the text has enabled me to devote more attention to these topics as well as to update information regarding other aspects of the human (biocultural) experience.

In keeping with requests from professors already using the text and other readers as well as from students, these additional updates ranged from a more in-depth look at the impact of colonialism and globalized capitalist enterprise on the biocultural experience of diverse peoples, including via the 'nutrition transition,' to a discussion of biocultural diversity in the context of the One Health framework linking human biocultural diversity with biodiversity more broadly, to deeper consideration of the impact assistive reproductive technologies have on population diversity, to an exploration of why some human groups have higher myopia (nearsightedness) rates than others. I also updated my discussion of the emergence of *Homo sapiens* in keeping with the latest science, paying more attention to the role of fire and cooking and giving more emphasis to the nonlinear and multistranded nature of our species' evolution. Beyond such improvements in coverage, I have refreshed many of my examples, so that they make sense to today's students (including those interested in healthcare careers: much that is contained here may be helpful in terms of the MCAT [Medical College Admission Test]). I have also developed, for every chapter, an extra case study that can be used to further facilitate focused discussion.

Although the additions and renovations made provide for a complete and current text, some gaps necessarily remain. This is in part so that the second edition's heft did not become too imposing or the length too long for a single quarter or semester's use. It also leaves instructors room to tailor lessons: plenty of territory remains open for supplementation.

These updates aside, the book's central aim has not changed. Therefore, and despite some aging phrases, the first preface (see below) still can supply a foundational understanding of this book's main messages and a map of how students will come to understand them. So it remains: please read on.

PREFACE TO THE FIRST EDITION

In Western society, we tend to think of biology and culture as totally separate and even opposing domains. This book takes a different viewpoint. This book's inquiry into human biocultural diversity takes a connectionist approach. It introduces the general principles of anthropology and specific biocultural topics by focusing on the synergistic interaction of biology and culture, or nature and nurture. In other words, this book demonstrates that we are, simultaneously and self-reinforcingly, biocultural creations.

Intended for entry-level courses in anthropology, medicine, human ecology, and other fields that investigate biology and culture, as well as interested readers, this book is unique in its holistic, systems-based approach, and in the way that it uses complexity theory and ideas from the emerging field of epigenetics to shift our thinking about biocultural diversity away from the traditional dualistic or either-or mode. Instead, readers learn to conceptualize human biology and culture concurrently: as an adaptive constellation that exists in relation to the broader environment in which humans live. Our adaptive biocultural capacity has helped to produce the rich range of human diversity seen today.

Human biocultural diversity has long been a key area of study within anthropology. Other fields speak of biocultural diversity, too; applied conservationists, for instance, reference all plant and animal life with the phrase, pointing to the role of human culture in relation to ecological balance and change. Humanity's impact on the earth system has in fact been so great that some scientists have declared we are now in a new geological epoch, the Anthropocene (*anthropos* means people). The human species is itself part of the earth system, and so humans, too, are affected by human cultural action. Indeed, the emergence of human culture complicated our species' evolution, in that the reactive process of natural selection now was conjoined with a proactive helper (culture); this led to greater diversity between human groups than otherwise would have been seen. It is this diversity—human biocultural diversity, and the dynamic, interconnected processes that bring it about—in which I am most interested.

The Biocultural Approach

Until now, many biocultural anthropologists have focused mainly on the 'bio' half of the equation, using 'biocultural' generically, like biology, to refer to genetic, anatomical, physiological, and related features of the human body that vary across cultural groups. The number of scholars with a more sophisticated approach is on the upswing, but they often write only for super-educated expert audiences. Accordingly, although introductory biocultural anthropology texts make some attempt to acknowledge the role of culture, most still treat culture as an external variable—as an add-on to an essentially biological system. Most fail to present a model of biocultural diversity that gives adequate weight to the cultural side of things.

Note that I said most, not all: happily, things are changing. A movement is afoot to take anthropology's claim of holism more seriously by doing more to connect—or reconnect—perspectives from both sides of the fence. Ironically, prior to the Industrial Revolution and the rise of the modern university, most thinkers took a very comprehensive view of the human condition. It was only afterward that fragmented, factorial, compartmental thinking began to undermine our ability to understand ourselves and our place in—and connection with—the world. Today, the leading edge of science recognizes the links and interdependencies that such thinking keeps falsely hidden. Building on the related campaign for a "new biocultural synthesis" (for example, Goodman and Leatherman 1998), this book takes culture seriously.

This book takes culture so seriously, in fact, that my use of the term 'biocultural' may seem generous in places where I apply it to traits that, on the surface, seem only biological. The same might be said for my use of the term in regard to traits that appear as if only cultural. Yet, our adaptive biocultural capacity operates constantly even when not obviously. This book invites readers to examine that fact. It challenges readers to look beyond shallow definitions of human variation focusing on traits, such as skin color, to include higher-order and more complexly derived variations, such as are seen in the diverse health profiles of various populations. In this book I strive to not simply describe our diverse biologies: I aim to examine how cultures create and maintain this variation. This includes addressing some of the misfits between modern lifestyles and certain features of the human body as it has evolved to date.

I use no hyphen in the term 'biocultural' because, instead of treating culture and biology separately, the book demonstrates how they really are connected. For example, the book shows that we humans make cultural interpretations of given biological processes and, in turn, manage and thereby modify these processes culturally, including through culturally infused forms of social organization. It describes a variety of ways in which our cultural modes of management affect our physical existence, such as when we settle in a particular geographic region, invent and adopt certain occupational roles, or subscribe to particular social structures, marriage and mating rules, or political systems. All such things alter our bodies in patterned, population-specific ways, much as do more obviously body-related practices such as circumcision or diet. Accordingly, in addition to genetic and related physiological diversity, I include in this book important anthropological subjects where the co-mingling of biology and culture are most obvious, such as kinship and gender. In support of this orientation, I use the term **culture** broadly, as is typical in US anthropology, to refer to the totality of each human group's shared, learned heritage. This includes social, political, economic, religious, cosmological, linguistic, health promotion, and other systems that, taken together, comprise our cultures.

In sum, from this book's perspective, **biocultural diversity** refers to all population-based human variation generated in or reflecting the dynamic, synergistic communion of biology and culture, neither one of which can function without the other. In other words, biocultural diversity, as I use the term, brings together topics that narrowly focused scholars might see as separate; it demands that we acknowledge and investigate their co-constitution. The result of such a unified approach is a book that offers a more comprehensive view on human variation than existing biocultural texts, truly engaging both sides of the biocultural coin and reflecting a holistic anthropological sensibility.

Roadmap

This book is organized into three sections that build incrementally on the ideas of dynamic systems, environmental pressure, natural selection, and adaptation. Chapters in Part I, "A Systems View of Human Adaptation," address our emergence as cultural beings through the process of natural selection and the biological diversity that resulted as humans spread out around the world. Part II, "Socio-political and Economic Factors," examines the varied adaptations and health profiles that were created around the world as an unintended consequence of the Agricultural Revolution and related transformations in how groups lived. For example, we examine the spread of infectious disease and resulting changes in the gene pool as well as the way health has become linked to wealth. The last part of the book, "Meaningful Practices," explores our expectations and experiences of our bodies, for instance in the ways in which our different cultures create bodies as male or female or in how standards of beauty lead some groups to skeletal modifications (see Figures 0.1 and 0.2).

Key Features

Health and Diversity

Many of the examples I use in the following chapters relate to health, and there are good reasons for this. First, health snapshots or 'epidemiological profiles' vary from group to group; they are a key indicator and mechanism of diversity. Second, varying health levels serve as an excellent index of adaptation. For instance, ill health suggests that a population is under stress or poorly adapted to its current environment. The focus on health helps students carry forward the initial lesson that biocultural diversity emerged as humans adapted to new or changed environments.

For instance, consider darker-skinned peoples now living in Britain. In comparison to many other places in the world, Britain is an environment with generally low levels of ultra-violet or UV light exposure: the island experiences late sunrises and early sunsets in winter, and often has overcast skies. These factors are compounded by the fact that British society has primarily

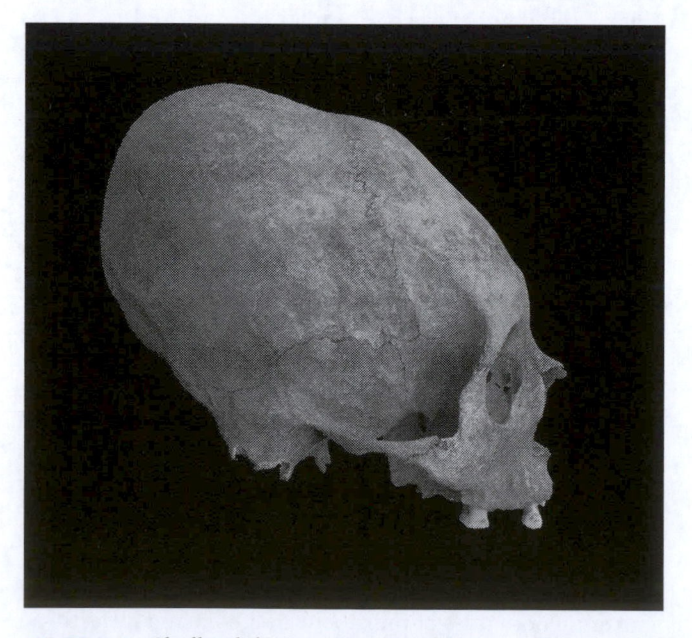

FIGURE 0.1 Skull exhibiting cranial modification. Photo courtesy of Valerie A. Andrushko.

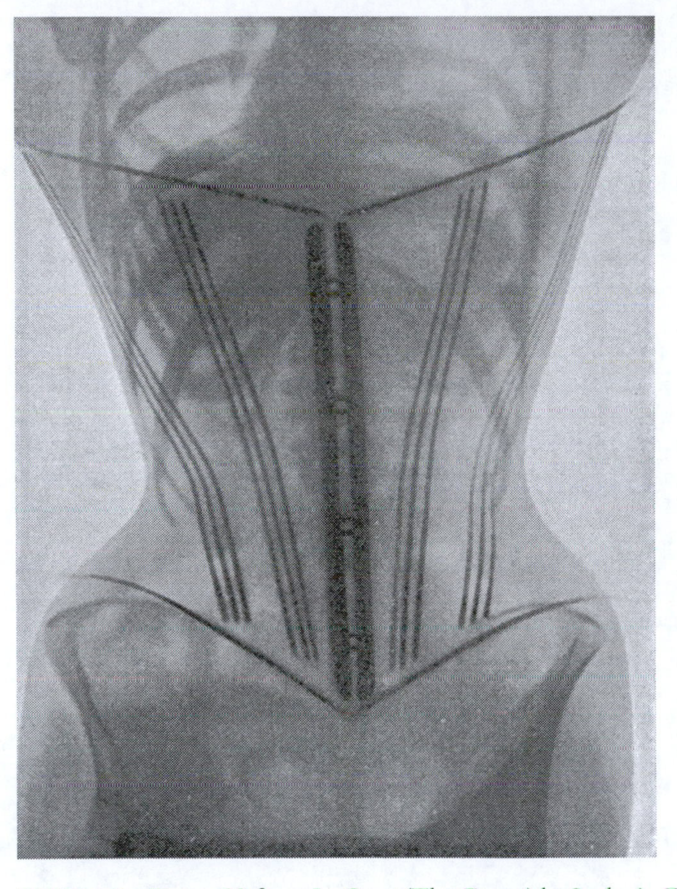

FIGURE 0.2 Figure 32 from *Le Corset* (The Corset) by Ludovic O'Followell, 1908.

indoor jobs and many sun-blocking high-rises. These features put darker-skinned peoples living in Britain at a higher risk for health problems such as rickets. In turn, lighter-skinned populations relocated to a higher-UV environment (for example, San Diego, California) are at higher risk for skin cancer, sunburn, folate deficiency, and other conditions. Without understanding the interrelatedness of culture and biology—without a comprehensive definition of biocultural such as this book offers—these 'facts' might be described, but their contingent qualities would remain unquestioned and our understanding of the diversity described would remain underdeveloped at best.

Learner Focused

The integration of "nature and nurture" into a complex systems framework is revitalizing the study of biocultural diversity. Questions asked and lessons learned by the vanguard of scholarship in this area can and should be brought into introductory teaching. To create a rich, upper-division learning experience that supports such an approach, this book:

- Teaches sophisticated concepts in jargon-free, reader-friendly language that renders them clear and accessible to readers from any discipline;

- Uses student-centered, current, and relevant examples from the past and present that bring lessons to life and demonstrate their here-and-now application;
- Keeps readers on track by focusing each chapter around answering three or so key conceptual questions that could easily be adapted for use as study questions or essay questions on tests.

In addition, each part of the book is introduced with a short overview steering readers toward the main points that the subsequent chapters discuss, and a concluding chapter highlights important connections.

Easily Supplemented

I developed this book over years of teaching biocultural diversity and with input from about two thousand students. Built around lectures fine-tuned over time, it is organized not only for reader ease but also so that it might provide other instructors the backbone for a similar course. Depending on the needs of end users, it could stand alone or be paired with supplemental resources.

In the old days, we might have assigned a reader to complement this text; today, using key terms from each chapter, instructors also can use library databases and the internet (see References) to compile a host of rich, supplementary, up-to-date material that might include both scholarly and popular articles; videos; web-based tutorials; images; podcasts; and so on. We can even engage students actively in creating their own course supplements. For example, using adaptations of Bernie Dodge's inquiry-oriented Webquest method, students can be sent to the internet on a topic- or problem-based treasure hunt. Of course, full-length books also are an option; some that I have used are in References, as are many of the short, topical readings and videos that I have assigned.

Accessible

This book builds on the scholarship of others. However, as this text is introductory, I have not burdened readers with a massive load of references. Whenever possible, I have chosen to cite works that are easy to find and to read; in addition to scholarly literature, I point readers to solid work written for the educated public and published in books, newspapers, educational websites, or public broadcasting sources.

In addition to minimizing in-text citations, the book is only lightly illustrated. This not only limits distractions but also helps keep the book's price relatively low. Accordingly, older photographs and artwork available via the public domain are included, not to suggest that the cultures pictured are dead when indeed they are not, but rather to ensure the book's affordability.

Broad Appeal for Teaching

Instructors with various kinds of expertise in an expanding number of disciplines are charged with teaching biocultural diversity courses. Because of its straightforward approach, this book is likely to appeal to anyone teaching such a course, not only biocultural anthropology specialists. The contents are suitable for other courses as well, from introductory courses in general and medical anthropology to courses that include a review of the rise of civilization and even human biology programs seeking to enhance their culturally informed course offerings. Because this book uses health as a key index of adaptation, it also will fill a niche in public health programs, which increasingly are committed to teaching cultural competence and offering courses that address health inequities and global health.

End Aims

In any course, this book will introduce the amazing—and necessary—diversity of humankind. By the end of the book, readers will be able to define and apply systems thinking, describe the key socio-cultural and political-economic reasons for human biocultural diversity past and present, and characterize some of the many ways in which the human being is, for better and for worse, an irreducible biocultural creation.

ACKNOWLEDGMENTS

This book summarizes lectures presented in an anthropology course called "Dynamics of Biocultural Diversity," which I've taught at San Diego State University (SDSU) since joining the department in 2005. Thanks are due to the multitude of students who've taken the course for their enthusiastic interest in the topic and, more so in this context, for their patient, constructive feedback on my methods for explaining and illustrating the fascinating multi-directional links between culture, biology, and human diversity. Without student feedback, this book would not exist. I also wish to thank colleagues at SDSU (particularly Brock Allen and Andrea Saltzman Martin, as well as others too numerous to mention) for providing specific guidance to me in developing the course, first for the lecture hall setting and more recently for online learners. I am extremely grateful for the lessons that they, and my students, have shared with me.

Thanks also are due to Jennifer Collier, for champion editorial oversight. Several anonymous reviewers and Erin Riley kindly provided thoughtful and constructive input on the manuscript and sections of it, and colleagues too numerous to list by name have aided me in my thinking through discussions of topics now reflected in the book's contents. Christopher Morgan graciously fact-checked some of the work, and Jennifer de Garmo helped hunt down illustrations. Dedicated graduate students and teaching assistants Samantha Coppa, Jules Downum, Elizabeth Herlihy, Sonia Khachikians, and Harrison White worked diligently to transcribe lectures, help with figures, copyedit and critique the manuscript, and sharpen the course from which this book grew.

For the second edition, I wish to add thanks to first edition users who provided invaluable feedback, and the hard-working contractors and staff at Routledge. I further thank my university for supporting me with the protected time necessary to craft this second edition. I am most grateful to my family, for putting up with yet another 'last' book project.

PART I

A Systems View of Human Adaptation

Conceptualizing biology and culture separately, as if each exists independently, may be convenient, but it dulls our understanding of who we are, how we work, and what our participation in the human enterprise accomplishes for us. Part I of this book lays the foundation for a comprehensive and unified understanding of our biocultural diversity—one that embraces the interdependence of biology and culture and all that they entail. Through Part I, readers learn to think holistically, and to see the world from a systems perspective. Part I also introduces students to 'adaptation': a complex process entailing multiple forms of feedback through which the theoretically separate domains of body, behavior, and environment work as one in effecting change.

Chapter by Chapter Overview

Chapter 1 cuts straight to the chase, describing the nature–nurture debate and then demonstrating how it is mistaken and misleading. Rejecting the dualist position, this chapter presents an alternative, unified approach grounded in 'systems thinking.' It explains and exemplifies the holistic viewpoint of anthropology, paying special attention to the concept of adaptation and the phenomenon of 'emergence.' We explore the concept of 'complexity' in relation to how it can inform our understanding of human nature.

Chapter 2 introduces evolutionary biology with a focus on genetic adaptation. After giving readers a good grasp of how genes and traits are passed on to the next generation, this chapter explains how natural selection at the population level works, and why certain forms of genetic variation are useful for a group's transgenerational survival or reproductive success from generation to generation.

In Chapter 3, we learn that the environment plays a key role in the ways that genes are expressed to begin with. This chapter introduces the concept of developmental adjustment, distinguishing it from natural selection and using that as a jumping-off point for an exploration of how genes are switched on and off via epigenetic mechanisms—biochemical transactions that affect gene expression. Given the human body's inherent plasticity, if exposed to divergent epigenome-altering events, even twins with identical genetic profiles can end up with surprisingly different physical characteristics. Information in this chapter keeps us questioning the relationship between nature (here, genes) and nurture (the environment, and all that it entails) while demonstrating how nature and nurture work together to shape a population's traits—both within and between generations.

Chapter 4 defines the term culture with reference to its adaptive function. Our capacity for cultural adaptation—once it emerged—has been crucial to humanity's survival. Culture's evolutionary emergence is explored with reference to the synergistic interaction of our ancestors' evolving capacities for motor skills, empathy, cooperation, imagination, and symbolic communication (for instance, through the language of ritual) as well in regard to tool use, control of fire, and cooking. 'Cultural relativism' and 'ethnocentrism' also are defined and discussed as contrasting standpoints for interpreting data.

In Chapter 5 we turn our attention to the spread and settlement of behaviorally modern humankind, which culture's emergence made possible. We examine how human migration around the globe fostered diverse biocultural differences between human populations, as well as how subsequent mingling may have dimmed or intensified these. We question assumptions about the biology of 'race' and demonstrate how geography in general and geographic clines in particular (geographically specific trait gradients) provide a much better explanation for the biological differences found within our species than the erroneous idea of biological race ever has done or can do. That said, making geographic, ancestry-based distinctions between subpopulations in certain limited circumstances can be useful; this also is discussed.

Central Lessons of Part I

At the end of Part I of the book, readers will be well-versed in systems thinking and able to explain the value of taking a holistic point of view or of adopting a unified approach in exploring biocultural diversity. In addition, they will be able to discuss and explain, using examples, how and why various forms of biocultural diversity came about.

Part I does not ignore the importance of human behavior in the evolution of human variation. However, its overall focus is on geographic or habitat-based adaptations and physical differences amenable to materialist scientific study, such as red blood cell structure, height, and lactase levels. This provides readers with a concrete basis for truly understanding the concepts presented. In this, Part I fosters in readers a deep grasp of the basics; it does so to ensure success as we study the more abstract forms of biocultural diversity that we address later.

1

ANTHROPOLOGY AND COMPLEXITY

This chapter prepares you to:

- Define biological determinism, discuss its shortcomings, and explain what people hope to accomplish when using it
- Describe anthropology with reference to holism (in contrast to reductionism) and the comparative method
- Define and differentiate ecosystems and complex adaptive systems, paying special attention to the concept of adaptation and the phenomenon of emergence
- Demonstrate how human groups may be studied as (and as part of larger) complex adaptive systems, focusing on adaptation and emergence

Why are we the way we are? Why are we all somewhat different? These may seem like straightforward questions, but they address two of the most important and complicated puzzles humans face—and their answers are neither singular nor simple. Begin to brainstorm about flavor preference, athletic ability, or intellectual capacity, for example, and you will find yourself awash in various biological and cultural explanations. Most people can be split into two camps here: one favoring biology and the other, culture. Which camp is right? Is it biology or is it the fact of having been raised in a particular culture that determines how we look, think, move, and feel?

Although most people, including most scholars, view biology and culture, or nature and nurture, as different, even opposing forces that work separately or against each other, the opposition is both clumsy and futile. Culture and biology are both part of a system in which multiple mutlidirectional feedback loops can lead to fantastic and unanticipated *biocultural* outcomes. This chapter aims to help readers really take hold of this idea by exploring what we know about systems, paying special attention to the concept of adaptation and the phenomenon of emergence in the process.

Where Did the Nature–Nurture Question Come From?

Biological versus Cultural Determinism

We are not the first to ask "Nature or nurture?" This question has many historical roots. For example, it was incorporated into scientific inquiry in the writing and work of Francis Galton, a productive and well-known nineteenth-century geographer, meteorologist, psychologist, and statistician. In 1869, Galton (cousin to Charles Darwin) asserted in his book *Hereditary Genius* that talent is physically inherited: genius simply runs in the family. Galton's scholarly peers berated him for ignoring the role of upbringing, so he bolstered his claims with another book, *English Men of Science: Their Nature and Nurture.* Galton chose this title purposively, explaining in the book that it supplied "a convenient jingle of words," nature and nurture giving the many elements that shape us "two distinct heads" (Ridley 2003, 71).

For Galton, nature was the more powerful head by far. In explaining all human characteristics as a result of nature (heredity or biology), Galton discredited the role played by upbringing (nurture, culture), diverting attention from 'nonbiological' causes of human diversity.

We know that **biological determinism**—the argument that biology determines utterly and completely one's capacities and characteristics—has problems. Calling certain arrangements 'natural' is a way to rationalize them rhetorically. It downplays, for instance, the vast amounts of hard work and sacrifice that records show really are put in by winning athletes such as Venus and Serena Williams, musical superstars such as Ludwig van Beethoven, and intellectual giants such as Albert Einstein. Positioning their achievements as resulting from inborn 'gifts'—calling them 'natural'—provides a self-serving excuse for the rest of us not to practice, drill, or work out. It diverts us from reaching our full developmental potential (Shenk 2011).

This does not seem so tragic when we think about it on the level of the individual or when the skills in question are rarefied. Yes, it is too bad that I am not the excellent soccer player or gold-medal swimmer (and so forth) that I could have been because I avoided dedicated training. However, when applied at the population level, biological determinism that leads people to place false limitations on their potential also can obstruct a group's progress or productivity. Worse, those in power can use biological determinism strategically, to justify their advantages and keep others oppressed.

If the genius it takes to fix cars, say, is hereditary in males but not females, then barring girls and women from auto shop classes can be seen as justifiable: to let them in would be futile and waste resources. Similarly, the poor might be kept from being educated based on an argument that they are inherently dim-witted. Certainly, there would be no need to examine underfunding in inner-city schools as a possible source of low achievement rates if those low rates were written off as due to biology. My examples may seem extreme but they represent real situations, present and past (see Figure 1.1). Biological determinism has even justified the enslavement of one race by another as part of the so-called natural order.

We know today, of course, that racist thinking has no scientific basis (see Chapter 5). The myth of 'giftedness' also has been debunked (Shenk 2011). Yet, biological determinism still haunts us—not only in the way it remains in use for justifying inequality but also in its antithesis or counter-form, cultural determinism. **Cultural determinism** argues that culture (rather than biology) determines one's capacities and characteristics. Early twentieth-century anthropologists in the United States, appalled by racism, popularized this position. It may have its appeal, but it, too, is one-sided. It, too, exists, in a way, as an outcome of Galton's "convenient jingle of words": with respect to our original questions—Why are we the way we are? Why are we all somewhat different?—Galton and his followers have led us to believe that the answer must be *either* nature *or* nurture, not both (see Ridley 2003). However, the forced choice approach is in error.

Nature and Nurture Work Together

Recent advances in philosophy and science demonstrate that determinism of either kind is wrong-headed. Rather than asking if nature *or* nurture is in charge, we should ask how the two forces work *together,* inseparably in a system, to produce the vast and wondrous diversity of human experience. Humans are neither simply biological nor simply cultural; we are both.

FIGURE 1.1 "Colored entrance" of a cinema in Belzoni, Mississippi, c. 1939. Photo by Marion Post Wolcott, Library of Congress Prints and Photographs Division, LC-DIG-ppmsca-12888.

To think otherwise is in some ways an artifact of language: separate terms label each of these facets of human existence, suggesting they are mutually exclusive. Any attempt to talk about them as blended ends up resorting to the inserted 'and' or, more efficiently, use of a silently hyphenated term, 'biocultural.' The assumption, even here, is that biology plus culture (or nature plus nurture) equals human experience; the two still are separate because of the additive characteristic of any compound phrase or word.

Nevertheless, with some effort, it is possible to get past this language barrier conceptually. Indeed, doing so—characterizing the inseparability of biology and culture while exploring human diversity—is the focus of this book. To that end, we must first learn a little bit about systems and systems thinking.

Holistic Systems Thinking

A **system** consists of two or more elements whose organization serves a common purpose or outcome, intended or not. A central feature of all systems is that, in listing their parts, something more like 'multiplied by' can take the place of 'and' or 'plus': systems are interactive or multiplicative, not additive. The term **synergy** can describe this kind of thing—when the combined effect of two or more substances or agents is altogether distinct from the separate effect of each ('syn' as a prefix means 'with' or 'together').

In **systems thinking**, we focus on productive interconnections: systems thinking highlights relationships and what emerges from them. Over 2,000 years ago, Aristotle summed up this point of view in *Metaphysics*: "The whole [that is, the system] is more than the sum of its parts." Indeed, and to push the mathematical metaphor, the whole is the *product* of its parts. What does it mean to say that a system derives from the relationships between its parts—not merely from the addition of these? Well, a cake is not just a pan filled with eggs, milk, sugar, flour, and butter. It is the thing that emerges through the combination of ingredients plus a cooking process. In other words, *the properties of a system can be neither explained nor determined by examining its parts alone.*

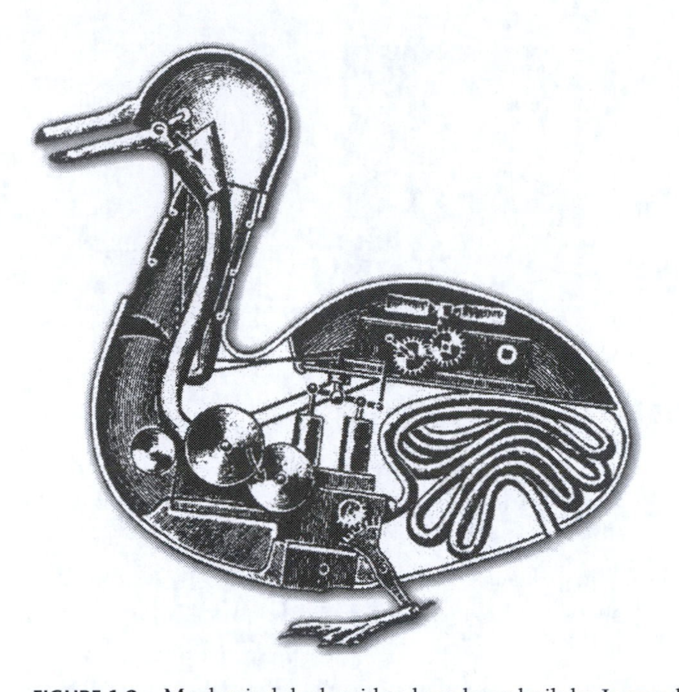

FIGURE 1.2 Mechanical duck, said to have been built by Jaques de Vaucanson; this imagined interior was drafted by an American observer, 1738–1739, France.

This tenet or principle of systems thinking is called **holism**. It often is in contrast with **reductionism**, the idea that an entire system can be explained by a single aspect of it. For example, explaining organisms such as humans (or ducks, say) as machines reduces them to their moving parts (see Figure 1.2). Doing so might be helpful in certain circumstances, but it cannot lead to a full understanding of human (or duck) experience. Not only does it fail to bring into consideration other dimensions of the whole human (or duck) being, but it also fails to position humans (or ducks) in a larger context—despite the fact that everything is part of something bigger.

Holism is anthropology's hallmark. This in part explains the breadth of the discipline's title: rooted in the ancient Greek terms *anthropos* (human) and *logos* (word), and by extension the ancient Greek suffix *logia* (spoken knowledge regarding), **anthropology** concerns the study of humankind. Anthropologists study humans across time and space; we study human systems and human interactions with the systems in which they are embedded (such as their physical environments). Humans affect these larger systems just as they are affected by them—for instance, developing complicated cultural strategies for finding water in arid, drought-prone climates, or evolving a greater genetically driven capacity for cold tolerance after generations of living in Arctic regions.

Ecosystems

One strain of systems thinking is the ecological or ecosystems approach, which gained popularity in the United States in the 1960s and 1970s as people began to realize in large numbers that some human actions (such as dumping pollutants in rivers) were harming the environment. **Ecosystems** include multiple interconnected populations, both **biotic** (living) and **abiotic** (for example, sunlight). Most biotic populations—that is, most **species** or breeding groups—are quite specialized, requiring particular resources, interactions, and conditions for survival. For example, a species may eat only a particular bug in a particular stage of its life cycle. Competition for resources, etcetera, limits the number of species that can rely on, say, that kind of bug; so, species are mostly fairly specialized. Each species generally occupies a very particular **niche** or does a particular job within the ecosystem, playing a particular role in maintaining the status quo—at least in theory. The total number of members of a given species that a niche or habitat can support without provoking its own collapse is referred to as its **carrying capacity**.

A new way of talking about the beneficial maintenance functions of various parts of the ecosystem uses the phrase '**ecosystem services**.' For example, bees provide pollination services, without which many plants cannot reproduce (see Figure 1.3). Maggots provide decomposition services, breaking down, for instance, fallen fruit, putting nutrients back into the ecosystem's soil. Microbes in streams, lakes, and oceans effectively serve as water filters; if they come to harm water can become polluted even without human dumping. Trees and other plants provide the ecosystem with oxygen; further, they release water into the air while absorbing potentially dangerous atmospheric carbon dioxide in support of photosynthesis, a process by which they convert sunlight into chemical energy forms such as carbohydrates. These carbohydrates not only fuel the plant's ecosystem-serving and survival-related activities; they also fuel those of species that eat the plants.

Speaking of energy, it is worth pointing out that an ecosystem's populations all are linked to each other through various ties involving *energy* flows, whether in the form of nutritional calories, the kind of energy the sun provides, or otherwise. For instance, squirrels rely on oak trees for acorns (calories), and oak trees rely on squirrels to spread their acorns so new trees can grow, and to provide, via their dead bodies as well as excrement, fertilizer that sustains the trees. Add in other populations, such as fleas, squirrel-eating wolves, grasses, a river, limestone—even people—and we have an ecosystem.

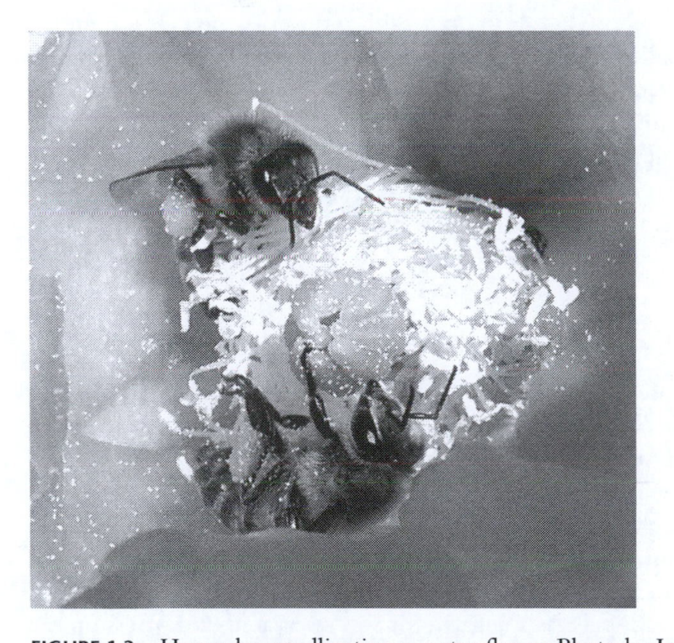

FIGURE 1.3 Honey bees pollinating a cactus flower. Photo by Jessie Eastland, 2011. CC-BY-SA-3.0.

Examples

One Health is a theoretical model emphasizing connections between animal and human health whereby the health (and health-related actions) of one influences the health of the other. The One Health concept exemplifies the application of an ecosystems-influenced approach on a grand scale. That said, its history precedes the rise of ecosystems thinking. Some scholars trace the One Health idea (if not the catchphrase) to the work of Rudolf Virchow, a prominent nineteenth-century physician from Germany who noticed connections between animal and human health, and recommended comingling veterinary and human medical training and practice. Virchow's views, particularly regarding public health threats that overtly entail animal–human linkages, such as rabies, caught on slowly. In 1964 Calvin Schwabe, a specialist in parasitic diseases that move between animals and humans, coined the phrase "One Medicine" to better promote collaboration between human and veterinary medicine. In the 2000s adherents pivoted away from emphasizing medicine (i.e., doctors) in the phrase, changing its last word—thereby shifting the focus—to health itself.

Today, One Health is promoted by health organizations such as the Centers for Disease Control and Prevention (CDC) and the American Veterinary Medical Foundation, as well as by overtly political and economic institutions such as the United Nations and the World Bank. This reaffirms health's broader systemic import. Indeed, some refer to the One Health movement now as 'One Health, One World,' a phrase introduced in 2004 (National Center for Emerging and Zoonotic Infectious Diseases 2016).

The One Health ecosystem is global: it's huge. Minuscule by comparison is the individual human **microbiome**. *Biome* is a suffix that comes up frequently: it refers to a biological community; *ome* means the complete whole of a given class. The term 'microbiome' refers, technically, to the pooled genetic material of the microbial organisms inhabiting a certain space—in this case, the human body. Each of our bodies harbors about as many microbes as it does human cells. The bulk of anyone's microbial community lives in the large intestine or colon but there are plenty of microbes living on the skin, in the mouth, and elsewhere within the bodily human ecosystem (Sender et al. 2016).

The human microbiome plays an important role in our health and well-being. Our microbes affect our nutritional and immunological status, our mental health, and even our physiology. For instance, microbes that have colonized our gut can help us with digestion; they can also evict harmful organisms that otherwise might make themselves at home. We drill down into such specifics later; for now, the main point is that human survival depends not just on taking in food, water, oxygen, and so forth but also on receipt of the essential services provided by the beneficial microbes of the human microbiome, so long as our bodies hold the proper mix of them. Our cells and theirs comprise a delicately balanced ecosystem.

Stasis and Change

Over time, ecosystems thinking has evolved to accommodate complexity (a concept I'll explain momentarily) as well as real change over time. However, when the ecosystems approach first gained steam, ecosystems were understood to endure 'as is' over the long term. People assumed that an ecosystem had a stable balance point at which species' population numbers, and the relations between populations (for example, supply and demand), held steady even as time passed.

A system in balance is in **homeostasis**; it has achieved a steady state or one of equilibrium ('stasis' means inactivity; *homoios* is an ancient Greek term for 'like' or 'similar'). In the original ecosystems model, change was generally seen as a negative; it was associated with stress on the system. For instance, if people overhunt squirrels, oak trees suffer because they lose a source of fertilizer and a mode of seed (acorn) dispersal. A damaging chain reaction ensues, with one population's demise leading to growth for another, which in turn creates stress for a third population, and so on. Unless something happens to restore balance, in this view, the entire system collapses. From this perspective, change is not good.

Complex Systems

Information and Adaptation

Another approach to systems thinking has developed, and this one sees change as an inherent feature of the system—not a negative disruption to it. This approach stems from complexity science, which emerged with full steam in the context of the late twentieth century's revolution in information technology. It has since infused ecological science, but for now, to keep things simple, I will introduce complexity theory as a different entity, which it was when initially invented.

Some systems are **mechanical systems** and therefore totally predictable. For example, a thermostat-regulated heating unit will switch on and off when the temperature moves below or above a certain set-point. This system detects one thing—temperature—and responds in one of two predetermined ways (turning on or off). Human systems, however, are not mechanical. They can evolve adaptively in response to changes in the environment. They bear closer resemblance to **complex adaptive systems**—networks of dynamically interrelated parts (**dynamic** means in motion, or active). Importantly, in complex adaptive systems, not only energy but also *information* can flow between parts, making creative behavior or surprising actions possible. For this reason, the complex adaptive systems framework provides a fruitful source of insight into human existence.

Following this framework, let us assume that complex adaptive systems are driven to try to stay organized, or stay together. Given a change in the environment in which such a system is embedded, simple interactions between its parts that strive to keep the system working can lead to something entirely new—to a novel pattern or property: to **emergence**. Many things that we take for granted today, such as our reliance on electronic messaging, or fast food and ready meals, began spontaneously as emergences. Emergence itself also can lead to unintended consequences, whether on a small or large scale, as its ramifications ripple out. For example, electronic messaging fosters flexibility in event planning that was not possible in the previous generation.

Another example is the way the spread of the automobile fostered an unintended increase in traumatic injury because of newly possible motor vehicle crashes. This was in addition to changing the landscape (see Figure 1.4). The civilian emergency medical and trauma services systems developed or emerged in turn, addressing the rising need for immediate medical attention that motor-vehicle-related injuries created. The health services system was dramatically altered, as was our entire employment and residential structure, by the introduction of the private car, which effectively made suburbs and office parks possible. Other initial domino or ripple-out ramifications included the emergence of anti-theft, dangling air freshener, road map, and passenger travel game industries. Our social and cultural systems changed radically with the automobile's emergence and those emergences that followed on from it: we adapted.

Changes like those just described not only illustrate the dynamic, ever-changing nature of complex adaptive systems, they also are **nonlinear**; that is, consequences do not necessarily unfold in a predictable manner but instead as surprises that lead to surprises of their own. Cars led to the drive-in cinemas and drive-through burger stands that took the country by storm in

FIGURE 1.4 Clark Avenue and Clark Avenue Bridge, looking east from West 13th Street, Cleveland, Ohio, 1970. Photo by Frank J. Aleksandrowic, Environmental Protection Agency, NARA record: 8452210.

the 1950s and 1960s, which then led to the modern fast-food industry. If someone had told the automobile's inventors that they would revolutionize the world's food industry, it is unlikely they would have believed it. Nor would Sheldon Cheney have predicted today's obesity and diabetes epidemics in 1947 when he opened the world's first drive-through hamburger restaurant (Red's) Route 66 in Springfield, Missouri (In-N-Out came next, in 1948). The emergence of these things fulfills what is sometimes called The Law of Unintended Consequences (see Supplement Box 1). What we do to survive these unintended consequences—to adapt to them—is likewise an open question.

SUPPLEMENT BOX 1: THE NONLINEARITY OF SOCIAL MEDIA

Social media—interactive computer-mediated technologies by which we can share information electronically with individuals or in networked online communities—have exploded. Such technologies date back to the 1970s but social media really got a foothold in the 2010s, as smartphones became ubiquitous and the internet evolved from being a read-only platform to one supporting and even relying on user-generated content.

Intended initially for use by individuals with above-board community-building intentions, social media has become a corporate tool too. Moreover, it can be hacked and hijacked. For example, starting in 2014, a political consulting firm (Cambridge Analytica) improperly obtained and used the personal data of many millions of Facebook users, to influence voter opinions.

There have been many instances of persons and organizations creating and using (sometimes via robots or 'bots') bogus social media accounts to spread untruths or promote certain positions. They may, for instance, 'troll' or sow dissent by posting or responding to posts regarding polarized or polarizing topics, including not only elections but also debated social issues and practices, such as vaccination. Fake accounts also can provide a clever marketing boost through the influence that lots of 'likes' and positive comments have on our purchasing patterns: for instance, bots may therefore generate more than two-thirds of vaping-related tweets (Martinez et al. 2018, 555).

Manipulation and outright duplicity were not part of social media's original promise. What happened? In a nutshell, the kind of emergent behavior that typifies complex adaptive systems led to unintended consequences.

Regarding the Cambridge Analytica scandal in particular but touching on these broader kinds of challenges more generally, *New York Times* reporters asked Facebook founder Mark Zuckerberg, "Do you feel any guilt about the role Facebook is playing in the world?" Zuckerberg's answer highlighted the principle of nonlinearity:

> That's a good question. ... We're doing something here which is unprecedented, in terms of building a community for people all over the world to be able to share what matters to them, and connect across boundaries. I think what we're seeing is, there are new challenges that I don't think anyone had anticipated before. If you had asked me, when I got started with Facebook, if one of the central things I'd need to work on now is preventing governments from interfering in each other's elections, there's no way I thought that's what I'd be doing, if we talked in 2004 in my dorm room. I don't know that it's possible to know every issue that you're going to face down the road. But we have a real responsibility to take all these issues seriously as they come up, and work with experts and people around the world to make sure we solve them, and do a good job for our community.
>
> *(as quoted in Roose and Frenkel 2018)*

In many ways we are all, at certain times in our lives, like young Zuckerberg sitting in his dorm room. Even the most apparently inconsequential actions we take can have huge, unanticipated ramifications—for better, or worse.

A complex adaptive system is one that alters itself, or adapts. Technically speaking, an **adaptation** is a survival-enhancing change in a system brought about in response to an atypical, stress-producing change in the environment. In contrast to the traditionally-conceived ecosystem or a mechanical system, which either breaks down in response to the pressure exerted through such change or tries to restore itself to some preset balance point, a complex adaptive system evolves in tandem with environmental pressures. It learns from its experiences.

One reason that adaptation can happen in complex adaptive systems (which, as scientists now understand, ecosystems really are) is that the complex adaptive systems model accounts for information flows, not just energy flows. In mechanical systems, energy flows are linear and their organization is static—recall the classic ecosystem relationship in which a squirrel relies on an oak tree for nuts, and the tree relies on the squirrel to fertilize the soil and plant its acorns, and 'twas ever thus. In complex adaptive systems, however, information flows can lead agents to instigate novel changes. For example, ants may change their behavior based on scent and sound signals. Humans may change theirs based on verbal or facial expression feedback from friends, bosses, teachers, or students. Such dynamics characterize complex adaptive systems. Complexity theorists focus their studies on the flow of information and the way in which new processes and structures emerge in response.

Delimiting Systems

Systems thinking is used to examine a variety of human phenomena at numerous levels of analysis. For instance, the state is a system, and so, too, is a city within it, as is the neighborhood, family or household unit, and individual body. Indeed, each of these systems is itself a system within a larger system and home to smaller, internal systems. The class for which you may be reading this book is its own system, but also part of the larger university system, and so on.

A system can be simultaneously a whole unto itself and nested into or otherwise a part of something larger. Where one system ends and becomes part of another is really a matter of how we package or ring-fence systems-within-systems conceptually, for discussion or scrutiny. For instance, the circulatory system can be studied on its own, or as an integral part of the human body system. The body in turn can be studied as part of a population grouping (as we mostly study it in this book).

In reality, however, treating any system as if self-contained is somewhat deceptive. Differentiating a system from its environment rests on a false distinction between the two. Boundaries like walls, or skin, or a village's borders, however real, always are permeable (and even apparently impermeable boundaries are worked around by other connections). Multidirectional links continue to exist; energy and information continue to flow. Indeed, *if linkages did not persist, changes in the so-called environment enveloping a system could not actually have an impact.*

Delimiting the system under study can help focus discussion, and that is a good thing. However, the contrived nature of any system boundary is all the more apparent when we take into account that where said boundary is set depends on the observer's standpoint. For instance, investigators who wish to study the mechanics of breathing will focus on the respiratory system, or perhaps the higher-level system of the body. Those interested in studying what is breathed—air quality—may take a neighborhood perspective or extend their study to weather systems if they find pollution coming from a far-away source. They may need to think globally because weather is a global phenomenon; likewise, the factories whose pollution the weather may spread could be organized at the level of the global political economy.

Getting bogged down in questions about where systems begin and end can distract us from their more important features (dynamic connectivity between the system's parts, adaptation, emergence) just like the nature–nurture debate distracts us from the jointly biological and cultural dimension of human experience. For instance, we could delimit a college class based on who is enrolled; alternatively, we could use the classroom's walls—but the class is not just a space plus students plus a teacher plus course materials: the class is what *emerges* when all these features or system parts interact. The classroom's physical layout, the personalities of participants, and even the time of day we meet are just a few of the many factors that come into play as information and energy flows. If the class is an online class, a key factor is the kind of distance education software that the school uses, which in turn determines to some degree the shape that learning takes.

When we envision the class as a system, it becomes clear how many things affect learning outcomes: whether students and teachers are timely, respectful, prepared; the temperature of the classroom; the state of the room's furniture and its styling; whether or not someone is eating lunch in the room, or snoring, or using social media on a large-screened device (in an online course, some factors might be the quality of the internet connection, where the student is logging in from, and so on). We can see how asking questions in class helps a teacher gauge how well the material has been understood. We can see how learning emerges, for better or worse, from the relationships between all the system parts. We can acknowledge other things that emerge, too: friendships, for example, and the spread of the flu.

Studying Human Systems

The Evolution of Anthropology

Anthropology, we have learned, concerns human systems, human culture included. During the late nineteenth and early twentieth centuries, when academic anthropology came into its own in Europe, anthropologists from there often made long, transdisciplinary field trips to lands where they could study native or indigenous non-European cultures from a variety of angles. Many spent a great deal of time trying to measure and document human physical diversity. For example, the 1898 Cambridge Expedition to the Torres Straits team even brought an 'olfactometer,' meant to assess smell (participants would not insert its tubes into their nostrils, however; Richards 2010). Anthropologists also took pains to record native languages and to create descriptive written accounts of indigenous social structures and cultures—**ethnographic** accounts. Specialists in doing this often also practiced **ethnology**: they made cross-cultural comparisons in order to understand the significance of the cultural variation that they had documented, as well as to test for universality.

Here, terminologically, it will help to recall that *anthropos* means people in general, and to learn that *ethnos* refers to one people or nation. In this, *ethnos* refers to the culture or design for living held to by a particular population. It indexes a given group's shared, learned heritage—including the systems this heritage supports ideologically: social, political, economic, spiritual, and so on. In fact, today ethnology is often termed 'sociocultural' or even just 'cultural' anthropology.

Many anthropologists today focus solely on sociocultural issues. Many focus, equally narrowly, on the physical or biological dimensions of humankind. Other specializations entail linguistic and, in the United States, archaeological anthropology. This 'four field' partitioning does not fit well with the original view of anthropology as the study of humans in the round—but it does make sense when seen in the context of academia's evolution (and see Chapter 5 regarding four-fold thinking).

From Fission to Fusion

The history of academia has been one of territory-staking, with each discipline carving off a precise aspect of human existence for study. Today, we are noticing problems resulting from the narrow thinking that such subdividing can foster. The twentieth century seems bound to be remembered as a century of disciplinary fission or fragmentation. It remains to be seen whether twenty-first-century academia can, in contrast, be remembered as a time of fusion or cohesion, in which holism pervades not only anthropology but other disciplines as well which, for reasons that should be clear to you by this book's end, is the hope of many.

The unified, transdisciplinary, multi-methods team approach now regaining popularity has led to more comprehensive, realistic, illuminating, and therefore relevant findings. The individual researcher, too, can strive for holism by crossing humanly created academic boundaries. The biological anthropologist interested in population-level differences in rickets vulnerability, for instance, can examine not only possible genetic differences relating to skin pigmentation but also breastfeeding and dietary practices and preferences; how the built environment, employment patterns, and latitude affect sunlight exposure; cultural views on tan skin; understandings about the risks of sunscreen, and so on. Importantly, these pieces of the puzzle cannot just be studied in parallel. A concerted effort must be taken to explore how they interrelate, and how changes in one part of the picture may lead to or derive from changes in another: a complex adaptive systems perspective should be applied.

Ethnocentrism Is Not Part of a Unified Comparative Method

We got onto the topic of (re)unification when I mentioned the fragmentation that occurred as academia—and anthropology—grew. Beyond the problems associated with the territory-staking, however, any discussion of anthropology's evolution also must address the deeply embedded biases many early practitioners had regarding the peoples studied. For example, many held fast to the theory of progressive cultural evolution, in which other cultures were cast as inferior versions of the anthropologist's. This idea has long since been soundly rejected and labeled 'ethnocentric.'

Ethnocentrism involves putting one's own culture at the center of an interpretation rather than using that culture's norms and values to make interpretations (see Chapter 4). We are ethnocentric when we take it for granted, based on our own cultural habits, that everyone marks birthdays with cake, or knows a particular Qur'an or verse. Frequently, we are ethnocentric because we assume that the way we do things is the best or natural way, or that everything can be understood from our own point of view.

Although anthropology warns now that ethnocentrism severely limits our ability to understand other cultures, many early anthropologists went into the field with an unabashed ethnocentric perspective. Technically, they did not have to: bias is not a prerequisite of the **comparative method,** in which cultures under study are compared with each other as a basis for generalizing about humankind. When anthropologists compare cultures they may do so intentionally and fair-mindedly as ethnologists—but they also may do it unreflexively, unthinkingly positioning their own cultures as ideal. This is generally what happened in the beginning. Early anthropologists often held their own cultures up not as fair comparisons, then, but rather as ideals. Today, however, anthropologists recognize that **reflexivity**—deeply examining our cultural (and personal) biases and motives to grasp how they influence our perceptions and conclusions—is necessary to minimize the limitations ethnocentrism puts on the ethnological endeavor.

Another way to minimize ethnocentrism is to actively participate in the culture under study—a common practice these days but not at anthropology's start. Early anthropologists often hired native interpreters to help them in the day-work of data collection and retreated at night to temporary camps or missionary quarters set back from the settlements of the cultures under study. By the 1920s, however, the need for long-term immersion had grown apparent. Rather than living nearby those under study, anthropologists now lived among them, learning their language, eating their food, and participating in as much of their daily lives as was possible. **Participant-observation** is the name for this approach, and it is now the classic method for creating ethnographic accounts.

All ethnographic accounts, old and new, taken together, comprise the **ethnographic record**. The **archaeological record**, by contrast, is made up of all that we know from what people left behind and how those objects were organized. It includes not only the artifacts of their cultural worlds, such as their houses and cooking tools and high- and low-status objects, but also (in a subset termed the fossil record) their physical remains, such as bones and coprolites (fossilized feces), which provide important biological data.

The ethnographic and archaeological records often are used in combination to form a better picture of the various lifeways enacted by human beings across space and across time, and to reconstruct humankind's evolutionary history. Throw in contemporary biological studies, and—assuming we can turn away from biological determinism, apply systems thinking, and steer clear of ethnocentrism—we can begin to apply the comparative method fruitfully as part of a unified approach to understanding human variation. We can begin to hypothesize productively about the origins, functions, and magnificent range of our biocultural diversity.

2

GENETIC ADAPTATION

This chapter prepares you to:

- Explain, with reference to genes and alleles, how traits can be inherited physically
- Describe basic sources of genetic variation
- Demonstrate understanding of the basic principles of natural selection at the population level, with a focus on expressed genetic variation
- Differentiate ecological from sexual selection and explain the value of sexual reproduction for assuring within-species variation
- Explain and exemplify balanced polymorphism

For the archetypal ecosystem in homeostasis, everything is perfect. Life is balanced and adaptation unnecessary. However, as we now know, apparent equilibrium depends on change, which is constant and inevitable—even when incremental.

Another word for change is **evolution**, and the kind of evolution this chapter deals with is genetic. **Genetic evolution** happens at the level of an entire population. It entails a change in the frequency of a given gene or genes in the **gene pool**—in the entire population's sum total of genes—from generation to generation.

More often than not, human genetic evolution occurs as an adaptive response to an environmental change that we humans in fact have instigated in a process termed, with increasing frequency, **niche construction**. The rise of this phrase, awkward as it may be, reflects a growth in researchers' awareness of species–environment interaction. It refers to all of the ways that species, simply by going about their species-specific lives, shape their environmental niches just as their niches shape them. It acknowledges the ways that species construct the worlds (niches) that they, and other species, occupy (see Odling-Smee et al. 2003).

A classic example of how niche construction processes can support genetic evolution entails the adaptive evolution of some resistance to malaria in human populations with increased exposure to the disease as a result of cultural practices fostering mosquitoes, which carry the parasite that causes it. We explore this in detail later; for now, it suffices to say that such mutual interaction between behavior and body demonstrates that we are truly biocultural beings. Genes alone do not determine our problems or potentials.

Of course, without genes we would not exist. Knowing this prepares us to ask: what is a gene, and how, exactly, does the gene pool change to accommodate (when it can) challenges to survival such as increased exposure to malaria? Ultimately, we seek an understanding of how and to what extent human genetic adaptation depends on as well as fuels biocultural diversity. To achieve this, we begin by building foundational knowledge of genetics and heredity.

Genes and Traits

Put simply, a **gene** is a linear sequence of deoxyribonucleic acid (**DNA**) that codes for (or, contains the recipe from which our bodies build) a specific protein or trait. DNA itself comes in long strings, and each string contains many genes or information storage units. DNA strings are generally wound into little wads or structured packages, called **chromosomes**. In sexually reproducing populations the chromosomes come in paired sets. They are inherited: half come from the male parent and the other half from the female parent. We inherit two copies of each gene, one from mom and one from dad (biologically speaking).

While those new to genetics might associate each gene with a specific bodily trait, such as nose size or singing voice, things aren't that simple. Most traits actually are **polygenic**; that is, they develop through the interaction of a collection of numerous genes ('poly' means many or more than one). Most traits emerge from the combination of proteins and biochemical reactions that a set of genes 'codes for' (and see Chapter 3).

Genes themselves can be **polymorphic**, meaning they can come in various forms, called **alleles**. For example, the 'mid-digital (mid-finger) hair gene alleles code for proteins involved in both hair growth and nongrowth. The gene entailed in cystic fibrosis, a generally fatal disease in which the lungs fill up with fluid and certain digestive processes do not work well, also has two alleles—one that codes for proteins involved in the disease and one that does not.

Some alleles are **recessive**, or easily dominated, and therefore less frequently expressed; others are **dominant**, making their own instructions take precedence over recessive alleles so that only they are expressed. This phenomenon—the dominant–recessive mode of gene expression—was described by the scientist-priest Gregor Johann Mendel even before we knew about genes per se. Examining peas in his monastery's experimental garden, he noticed dominant and recessive inheritance patterns in color and texture.

A **homozygous** gene pair, or person, has two alleles that are the same for a given gene; a **heterozygous** person has two different alleles for the gene in question (remember, we each have two of every gene because of how humans reproduce). A person can be homozygous when it comes to one kind of gene, but heterozygous for another.

Why distinguish whether a given allele has been inherited from both parents (homozygously) or not (heterozygously)? For one thing, co-occurring recessive alleles cause their associated trait to be expressed. So, if a child is heterozygous for cystic fibrosis—if the child received the recessive allele that codes for cystic fibrosis from one parent and the dominant allele that does not from the other parent—the child cannot develop the disease. Alternately, if the child is homozygous, having *two* recessive alleles, cystic fibrosis is inevitable. If both parents are heterozygous and so carry one copy of the recessive gene, the child has a 25 percent chance of inheriting two recessive copies.

To better grasp this, consider Figure 2.1. It maps out inheritance using the four-square graphic form known as a **Punnett square**, after Reginald Punnett, who famously devised it. Punnett squares are templates used to assist in the calculation of possible allele combinations resulting when two individual organisms reproduce.

Of course, most traits are not **monogenic:** most are not traceable to the alleles of just one gene; and most are not ruled by simple dominant–recessive relations. Most traits have complex, polygenic origins. On top of that, the cultural contexts in which we live—in which we are nurtured—directly

		Mother = Rr	
		R	r
Father = Rr	R	RR	Rr
	r	Rr	rr

FIGURE 2.1 Inheritance of a recessive (*r*) trait.

affect gene expression; this must be taken into account for a full understanding of our genetic dimension. Still, by thinking about clear-cut (dominant–recessive, monogenic) examples first, we can begin to understand the basic mechanisms of heredity, such as how genes are passed along. Genes are not the full story, as we shall see, but understanding heredity, and its relationship to genetic variation, provides a good first step in deciphering the puzzle of human biocultural diversity.

Evolutionary Change

In the absence of environmental changes and when mating occurs at random, genes will be reproduced in the next generation at a steady rate. This fact can be referred to as Hardy–Weinberg Equilibrium, after the fellows who first wrote about it. However, in reality, no population exists in such conditions except in the lab. When the rate of a given gene's reproduction in the next generation therefore shifts, genetic evolution has occurred.

General evolution can happen in a gene pool due to simple random chance, and this type of change is called **genetic drift**. For instance, genes are shuffled during sexual reproduction, and sometimes this randomly leads to changes in gene frequencies. Random events, too, can affect gene distribution, such as when a tree in the woods is knocked over arbitrarily (say, by a lightning strike), and the falling tree kills a badger that happened to have been sheltering below it. If this unlucky badger had not yet reproduced, its chance removal could have an impact on gene frequencies in the next generation. Now, drift may not add up to much with a large population, but it can have a noticeable effect when a population is small. If there are only ten badgers, then losing one is quite significant; a big shift in the next generation's gene pool can result. If there are thousands of badgers, the untimely loss of one poor animal will not have the same kind of impact.

A second mechanism for changing a gene pool is **gene flow,** which is when genes migrate into one gene pool from another, for instance due to one population moving into or through another population's territory such that some interbreeding occurs. One group might dominate the other, assimilating its gene pool; or the two groups could just interbreed a little bit and then separate again (maybe even splitting this time along new lines). This is what scientists today think happened between various archaic forms of humankind, so that *Homo sapiens*' eventual ascendance was not the outcome of a linear, progressive series of changes happening to one ancestral line but rather a process informed by many chance encounters between various separate *Homo* population groups that interbred, exchanging genes (a point we'll return to in Chapter 4).

The US gene pool has been greatly affected by the gene flow immigration entails. Take American eyes, for instance (the Americas comprise two whole continents, but I use the term 'American' in the lay sense to refer to people in the United States). American eyes, generally quite dark prior to the European invasion, became much lighter by 1900, when about half the population had blue eyes. By the 1950s, however, only about one-third of the US population had blue eyes; today, fewer than one in six people does. These changes seem to have been caused by the flow of genes across national borders; other recent cultural changes, for example in mating and marriage patterns, have accelerated the process (Belkin 2006).

Mutation is a third way change gets introduced into a gene pool. When genes are duplicated for reproduction, alterations can occur; a mutation is simply a miscopied gene. Miscopying often happens as the result of some kind of exposure, for instance to a toxic chemical, radiation, or a particular virus. Culture can give the biological process of genetic mutation a boost: for example, although asbestos is found in nature, the asbestos fiber products that eventually triggered mutation-based cancers in so many factory and other workers in the twentieth century exist as a result of cultural invention. Similarly, our chances for exposure to solar radiation have been enhanced by cultural practices that have depleted the earth's protective ozone.

Not all mutations lead to cancer, of course, or allow fatal recessive diseases to persist in a population even when they kill homozygous individuals before the individuals reproduce and pass on their genes themselves. Most mutations that occur during reproduction are silent or neutral, meaning that they have no immediate impact on an offspring's survival and reproductive capacity. Some mutations, and sometimes, too, genetic drift or flow, actually increase a group's or species' chances for survival and proliferation (remember, a 'species' is an organism type that can reproduce itself, for instance by mating and giving birth to fertile offspring).

The Role of Natural Selection in Evolution

Natural Selection Acts on Existing, Expressed Traits

This leads us to the key mechanism of genetic evolution: natural selection. **Natural selection** is a process by which the genes for expressed traits that, under certain environmental conditions, happen to give an organism an adaptive advantage for survival (at least to the point of reproductive success) become, in this way, more abundant in the next generation's gene pool. In other words, the survival advantage that these particular traits confer allows their bearers to produce a larger number of healthy offspring. More viable offspring—offspring that themselves live on to reproduce—mean that more of the genetic material that gave the first generation a leg up will exist in the next generation of the population's gene pool, and in the next, and so on, until the environment changes yet again. Successful reproduction is the coin of the realm of natural selection.

Charles Darwin, the scientist popularly associated with natural selection, outlined its mechanisms most famously in his 1859 book *On the Origin of Species*. For Darwin, of course, genes did not exist; he spoke instead of traits themselves. This reminds us that observable, tangible traits—expressed traits—are what natural selection preserves. The advantage provided by expressed traits given a new environmental pressure—that is, **fitness**—is measured in terms of *lower mortality rates*, which means those with that trait are less likely to die in the face of that pressure than those without it, and *higher fertility rates*, which means those with that trait produce a relatively higher number of viable offspring for the next generation. Darwin called this "descent with modification" (Darwin 1859).

The following hypothetical example illustrates this kind of occurrence (it is far too simple to represent reality, but my goal here is to get the principles across; we deal with reality later; and see Chapter 3). Let's say that people in your world have only purple or green skin. For a long time, this variation just existed, much like hair color varies in our world. However, now, because of some change in the environment—in, say, the culture or climate—having green skin provides a selective advantage. Green-skinned people reproduce more.

Whether one has green or purple skin depends (in this just-so story) on only one gene with one dominant and one recessive allele, and the allele for green pigmentation dominates. You, fortunately, have green skin. Here's the interesting part: whether your greenness is expressed because you are homozygous for the trait or because you have one copy of the dominant allele makes no difference according to the law of natural selection. It only matters that greenness was in fact expressed or made manifest in your skin. That—not the genes themselves—saved your hide so that you could later proliferate. The genes came along for the ride when you reproduced; in this way, you passed those genes along to the next generation and experienced reproductive success. Note here that reproductive success means not just having more babies, but having fit ones—babies that can survive to reproduce again, as those with green skin will do in this admittedly simplistic scenario.

Natural selection's focus on manifest traits allowed Darwin and others to recognize it despite their lack of knowledge regarding genetics. It also points us to two new vocabulary terms: genotype and phenotype. Your **genotype** is the set of genes you carry; your **phenotype** is the manifestation or measurable expression of your genes 'in the flesh.'

To be clear: your genotype is not the main focus of the selective process. Rather, your phenotype takes center stage. Remember, a person who carries one recessive (purple) allele has no less chance of reproducing than someone homozygous for the dominant (green) one because the heterozygote still is green skinned. It is only when the recessive allele is homozygous and therefore expressed, as purple skin, that a person does not live to pass his or her genes on. It is the phenotype, not the genotype, that gives the green folks the advantage. Moreover, natural selection works on *existing* phenotypic variation. A trait must exist already if it is to be selected for.

This points to the fact that, while the general phrase 'genetic evolution' does cover the selective process just described, the more exact title for it is 'genetic adaptation.' The phrase **genetic adaptation** differentiates the particular kind of evolution brought about by natural selection from that brought about by drift, flow due to migration, or mutation alone. Those mechanisms just introduce change. In the kind of genetic evolution that occurs with natural selection—genetic adaptation—certain *already existing* phenotypic or expressed traits become situationally more (or less) advantageous in terms of that group's continued survival (see Figure 2.2).

Environmental Pressure and Ecological Selection

Ecological selection—genetic adaptation in response to a change in the environment that makes a given trait advantageous—is a mainstay of natural selection. In fact, and in part for this reason, the phrase natural selection often is shorthand for it.

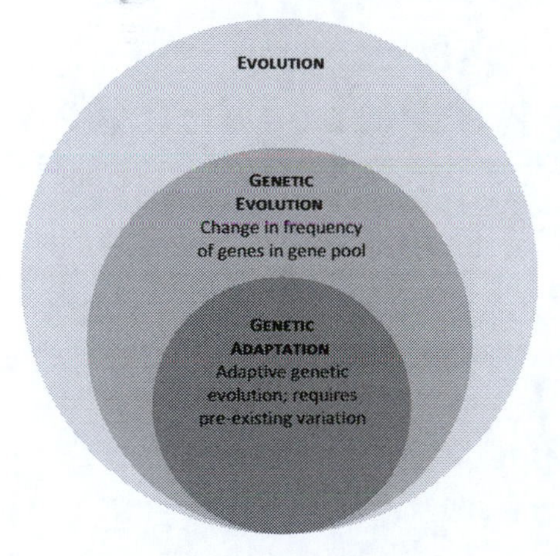

FIGURE 2.2 Conceptual model of genetic adaptation. Genetic adaptation occurs when the forces of natural selection lead to increases, in the next generation, of the frequency of genes supporting advantageous manifest traits that already existed within the population. Those traits were generally introduced earlier through simple genetic evolution, for instance via the mechanisms of gene flow, drift, or mutation.

Let's imagine that a new **pathogen**, or germ, enters and thus becomes part of our environment. Imagine that it is a lethal disease-causing germ. People who happen to be endowed with some pre-existing genetic trait that makes them less susceptible (or more resistant) to the pathogen will have a better chance of survival than will those who do not have the protective gene. Naturally, these protected people will have more children than their dead compatriots, leading to an upward shift in the frequency of the adaptive gene in the next generation. Should the same pathogen sweep through again during the next generation's lifetime, the death toll will be lower: a higher proportion of the population will be resistant.

The imaginary pathogen is but one example of **environmental pressure**. This category includes anything in the environment that reduces a population's ability to function or puts a damper on its potential, for instance by increasing **morbidity** or sickness, and **mortality** or death. Disease, deprivation (through drought, famine, and so forth), disaster (floods, fires), and an increase in predators are all environmental pressures.

To better understand how ecological selection works, we might go right now to collect some dry beans (or jelly beans or buttons)—their colors do not matter, so long as there are a variety of colors, and so long as some match the color of the desk (or table, or whatever) on which I'd like us to imagine we've put them for this experiment (see Figure 2.3). Variety in color represents pre-existing variation in this population's gene pool. Once the beans have been collected and laid out, let us assume that the bean population has lived in peace on the desk (or wherever) without a predator for many generations.

The beans that match the desk are to some degree camouflaged. Right now, that provides no advantage; but what if a predator came onto the scene? What if the local bean-eating bird population got so large that some birds moved into the room? This could create environmental pressure on the room's bean population. Bean color could now become very important. If the birds can see some colors of beans better than others, those beans will be eaten faster. The beans more likely to be eaten would therefore be less likely to contribute their genes to the next generation. Beans of their color would become exceedingly rare as selective pressure (hungry birds) removed them. By contrast, beans the color of the desk would have a selective advantage because they can better go unseen; their color would be selected for.

Readers will know by now that new colors are not created in response to hungry birds; the colors pre-exist. They already are there. The bean population's total color scheme changes as

FIGURE 2.3 Varieties of beans. Photo by Debora Cartagena, Centers for Disease Control and Prevention.

some of those with a disadvantageous color are eaten off. It is true that a bean in a new color may emerge through a random mutation, but random mutations occur independent of environmental pressures. A randomly introduced color may be selected for after it arrives; but random mutations are not a responsive development (regarding responsive genotypic, phenotypic, and epigenetic change, see Chapter 3).

SUPPLEMENT BOX 2: PRE-EXISTING TRAITS, NICHE (RE)CONSTRUCTION, AND GLUTEN-FREE FOOD

Until 10,000 years ago—that is, for most of human evolutionary history—there was no such thing as domesticated wheat. The amount of gluten in the human diet was effectively nil. When we domesticated wheat (and rye and barley), however, gluten consumption increased for those dependent on the grain.

The addition of wheat to the diet leveraged an until-then under-exploited, pre-existing genetically derived capacity. That is, most people could digest gluten with no problem: wheat would not likely have been domesticated if that were not the case, because it has little else to offer us humans than its food value. Our gluten-digesting capabilities were simply there but latent prior to that time.

Those general capacities notwithstanding, gluten exposure did trigger, in some individuals, a terrible response. Gluten activates celiac disease (CD) in those genetically predisposed to it. Individuals with CD become malnourished, and they are susceptible to growth problems, chronic constipation, abdominal bloating and pain, iron-deficiency anemia, fatigue, bone or joint pain, depression, anxiety, seizures or migraines, skin rashes, infertility, and so on: in other words, they have lower fitness when exposed to gluten.

Industrialized processed food production, which started in earnest about sixty years ago, intensified the issue. Processing foods industrially entails many steps that a home cook never would take, including the use of additives. We all know about preservatives and added sugars; but gluten, the main protein in wheat, also makes its way into many foods that wouldn't normally include it: it serves well as a cheap thickening agent and filler. Soy sauce, peanut butter, chocolate, sausages, potato chips—all may have some gluten in them. In short, due entirely to changes in our provisioning system eating has become a major problem for those who cannot tolerate gluten. The flourishing gluten-free food market that recently emerged did so partly in response to this.

That said, even if what scientists term the CD genotype was harmless nearly forever—that is, until the point at which we began to eat wheat regularly—once it became harmful why didn't natural selection weed it out? In exploring this evolutionary paradox, Kayla Morrell and Melissa Melby (2017) offer three hypotheses, each entailing unintended trade-offs in fitness.

The basic idea behind each hypothesis is that the gene pool includes CD genes because they do something good for us as well as underlying CD (which is wholly humanly triggered). In other words, CD genes control or are linked to at least two traits, only one of which is expressed among foragers (that is, the still unidentified helpful one). The transitions entailed in the Agricultural Revolution and in the industrialization of our food supply mediate our susceptibility to the CD genotype's otherwise latent negative impact.

The first hypothesis regarding why a CD genotype might be helpful focuses on this transition. Shifting from a foraged diet to an agricultural one, heavy on domesticated cereals, would entail moving away from iron-rich foods to iron-poor ones. The genetics behind CD might have originally protected foragers from hemochromatosis, a disorder in which too much iron is absorbed into the body; this is a potential problem when one's diet is iron rich. Not coincidentally, Basque people, whose diet has long been uniquely high in iron, have more tendency toward the ostensibly hemochromatosis-halting CD genotype.

Domesticated cereals are not only iron poor, making genetic protection from hemochromatosis not worth the CD risk; they also are associated with dental caries or decay because they stick to the teeth, where they break down into simple sugars. In a second hypothesis, the CD genotype is linked with protection against tooth decay, which would have been a much greater problem for farmers than for foragers. Individuals with some genetic dental protection would have had greater reproductive fitness, and thus their (CD) genotype would have contributed more to the next generation's gene pool than those who were unprotected—but the grain-based diet that made the mutation useful also could, if the grain was wheat, trigger celiac disease.

A third hypothesis holds that the CD genotype provides some sort of protection against bacterial illness. Indeed, the CD genotype may have received a boost when it helped populations survive a bacteria-based epidemic in some parts of Europe 1200–1700 years ago. The CD genotype may in fact be protecting us against other bacterial infections even now.

Each hypothesis has support. Whether data collected and analyzed in the future will disprove one of the three, or more than one, or maybe none, remains to be seen. In the meantime, the important point to remember is that without a change in diet, the problem of celiac disease could have remained latent forever. Its existence proves the biocultural nature of human experience.

Other Kinds of Selection

Natural selection via environmental pressure works like a filter on variation that is already there in the gene pool due to mutation, drift, or flow—or a prior instance of natural selection that favored a trait long ago for reasons unrelated to contemporary conditions (see Supplement Box 2). Ecological selection is natural selection's workhorse. It is, however, aided somewhat by another mechanism: nonrandom mating. **Nonrandom mating** is exactly what it sounds like it is: mating and, subsequently, reproduction limited by factors determined, whether consciously or not, by the population in question.

Among human populations nonrandom mating occurs when kinship or class-related rules or ethnic boundaries limit who can build a family with whom. We see nonrandom mating, for instance, when a person of one ethnicity or religion prefers a person of that ethnicity or religion to all others as a mate. We also see it when certain physical features are deemed attractive, perhaps because they index good health or some other culturally desirable characteristic (for instance, financial standing, moral standards, loyalty to the group; see Chapter 12).

Nonrandom mating happens even among nonhuman animals, although only humans would conceptualize it in the terms we are using in this discussion. Female peacocks (pea hens) are more likely to bear offspring with male peacocks that have bigger, brighter tails. Female lions are more likely to bear offspring with male lions that have fuller, glossier manes. The trait that attracts one to another is often termed an **index trait**; it is not important in itself, but rather for what it indexes or symbolizes (in this case, better health, including healthier sperm, which helps ensure the viability of the next generation). Note, too, that traits such as scent can be as important as visual indicators.

Another label for nonrandom mating is **sexual selection**. This term highlights the contrast between this kind of selection, which organisms themselves have a hand in, and ecological selection, or the kind brought on factors beyond the organism itself—for instance, drought or a predator's evolved ability to run faster or bite harder or even outwit its prey. It is important to remember that sexual selection refers to nonrandom mate choice, not to sexual reproduction; the latter is just a way of making more individual organisms—of reproducing.

Speaking of reproduction, **artificial selection** entails the purposeful breeding, by humans, of plants or animals to produce in their offspring particular traits, such as bigger corn kernels, sweeter oranges, faster race horses, or better-nosed hunting dogs. Darwin himself brought observations of human plant and animal breeders' manipulations to bear in his formulation of natural selection. He figured that if "feeble man" could invoke generational changes in domesticated crops, stock, and pets through manipulative breeding, there might be "no limit to the amount of change, to the beauty and infinite complexity of the coadaptations between all organic beings, one with another and with their physical conditions of life, which may be effected in the long course of time by nature's power of selection" (Darwin 1859, 109).

Because human interference was—and is—so explicit in plant and animal breeding, artificial selection provides a fine contrast to natural selection, and thereby to ecological and sexual selection (see Figure 2.4). It is worth considering, however, the ways in which this dualistic mode of classification can misguide us, much like the mind–body scheme that places culture and biology in contrast to one another rather than as two sides of the same coin.

Humans are part of nature, so to call humanly orchestrated selection artificial is illogical. Moreover, so-called natural selection also can have a human handprint. For instance, humans can bring on environmental pressures like those implicated in ecological selection through various means, such as when certain agricultural practices led to conditions that fostered the natural selection of the sickle cell trait as protection against malaria (more of which later). Further, the line dividing artificial selection from nonrandom mating based on conscious criteria is fuzzy, particularly when arranged marriage is considered.

Problems with commonly accepted terminology and related assumptions notwithstanding, if we are to engage in intelligent discussion regarding human diversity, we do need a vocabulary from which to start. We do need to know what is taken as common knowledge. Concurrently, we need not accept common knowledge just because it is common; classification schemes can and often do fragment processes and outcomes that in real life exist holistically. Only if we keep this in mind—only if we refuse to assume that nature is somehow separate from human

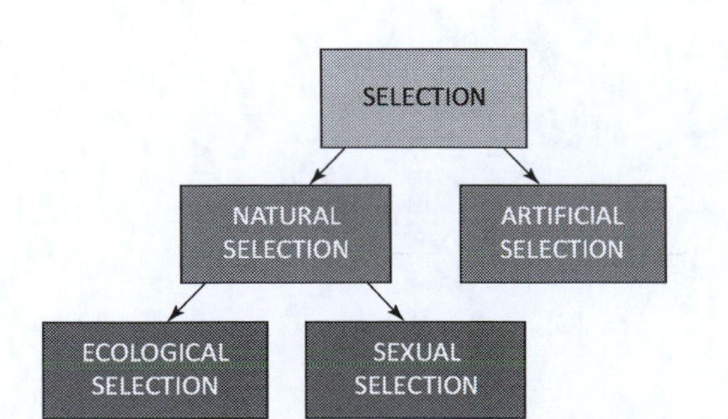

FIGURE 2.4 Natural (ecological, sexual) and artificial selection.

existence and instead continue to look for connections between (or reunify) dichotomous constructs such as nature–biology and nurture–culture—will our exploration of human diversity be productive.

Co-Evolution and the Importance of Within-Population Variation

Remember Darwin's reference to "coadaptations between all organic beings, one with another" (Darwin 1859, 109)? This reminds us that adaptation does not occur in a vacuum; all populations are continually and conjunctionally adapting—or failing to do so.

Whenever a gene pool with some variation meets an environmental pressure, natural selective processes go to work, and genes that provide some kind of advantage will occur more frequently in the next generation. If there is little or no variation in the gene pool, such that no subset of individuals has an in-built genetic advantage, the environmental pressure may extinguish the entire population. This could happen, for instance, if a new and lethal pathogen arrives on the scene, or if an existing pathogen evolves a fatal strain. A population in which genetic variation is not maintained is at a distinct disadvantage in a world that is constantly producing new environmental pressures.

This idea also is known as the **Red Queen Hypothesis** after a character in Lewis Carroll's *Through the Looking-Glass* (1871), the sequel to *Alice's Adventures in Wonderland* (1865). At one point in the story, in a garden of talking flowers, Alice meets the Red Queen. They begin to walk. Soon they find themselves trotting along—and getting nowhere, to Alice's but not the Red Queen's surprise. The world around them has begun moving. In the Red Queen's world, learns Alice, you run just "to keep in the same place" (see Figure 2.5). Of course, neither the Red Queen nor Alice remains in the same spot literally, because the spot itself differs over time, as do they: the illusion of standing still—represented in the construct homeostasis—rests in fact on constant change. A gene pool in an evolving environment must produce variation from generation to generation just to survive the changes that surround it.

FIGURE 2.5 Alice and the Red Queen running in place. Illustration by John Tenniel for Chapter 2 of *Through the Looking-Glass* by Lewis Carroll, 1871.

Happily for us humans, like so many living creatures, we reproduce sexually. In sexual reproduction genes are shuffled around, with half coming from one parent and half from the other. This means that every offspring can receive a slightly different combination of genes. In other words, sexual reproduction ensures variation. By contrast, some gene pools do not reproduce sexually and do not produce variation; examples include many single-celled organisms that reduplicate by splitting themselves in two.

If a population reproduces asexually, through **cloning**, each generation is, at least in theory, an exact duplicate of its precursor. This can leave a population vulnerable to extinction because there is generally no subset that might exhibit a genetic advantage. Scientists have learned recently that clonally reproduced species do have a few means of varying themselves. For one thing, mutations can occur as genes are split. For another, external contingencies, such as a temperature shift, can invoke particular adaptive changes through means that scientists are only beginning to explore (see Chapter 3). Still, with clones, what we mostly have is a whole lot of faithful facsimiles.

Although cloning does not completely condemn a group to consist of absolutely identical beings, sexual reproduction *ensures* variation. Even without mutation and flow, simply through the gene rearrangements and recouplings that occur as part of the sexual reproduction process, there will always be some variation among individuals that can help to insure against catastrophic extinction in the face of a potentially fatal environmental pressure. In this context, human diversity is good.

Natural selection is always at work somewhere. However, human evolution in action is hard to detect because each human generation lives such a long time and because our species is now so easily able to move from place to place and intermingle (causing additional change via gene flow, not genetic adaptation). By contrast, it is easy to see natural selection at work in creatures with quicker generational turnover, such as fruit flies or certain germs. We also see it in pathogens that develop a **resistance** to antibiotics: a dose may kill most of a given germ the first time you take it, but those germs genetically endowed to resist the drug then produce many more offspring than those killed off before reproducing, and soon the genes that confer antibiotic resistance predominate in the germ's gene pool.

Examples of Natural Selection

Another creature that has lent itself well to the study of natural selection is the peppered moth, *Biston betularia.* This moth is found in many places but most famously in the forests of England. Two hundred years ago, the majority of peppered moths living in and around Manchester were lightly colored, which camouflaged them from their predators when they alit on the light-colored trees and lichen in the area. With the onset of the Industrial Revolution, however, the trees (like other bits of England) became dark with soot; much of the light-colored lichen died off. The lighter moths were no longer hidden from sight.

Meanwhile, their darker kin now could not be seen by predator birds as well as before, and so those moths gained the advantage. Very swiftly, the population became darker. This process is termed **industrial melanism**, after the industrial processes that darkened the environment and the melanic pigment expressed in the darker moths. Safer now than ever from the birds that otherwise would have eaten them, the darker moths flourished. The lighter moths, now more easily seen, were pecked off in larger numbers. Natural selection worked as a filter on the existing color variation. Had there been no genetic color variation, the entire population might have been wiped out (see Figure 2.6).

This kind of genetic adaptation is termed **microevolution,** because it occurs on a relatively minor scale (the kind of genetic adaptation that brings about whole new species is **macroevolution**).

FIGURE 2.6 Peppered moths. Photos by Olaf Leillinger, 2006. CC-BY-SA-2.5.

Examples of microevolution via natural selection in people may be hard to find at first, but they do abound.

One classic example is the evolution of **lactose tolerance**: the ability to digest milk sugars (lactose) until late in life. For most of human history, this ability disappeared in late childhood as the genes responsible for creating lactase, the enzyme that breaks down lactose, turned off. In typical foraging or hunter-gatherer populations (the norm for most of time) lactase is unnecessary after breastfeeding is done; milk from animals is not a food source, and so someone who cannot digest milk later in life is not disadvantaged in comparison to someone who can.

In populations that began to keep milk-bearing animals, however, those who could digest animal milk were generally better fed than others—and had more offspring as a side effect. Eventually, those groups evolved, genetically, so that the majority could digest milk until later in life. This evolution occurred in Northern Europeans as well as in several cattle-herding groups in North Africa. The rest of the world's population had no selective need for long-lasting lactose tolerance and so lactose tolerance is still the exception, not the norm.

Likewise, some scientists think that the ability to digest starches easily was selected for when agriculture was invented and we began to rely more on grains (see Chapter 7). Amylase, an enzyme that breaks down starch, is secreted in our saliva. People who come from agrarian stock have more copies of the gene that codes for this enzyme than do those who come from hunting or fishing populations. Like the ability to digest milk, the ability to digest lots of starch quickly and easily may be a relatively recent adaptation (Wade 2010).

Balanced Polymorphism

As the digestive examples suggest, potentially adaptive traits can be around in a portion of a population for a long time without consequence. They persist here and there because they do no harm. However, once the environment has changed (for example, once people started raising dairy cows or grain) certain traits may confer a sudden advantage, and so their frequency quickly escalates. The example of the peppered moth illustrates this process. Other changes in the physical environment, including those related to migration, for instance, from a hot to a cold place or a bright one to a dim one, also can instigate genetic adaptation.

Sometimes, a potentially harmful gene variant can even be selected for. This is what has happened in some regions where malaria is a problem, including parts of Africa and Asia. **Malaria** is a disease caused by parasites belonging to the genus *Plasmodium* and carried by the *Anopheles* mosquito. However, malaria is more than a 'naturally occurring' disease threat; it exists as an **anthropogenic** or humanly generated condition.

Changes in human agricultural and settlement patterns (that is, in human niche construction) have fostered population growth among malaria-spreading mosquitoes and so malaria parasites. For instance, deforestation—clearing land of trees and undergrowth for farming—harms certain bat and bird populations. With fewer of these predators around, more mosquitoes survive to reproduce. Deforestation also reduces plant life that preserves topsoil, so more standing water is available—and that is where mosquitoes reproduce. Irrigation ditches, too, provide lots of lovely standing water, further bolstering *Anopheles* population booms (see Figure 2.7).

FIGURE 2.7 Standing water created through irrigation practices. Photo by Lynn Betts, USDA Natural Resources Conservation Service.

The disease caused by the *Plasmodium* parasites that the *Anopheles* mosquitoes can carry entails cyclical high fevers, headaches, and often death in humans. Indeed, malaria is one of the top ten causes of death worldwide, killing one to three million people annually, mostly children.

In a nutshell, malaria parasites enter the body via the mosquito bite, spend some time in the liver, and then invade the red blood cells. Red blood cells, which carry oxygen throughout the body in their **hemoglobin** molecules, are normally round or doughnut shaped. To the malaria parasite, healthy red blood cells are lovely, fat, round nests. They live there and reproduce, damaging and bursting the cells, after which they invade fresh red blood cells as part of an amplification process. They always need healthy red blood cells to proliferate.

There is genetic variation in our red blood cells. One of the alleles for their hemoglobin molecules, which transport oxygen, causes fragility in the structure of the cell: such cells often take on a crescent or sickled shape as they cave in (see Figure 2.8). These frail cells are not appealing to the malaria parasite: they do not provide good nests. So, in an interesting twist, the more sickle-prone cells you have, the less vulnerable you are to malaria.

Given this scenario, one might expect the sickle-cell allele to have been selected for and the typical version to have been selected against over the generations in areas where malaria is endemic or a constant threat, such as where humans have introduced agricultural and other practices supportive of *Anopheles* breeding. Not so, however. Why? Because too many sickle-prone cells will kill a person. Sickled cells tend to get stuck in capillaries (the body's smallest blood vessels), jamming up on each other like logs in a narrow stream, leading to a disease known as **sickle cell anemia**, or sickle cell disease. This disease entails, among other things, crisis periods in which individuals suffer great pain, as well as early death. So we are caught on the horns of a dilemma: people with 100 percent round red blood cells are good hosts for malaria and more likely to die from it; people with 100 percent sickling cells will not die from malaria—but they will die from sickle cell anemia.

Sickle cell disease is sometimes called a recessive disease—one contracted when an individual inherits two copies of (is homozygous for) the sickle cell variant of the hemoglobin gene. In truth, however, sickle cell disease is not fully recessive; heterozygotes who carries the trait in just one allele still can show symptoms because some of their red blood cells, but not all, will sickle. Importantly, though, such individuals are likely to be healthy enough to reproduce, thus

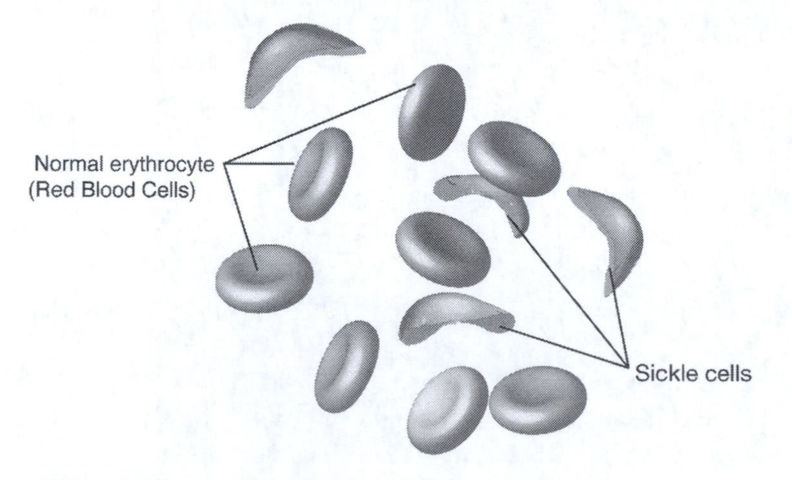

FIGURE 2.8 Round and sickled red blood cells. Illustration by Darryl Leja, National Human Genome Research Institute, National Institutes of Health.

contributing genes to the next generation's gene pool. Moreover, when in a heterozygous pair, the sickle cell allele provides just enough protection from malaria to maintain itself. This is why the heterozygous condition persists.

Given the anthropogenic threat of malaria—that is, in concert with a culturally created situation—natural selection promotes a balance between the round and sickling variations of the hemoglobin allele. The fancy label for this situation is **balanced polymorphism**. In balanced polymorphism, the pressure for one form of the allele is balanced or offset by the pressure against that form. An allele that itself is deleterious or harmful can be kept in a population's gene pool in this fashion because the hazard it entails is offset by the way in which it protects the organism from another problem.

Where malaria has been tamed through cultural practices and ceases to pose a risk to human health, the sickle cell allele will offer no benefit and so could be cleared from the gene pool eventually. Such an adaptation, in which human behavior begets a change in the status of malaria that in turn begets a change in the frequency of the sickle cell allele, demonstrates clearly the interlinked nature of biology and culture.

We will close this chapter on the malaria–sickle cell example, which amply illustrates important concepts in genetics and in regard to heredity, and which, moreover, makes manifest the value of the adaptive processes that produced diversity in the distribution of the sickle cell trait to begin with. If populations exposed by their own agricultural and settlement patterns to malaria hadn't in turn evolved, via natural—more specifically, ecological—selection to feature a high frequency of sickle cell alleles, these populations might not have persisted. The synergistic intersection of natural and cultural forces within the human body underwrites the global distribution of biocultural diversity.

3

DEVELOPMENTAL ADJUSTMENT AND EPIGENETIC CHANGE

This chapter prepares you to:

- Define developmental adjustment and distinguish it from natural selection
- Identify, define, and distinguish the genome and the epigenome; discuss both in relation to the differences between as well as within species
- Describe how developmental adjustment and epigenetic events affect gene expression, or help organisms translate a genotype to a phenotype
- Delineate the paradigm shift model of scientific revolution; show how it applies in regard to epigenetics

In the preceding chapter we learned how human variation supports the survival of human populations over time. We distinguished the genetic makeup of the individual—the **geneotype**—from the physical manifestation of genes in the flesh—the **phenotype**. Although we often think of visible phenotypic features such as hair, skin, and eye color when considering our genes' concrete expression, many phenotypic features are invisible. Take, for instance, how much hemoglobin (an oxygen-binding protein) a given person's body can produce, or the average shape of the red blood cells carrying the hemoglobin. We manifest our diverse phenotypes both outside and inside.

How do genotypes become phenotypes? We know that most genes, in effect, contain information in a code that can be translated, by other cellular functions, into a particular protein, and we know that proteins are physical building blocks. Yet getting from genotype to phenotype involves more than just building bodies up from a series of genetically predetermined proteins. The incarnation of diverse populations is no simple matter. Given new scientific discoveries, seeing DNA as a blueprint carved in stone no longer suffices.

In Chapter 2 we learned that the classic dominant–recessive situation, in which the genetic instructions in the recessive allele remain suppressed or unexpressed, actually is quite rare. We also learned that, although understanding dominance and recession is a first step to understanding genetic adaptation, many if not most traits are polygenic (based in many genes, not one); and many alleles are co-expressive (that is, members of the inherited pair both affect the phenotypic manifestation of the relevant trait).

In this chapter we see that extra-genetic factors also come into play, affecting phenotypic expression. Figuring out just what the relevant extra-genetic factors are and how they make their impact provides a fuller understanding of how we turn into who we turn into as we develop, and how the processes entailed can differ across populations.

Developmental Adjustment

A **developmental** process is one that occurs while an individual organism is growing or maturing. **Developmental adjustment** refers to a long-term, irreversible change in an organism's growth or biochemical processes during development that occurs in response to environmental conditions. It is made possible by our physical **plasticity**: our body's ability to adjust certain growth outcomes, to some degree, in ways that are adaptive to local environmental conditions.

Some alternate terms for developmental adjustment are 'developmental adaptation' and 'developmental acclimatization.' **Acclimatization** (or **acclimation**) can refer to shorter-term, reversible self-protective changes, such as the seasonal tan that some summertime beachgoers develop, so for clarity I avoid using acclimatization in a developmental context. Likewise, I prefer developmental 'adjustment' to 'adaptation' because, although a developmental adjustment is adaptive in a general sense, adopting the term 'adjustment' instead helps avoid confusion: genetic adaptations are inheritable and occur at the population level, affecting the next generation's gene pool. In contrast, developmental adjustments happen in an individual's lifetime rather than as a generation-to-generation shift, and they are not directly passed along to any offspring (see Table 3.1).

For example, a boy growing up in a location at a high altitude, such as in the Peruvian Andes, develops a larger chest and greater lung capacity (and more red blood cells and capillaries) than he would if raised at sea level. His body responds to the lower concentration of oxygen at high altitudes, as bodies will do, by effectively growing bigger lungs and so forth. If, however, when he is older, our boy moves to a low altitude and has children, they will not automatically have large or so-called barrel chests just because he (their father) does. Their chests will develop at an altitude-appropriate rate. Similarly, identical twins raised apart, one in Tibet or Peru at a high altitude and one in, say, San Diego, which is more or less at sea level, will end up with different chest phenotypes despite their genotypic equality.

Now certainly the range for plasticity underlying developmental adjustment is itself genetically determined to some degree. Moreover, our genes delimit a certain point past which growth is not possible. Think about the distribution of height that we would see if we lined up all the students in a class. While there will be an average, there also is an upper limit. Similarly, there is a range to chest growth at high altitudes. In a given high-altitude population, some individuals will be born with DNA that supports more expansive chest growth than others. These individuals may be better fitted to the high-altitude environment through this genetic endowment, and

TABLE 3.1 Genetic adaptation versus developmental adjustment

	Genetic adaptation	*Developmental adjustment*
Mechanism or cause	Natural selection: Lower mortality and higher fertility rates in subpopulation with advantageous trait	Exposure: Physical plasticity responsively optimized during development
When realized	Next generation	This generation
Directly heritable	Yes (Innate)	No (Acquired)

so survive to have more children than low-growth-potential individuals. They will pass along their high-growth-potential genes at a greater frequency, thereby altering the gene pool. Yet the accomplishment of extra chest growth, however extensive, is not itself passed on to progeny. It depends on developmental circumstances, not on genes per se (see Supplement Box 3).

Franz Boas famously documented this in 1912 for immigrants who came to the United States from Europe, using the plasticity findings to argue against the racist version of biological determinism predominant at that time (Boas 1912). The accepted view then was that humanity consisted of a handful of 'permanent forms' or unchanging racial types. As Gravlee et al. note, "Boas's immigrant study is significant because it treated this assumption as an empirical matter" (2003, 331), systematically testing rather than simply accepting it. Boas's most important finding concerned head size, at the time considered a stable, permanent aspect of race. Boas's demonstration of small but significant changes in head size for immigrant populations in just one generation shook this faith in a way that "was nothing short of revolutionary" (331).

SUPPLEMENT BOX 3: NOT A TALL TALE

Height is often subject to developmental adjustment. Barry Bogin, who has studied height for many years, reports that while in 1850 Americans were the tallest people in the world (with men averaging 5'6" in height and women 5'1") they now are in third place, with men at an average of 5'8" and women 5'2". The Dutch, whose men were among the shortest in Europe in the 1850s, are now number one, with their men averaging a height of 5'10" and women about 5'5".

Bogin asks, "So what happened? Did all the short Dutch sail over to the United States? Did the Dutch in Europe get an infusion of 'tall genes?' Neither" is his answer (1998, 41). The change had more to do with nurture (the culturally created environment) than nature (genes). Living standards improved in both the United States and the Netherlands, but they improved more, on average, for the Dutch, and in the process height in the Netherlands increased more than height in the United States. Height is, for sure, due in part to genetics, but environmental factors also affect developmental achievement, for example leading our bodies to grow parts larger or smaller as needed. In undernourished children, body growth is reduced by circumstances: when raw materials found in food are in short supply, these children cannot build as many cells as they otherwise would. This results in shorter adult stature—but also better chances for survival than would have ensued if scarce food resources were 'wasted' on growing a tall body.

Bogin's data also show the reverse. Maya refugees to the United States during the 1980s found security in their new country, including abundant food and treated drinking water. After just a short while living here, their children were bigger: first-generation immigrants were 2.2 inches taller than their mothers and fathers. Similar but less dramatic increases were shown for immigrants from Europe in the later 1800s and early 1900s.

Epigenetics

In the past, it was typical to paint the contrast between developmental adjustment and genetic adaptation in black-and-white as distinct phenomenon. However, in recent years, scientists have realized that extensive overlap exists. It is not only that, as noted, the boundary of one's plasticity has a genetic dimension: assumptions about DNA's directive role and its role in separating species have been vastly too simplistic.

Different Species Are Not So Different

Problems with late twentieth century assumptions about DNA actually go back for quite some time. For example, in the 1970s scientists discovered that near to 99 percent of the DNA in humans and chimpanzees is identical. Despite this fact (or perhaps because of it), in 1990 the scientific community began a quest to map every single gene in the human **genome**—the entire stock of genetic information carried by the species. Thirteen years later, the project was complete. Concurrently, other scientists were studying the genomes from other species—and the assumption that more genes mean more complexity was debunked. While the human genome comprises about 21,000 protein-coding genes, *Caenorhabditis elegans,* a tiny and essentially brainless but much studied worm, has about 20,000; moss (*Physcomitrella patens*) has c. 28,000; wheat (*Triticum aestivum*), c. 95,000. Returning to mammals, including say the mouse (with c. 20,000 genes), we find that the genes themselves are, in addition to being similar in number, not far different in kind (Milo and Phillips 2016). How, then, does each species attain such different phenotypic results (Figure 3.1)?

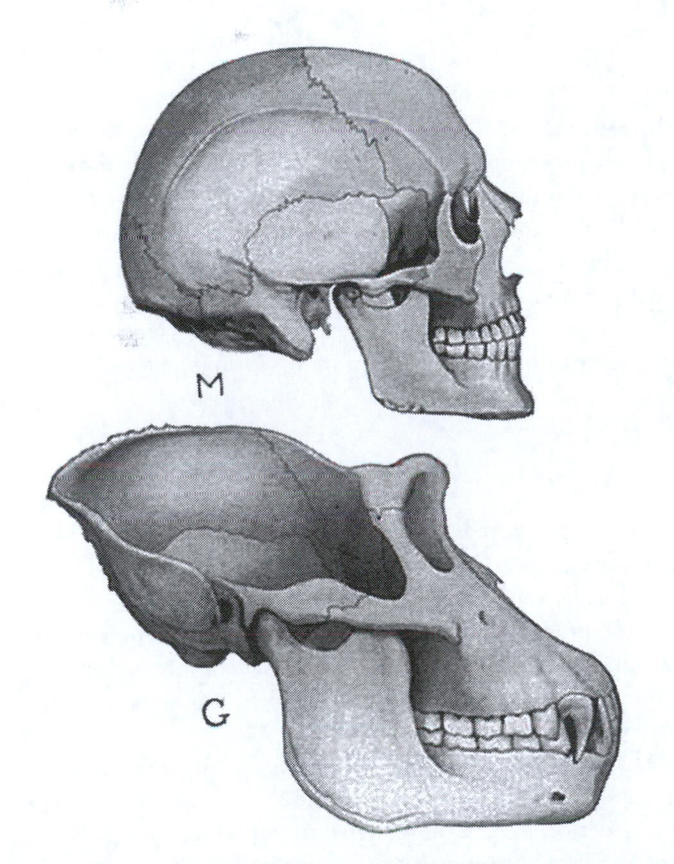

FIGURE 3.1 Side view of human skull (M) and gorilla skull (G). Alfred George Fysh Machin, *The Ascent of Man by Means of Natural Selection* (1925): Longmans Greene and Co, London.

Ridley (2003) helps explain by pointing to the difference between *David Copperfield* by Charles Dickens and J. D. Salinger's *Catcher in the Rye*. Using mostly the same words, Dickens and Salinger have produced two very different books. The difference, says Ridley, lies in the pattern and order of word use. Likewise, gorillas and human beings have and use many of the same genes, but each in a different pattern and order. So, for example, gorilla heads and human heads are visually distinctive because gorillas grow their jaws for a longer time and their craniums for a shorter time than human beings do. Their jaws protrude more than ours do and our craniums are higher (see Figure 3.1). As Ridley says, "The difference is all timing" (34).

Consider human and gorilla skeletal structures as a whole now, and, for added effect, let's throw in the skeleton of a zebra, say. Each species has not only a skull, whose constituent parts grow more, or less, depending on genetic programming, but also a rib cage, four appendages (we call ours arms and legs), a backbone, hips, and so on. The pieces are in some ways quite similar; the difference has to do with where a bone curved in the growth pattern, how long it was allowed to grow, and so forth. The bones in the human tail, for example, stop growing pretty early in comparison to the tail of the zebra. Now bring into our comparison chickens, say, and pythons. Each has thoracic (chest) vertebrae; but the number of vertebrae—the extent to which the thorax is directed or allowed to grow—differs so that in chickens it is limited and in pythons it is extensive; that is, a python's thorax extends all along its snakey body. The illustrations in Figure 3.2 help make the point clear: the same kinds of limb bones follow very different growth patterns depending on which creature owns them.

All Genes Are Not Alike

Bone growth timing (and the timing for growth of other bodily components) is regulated in many ways. Some gene regulation is, literally, genetic; that is, instead of or in addition to coding for proteins, some genes act as regulators, signaling a production schedule to the dedicated protein-coders. Regulators tell protein coders to turn on or off, for example.

Broad acceptance of this fact has been a long time coming. The original definition of DNA as a blueprint directed attention to its protein-coding bits and away from genes or parts of genes that could not be associated directly with proteins. The latter often were cast as mere fillers. However, today's scientists recognize gene regulation as where the real action lies. Scientists also now know that, in addition to receiving signals from other genes, protein-coding genes receive work (and stop-work) orders from beyond the thread of DNA that contains them.

Epigenetic Gene Regulation

The term **epigenetics** was first promoted by Conrad Waddington in the early 1940s. He defined it as the study of interactions between genes and their environment that "bring the phenotype into being" (1942). By the late 1950s, he was using a landscape metaphor to help others understand the systems perspective he promoted. Although the multidimensional "epigenetic landscape" Waddington described was microscopic, his vision resonates exceedingly well otherwise with what we learned about developmental adjustment and how the environment (given or culturally created) affects the phenotypic expression of a genotype.

A newer definition—one that drills down even deeper into the mechanisms involved—defines epigenetics as the domain of science focused on the **epigenome**: the layer of biochemical interactions that surround our genes or, as Cassandra Willyard puts it, decorate them (2017). When we think of this layer of biochemical interactions as enveloping, layered upon, or wrapped around the DNA, and recall that *epi* is Greek for upon or above (the top layer of

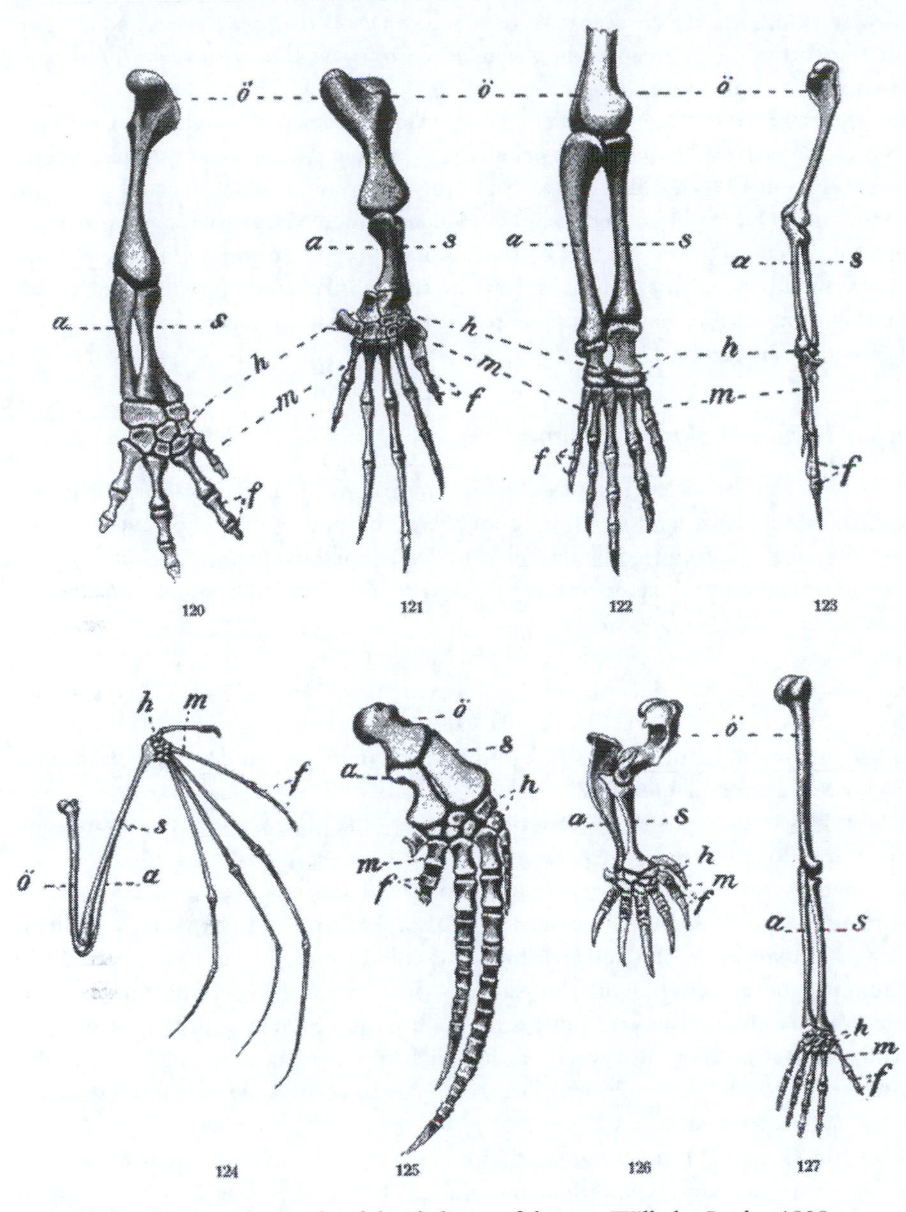

FIGURE 3.2 Comparative study of the skeleton of the arm. Wilhelm Leche, 1909.

skin is the 'epidermis'), the term 'epigenetics' makes good sense. It must be said that its definition in the 1940s, when we knew so little about genes themselves, was not so narrowly focused. Nevertheless, the epigenome and related systems seem crucial to our understanding of diversity.

Genetic expression is regulated when **epigenetic events** turn genes on or off or amplify or dampen their activity. Although the biochemistry behind epigenetic events is beyond the scope of our inquiry, it is worth noting that much of the action (or lack thereof) happens through a process by which chemicals such as methyl groups tag or mark DNA and related structures. Think of these tags like the sticky notes you insert in your textbooks to flag certain passages. The notes lead you to pay more attention to those tagged passages. Likewise, you may use a paper clip to bind together and limit your access to pages a teacher tells you not to read. To push the analogy, if your book is a chromosome or string of DNA, the sticky notes and clips are epigenetic tags, regulating the degree to which you attend to or engage with various sentences or passages—that is, to certain genes.

Beyond directly applied tags, some epigenetic events come into play through **RNA** (ribonucleic acid), DNA's self-generated biochemical helpmate. RNA once was thought to simply fetch and deliver the proteins that DNA called for, so that bodily parts could be constructed according to plans contained in the DNA. However, some RNA takes a more active role in the process of regulating our protein-coding genes so that collecting proteins and building with them either starts, stops, quickens, slows, or just continues. Data on this existed as early as 1974 but we are only now attending to it: until recently most scientists would not question accepted assumptions about RNA passivity (Willyard 2017).

Environmentally Induced Epigenetic Events

Waddington's early definition of epigenetics as gene–environment interaction that leads to a phenotype's emergence and his landscape metaphor were quite perceptive. Epigenetic events and the changes they foster can indeed be triggered by environmental factors, such as food we eat or toxins we get exposed to. The environments in which we live make a direct difference to gene expression.

For example, by fiddling with epigenetic tags on larval bee DNA, Robert Kucharski and colleagues turned generic larvae into queens rather than letting them develop into workers, which they would have done naturally. In real life, only larval bees fed royal jelly in the hive turn into queen bees, as this nutriment naturally affects tags implicated in the queen development process (see Genetic Science Learning Center 2013).

In another now classic example of **epigenetic change**—the phenotypically relevant, epigenetically driven modification of how a gene is expressed—mice prenatally exposed to a particular toxin exuded by certain plastics (bisphenol A or BPA) suffered over-expression of the *agouti* gene—a gene that we humans have a version of also. As Dana Dolinoy, Randy L. Jirtle, and other colleagues involved in this research found, this led to obesity and a propensity for cancer. Some human cancers seem to work this way also: they occur when certain epigenetically affected genes don't turn themselves off in due time. In any case, returning to the mice, over-expression was damped if mothers also were fed diets rich in nutrients such as folic acid. The supplemental diet provided certain molecules that in effect turned back down the *agouti* gene's volume (see Genetic Science Learning Center 2013).

Not all epigenetic change can be reversed at all or completely, and findings from nonhuman animal research do not always have human parallels, but this finding is indeed suggestive. Great promise lies in the possibility of interfering therapeutically with the epigenome's regulating mechanisms. Scientists hope the switches or tags that cause disease can eventually be manipulated by doctors—or through changes in diet and lifestyle that in turn can alter a person's exposure-linked risks.

Other examples of the environment's impact on genetic expression abound. Take, for instance, so-called animal instincts. It is said that a duckling instinctually follows its mother after being hatched. Its instinct, however, if we can still call it that, really is to follow the first moving body it sees during a certain window of time in its early life. In most instances, the body the baby duck sees will be its mother; her image will be imprinted, so to speak, in its mind. However, if the duckling first sees a different moving object, such as a person, it will assume that that person is the one to follow (see Figure 3.3). There is another interesting thing about ducklings—they

FIGURE 3.3 Young whooping cranes complete their first migration following an ultralight aircraft piloted by Operation Migration, www.operationmigration.org. Photo by Tim Ross.

are 'naturally' attracted to the call of other ducks. However, it turns out that this is only because such calls are the first calls that they hear: they hear their own calls while they are still in their eggs. If their vocal cords are cut before they hatch, they will not have a preference. Their 'instincts' thus depend quite clearly on environmental cues (Ridley 2003, 152).

People, too, are affected by early exposures. For instance, infants learn flavor preferences through breast milk and even prenatal exposures (Mennella et al. 2001). Certain early life conditions even can lead to problems in adulthood, such as cardiovascular disease (CVD). In fact, early exposures underwrite a portion of diseases typically thought of as racially determined, such as hypertension, diabetes, stroke, and coronary heart disease. Emerging evidence suggests these all occur at heightened rates among African Americans due not to some spurious genetics of race but rather to epigenetic changes tied to stressors in the socio-economic environment in which race is lived (Kuzawa and Sweet 2009; and see Chapter 9).

The Gene-Gift Hypothesis

There is another side to environmental exposure: 'talent.' Children who live in environments necessitating they run often and far—such as Kenyan children from the Kalenjin tribe—develop into excellent runners. Kalenjin children, as a practical solution to their remote location, run an average of eight to twelve kilometers every day, to and from school, to and from the river to bathe, and so on. They have a long-time tradition of long-distance running related to their cattle economy, as well as a culture that fosters what psychologists call a strong "achievement orientation." In a context where other careers are limited, and running already is common practice, and people grow up with a desire to do well or even better than others, the lucrative lure of the global athletic arena draws many children into training. Indeed, coaches can adopt a sacrificial attitude, pushing young runners to the limits of endurance and promoting the cream of the crop. It should therefore come as no surprise that Kalenjins and other Kenyans do so well in long- and even some middle-distance running competitions (Shenk 2011, 102–105).

Some might like to write this off to genetics only. David Shenk refers to this as the "gene-gift hypothesis." Yet professional athletes themselves know better: a standing in-joke regarding how to hobble the Kenyans involves buying them school buses. (Shenk 2010, 105). This would alter their developmental exposures, affecting their average phenotypes so that the Kalenjins would run more like members of non-running populations.

Jokes aside, the Kenyan success seems to hinge on **deliberate practice**—systematic, focused, practice undertaken purposefully with the conscious aim of improving one's technique, timing, and so on. The same is true for the Jamaicans, although where Kenyans dominate in distance and endurance running Jamaicans are known for sprints.

The 'you can do it' implication (just practice!) may seem like distinctly American cultural claptrap. Testimony provided by superstars regarding their definite dedication to practicing may be written off as anecdotal by those who wish to tie this all to genes. Plus, it is indisputable that coming from a population that, on average, has longer legs, say, will be an advantage in certain sports; being super-tall helps in basketball; large hands benefit swimmers. Nonetheless, in study after study, data indicate that, such things being equal, and particularly given cultural support and good equipment, time spent truly engaged in a given activity (studying included) will pay off.

Just as for duckling imprinting and human food preferences, the epigenetic mechanisms underlying the manifestation of running prowess in Kalenjins (and Jamaicans) have yet to be established. However, as David Shenk has noted in his discussion of running, based on extant science it seems clear that a genetic response, triggered epigenetically, leads to increases in certain structures within and around skeletal muscle cells that affect their stamina, given running's exertions. Likewise, muscle fibers grow in size and strength due to certain kinds of responsive "upregulation." Moreover, muscle fibers can change themselves from 'fast twitch' to 'slow twitch' as required (a higher proportion of slow fibers benefit marathon runners and cyclists; more fast twitch fibers benefit sprinters). While our plasticity does have pre-set limits, our muscles, says Shenk, "are designed to be rebuilt" in response to training (or inactivity): "At any given time, each muscle is adapted to a status quo of activity and exertion" (2011, 310, 312). Noting that one group of star athletes has more fast or slow twitch muscle fibers than another therefore means little; the distribution reflects their hard work and dedication to training rather than being a genetic 'gift' that allows them more wins. Again, some skills (traits, capacities) seen as natural endowments or inborn talents are not simply given but largely earned, through the hard, repetitive labor entailed in deliberate practice.

Categorical Differences

Epigenetic events may be behind most developmental adjustments such as those occurring at high altitudes. Like developmental adjustments, they are not by definition passed on to offspring. Yet, there are some key conceptual differences between the two constructs. For example, some epigenetic changes wrought by epigenetic events may be reversable while developmental adjustments are not. Also, while developmental adjustment always occurs as an aid to survival, some epigenetic change is patently maladaptive; think of cancer here. Further, epigenetic events can occur well after development is over; that is, they can make a phenotypic difference even in mature adults.

Take London taxicab drivers who, at least prior to computer-assisted navigation, had to memorize vast quantities of geographic knowledge to be licensed: there are 25,000 streets intersecting at various angles within six miles of Charing Cross Station alone, and many dead ends. In learning London's layout, drivers changed the shapes of their brains. We know this because

Eleanor Maguire compared experienced cab drivers' brains to those of a non-driver population. The cab drivers' brains had notably larger posterior hippocampi. Brain related variation existed not just between populations, but within the population under study, also: the longer a driver had been in the business, the larger his or her hippocampus was. The latter fact helped Maguire rule out the possibility that people with outsized hippocampi were simply attracted to the cab driving business to begin with (Shenk 2011, 34–35). Further research by Veronique Bohbot and others has shown that dependence on computer-assisted navigation is in fact associated with hippocampus shrinkage; populations that commonly rely on GPS or 'sat nav' may thereby be placing themselves at risk for relatively more cognitive problems later in life than those who still use spatial cues to get around (Raymond 2010).

To summarize: Genes are switched on or off, silenced or promoted, by the environment via the epigenome. In other words, so-called instincts, skills, and certainly our preferences are shaped, this way or that, depending on environmental conditions and what they do to our epigenetic and, thus, our genetic systems. As Ridley notes, genes are "not just units of heredity... they are themselves exquisite mechanisms for translating experience into action" (2003, 275); they are "the servants of experience" (280).

Transgenerational Modifications

The changes so far discussed in this chapter happen in a lifetime. They are developmental and/or epigenetic and not inheritable genetically—at least not in any direct sense. Having said that, maybe we haven't seen inheritance because we've been looking in the wrong place for it, or at the wrong scale.

Long before Darwin and others described the laws of natural selection, many people believed that changes brought on through experience in one generation could be passed to the next. For example, according to this model, if a person works hard to build up an extremely fit and muscular physique, that person's children, too, may be strong and fit. Or giraffes that stretch their necks by reaching upward will have babies with longer necks than will giraffes that do not stretch—so this theory goes. The idea that acquired characteristics can be inherited is called **Lamarckism**, after Jean-Baptiste Lamarck.

When systematically tested, the idea that acquired characteristics can be passed along falls short. For instance, a blacksmith who develops arm muscles through hard work at the forge may have children who end up with strong arms, too—if they also take on blacksmithing (see Figure 3.4). However, the child of the blacksmith who instead goes off to work as a movie director will not. It is true, of course—as we saw for chest growth in higher altitudes—that any inbuilt propensity the blacksmith may have had for greater arm muscle development than normal could have been inherited—but the blacksmith's children's muscle development itself is not a foregone conclusion. The blacksmith's acquired strong arms cannot themselves be inherited.

The Lamarckian theory of inheritance fell flat for more than just a lack of evidence. Its competitor, the theory of natural selection, easily and cleanly explained how characteristics were passed along without anyone having to acquire or work at developing them. Take, for example, the evolution of those long necks in giraffes. According to natural selection, those born with slightly longer necks than others are at an advantage when leaves are high up. Being better fitted to an environment where leaves are high, they simply have more offspring than those that, with shorter necks, cannot reach to eat as well. Although intergroup variation still will exist, given this advantage giraffes in the next generation will be, on average, slightly taller. Active self-development is not a part of a Darwinian selective process, which works on pre-existing variation within a population.

FIGURE 3.4 *Village Blacksmith.* Rudolf Epp, steel engraving after an original painting, c. 1890.

Epigeneticists may help rehabilitate Lamarck's reputation, however. Most epigenetic tags are stripped off when DNA is reset prior to embryonic development, meaning they cannot be passed on. Yet sometimes tags stay—even in the absence of the original environmental trigger. An epigenetic change that remains in a man's grandchild would provide evidence of Lamarck-like epigenetic inheritance, but because paternity can be uncertain it is only when a change gets passed on to a woman's great grandchild in the absence of the initial stimuli that we can be sure that we are seeing a true transgenerational epigenetic modification: her child (as a fetus) and even her grandchildren (as that fetus's germ cells) could have the trait by dint of direct, prenatal exposure in the womb. Her great grandchildren could not, so if it showed up that far down the lineage, we would have to tip hats to Lamarck.

Research on transgenerational epigenetic modification is in its infancy, but scientists have documented some apparently epigenetic transgenerational persistence in mice, fruit flies, and plants. Research in humans is harder; it often depends on proxy measures and seldom are there four generations of data. For example, Marcus Pembrey and colleagues examined historical records from 1890, 1905, and 1920 for an isolated community in northern Sweden. They correlated men's smoking and food supply in mid-childhood (that is, prepubertally) with grandsons' mortality risk (Pembrey et al. 2006; see also Whitelaw 2006).

Vocabulary Shifts

The shift in thinking demanded by conceiving of epigenetic inheritance or even of mere epigenetic change is leading to a new way of talking about human variation. For a start, given our expanding understanding of epigenetics some thinkers have wondered about replacing the term 'developmental adjustment' with 'epigenetic adjustment'—but this assumes that all developmental adjustment is epigenetically controlled, and we don't yet have proof of that. It also could be confusing in that some epigenetic changes are reversible or maladaptive while developmental adjustments, by definition, are not. Likewise, some epigenetic change happens later in life. Therefore, we must use the phrase 'epigenetic adjustment' carefully, for instance in reference to specific developmental adjustments that we know are epigenetically driven—not in reference to developmental adjustment or change in general.

Complementing the broad category of epigenetic change, some have added another to describe the possibility of epigenetic inheritance: transgenerational epigenetic change (see Table 3.2). It would be convenient to twin transgenerational epigenetic change with genetic adaptation or even to label it as 'epigenetic adaptation'; only, as with single-generation epigenetic change, many of its effects may be maladaptive in the long run. That is, transgenerational epigenetic change is not necessarily beneficial to fitness, particularly if the environment has changed. For instance, being able to survive on less energy is helpful in a time of famine (when food supply is low). However, if the trait allowing it is transgenerationally, epigenetically inherited, and if by the time that happens the famine is over and food has become abundant, this may lead to over-storing fat; that is, it could lead to obesity and related ills (see Chapter 12).

We still have much to learn about how epigenetic change works, and how and under what conditions it persists over time. The epigenetics chapter that all next-generation human evolutionary biology textbooks will include has yet to be definitively written.

TABLE 3.2 Genetic adaptation, epigenetic change, developmental adjustment, and transgenerational epigenetic change

	Mechanism or cause	*When realized*	*Innate or acquired*	*Always adaptive?*
Genetic adaptation	Natural selection: Lower mortality and higher fertility rates in subpopulation with advantageous trait	Next generation	Innate	Yes
Epigenetic change	Phenotypically relevant modification of gene expression caused by environmentally generated biochemical tagging of DNA (may be reversible)	This generation	Acquired	No
Developmental adjustment	Exposure: Physical plasticity responsively optimized during development, sometimes via epigenetic mechanisms (see 'Epigenetic change')	This generation	Acquired	Yes
Transgenerational epigenetic change	Epigenetic change that resists stripping during embryonic development and persists into the fourth generation in the absence of its initial stimulus	Great grandchildren (i.e., children not exposed via the womb)	Innate but on, not in, DNA	No

Paradigm Shifts

A **paradigm** is a theoretical framework that guides scientists in deciding what questions to ask and forms a lens through which scientists interpret data. Questions regarding Lamarck's repudiation notwithstanding, the exciting discoveries in epigenetics have heralded a radical change in scientific thinking—a **paradigm shift** regarding the relationships between species as well as concerning developmental change. Further, the debate over whether our genes or our environment contribute more to human biocultural diversity has shifted to emphasize discussion of the mechanisms by which genome, epigenome, and environment interact.

This is not the first time science has experienced such a transition. Some classic examples of changes in scientific viewpoint include when science came to accept that the earth is not flat, the sun does not revolve around the earth, and the earth is much, much older than we originally thought it to be. The revolution begun when scientists proposed that the origin of speciation lies in a process called natural selection and that humans and chimpanzees as well as mollusks and fruit flies and even peach trees all share a common ancestor provides another good example of a jarring transition. All these ideas were incredibly revolutionary in their times even though today they are axiomatic. Although data to support them were there, these ideas could not be accepted until a critical tipping point was reached.

Thomas Kuhn, a physicist and historian of science, described the radical nature of such transitions in *The Structure of Scientific Revolutions* (1996 [1962]). Prior to every revolutionary transition—every paradigm shift—scientific practice is what Kuhn called "normal." What he meant was that it is unremarkable, driven by a shared, paradigmatic theory about the world. During **normal science**, most scientists in the community dismiss any contrary results or findings as due to faulty measurement or some kind of mistake in the analysis. A piece of dust must have been in the equipment, or the calculations indicating that, say, the earth is round, must have been in error. Investigator bias, too, may be suggested. Aberrant results are explained away or even ignored so as not to interfere with the daily work of science, and so it goes, until there are just too many anomalies or contrary results to rationalize away. Then, crisis ensues. Core beliefs are shaken as assumptions about the world reveal themselves as wrong.

So, science is culturally conservative. Yet, when well-enough challenged, paradigmatic or standard views and normal scientific practices do change. Such change occurs radically and abruptly rather than gently and gradually. In other words, science does not progress at a gradual pace but moves forward in bursts as the growing unrest regarding contradictory data spurs what Kuhn terms a "scientific revolution." The old paradigm then is relatively quickly overthrown, and a new paradigm takes its place. Disciplinary purviews are redefined; sometimes whole new disciplines or sub-disciplines emerge. Soon, however, things settle down and another phase of normal science (and related conservative thinking) ensues. Researchers carry on, now seeing the world through the new theoretical paradigm or lens. This is what recently happened when epigenetics arose. We now take human biocultural diversity to be a much more fluid and contingent product than scientists previously thought.

This is an advance, but it still bears noting that all theoretical paradigms bias research, no matter how much better than previous paradigms they may be. In his book, Kuhn refers to the widely known duck–rabbit image to convey this. Regarding the image (see Figure 3.5), Kuhn argues that, in a nutshell, if one has been told (or takes up the theory) that it is a duck, then a duck is what one will see. Those expecting to see a duck ask duck-type questions; they may seek to measure bill size, or buoyancy, or feather softness. Conversely, those who approach the image with the theory that it is a rabbit ask questions relating to fur thickness, ear length, or maybe breeding habits. Duck-theory people fail to attend to the image's rabbitness, and rabbit-theory

FIGURE 3.5 Duck-rabbit. J. Jastrow, 1899. "The mind's eye." *Popular Science Monthly*, 54: 299–312.

people do not see its duckness; all will fail to ask questions that fall outside of the box for them. Therefore, although people holding either theory in exclusion may do some excellent research into either ducks or rabbits, as the case may be, they will never be able to truly and fully accommodate the data. By highlighting the duck–rabbit figure, Kuhn helped readers to have the experience of shifting paradigms and to appreciate the degree to which paradigms, for better and worse, set our agendas.

This is not to say that Kuhn dismissed the notion of progress through science. Clearly, science does advance and discoveries are additive and even multiplicative. Who reading this book would not prefer today's treatments for, say, smallpox or cholera over the treatments available in the 1800s? Kuhn is not describing the simple replacement of one set of biases for another. Rather, he is pointing to the conservative disposition of science in general. The willing and sometimes automatic adherence of scientists to shared paradigms can muddy our view of the laws of nature, including those relating to heredity and evolution, and to human variation. Science is, after all, a cultural product.

What Makes a Square a Square?

We began this chapter asking how genotypes are translated into phenotypes. Sure, the proteins that genes code for are built and assembled into body parts or process-supporting biochemicals. However, like baking a soufflé, turning a recipe into a body is a complicated and contingent act. As we have now seen, the phenotype is in fact the result of the concerted interplay of genotype with epigenome and environment. Put into an equation: phenotype = (genotype + epigenome) × environment.

Another relevant puzzle—and one that is implicated in the entire content of this book—is whether nature or nurture contributes more to who we are, or to the ways in which populations bioculturally differ. However, that question leads to a dead end. Just as it is illogical to ask which contributes more to a square's area, the width or the height, so, too, it is nonsensical to pit nature against nurture or vice versa and search for a victor. Like the square's height and width, nature and nurture work together to produce the wondrous diversity we see in humanity. Nature (biology, genetics) *and* nurture (culture, the environment in which we are conceived and raised) both work together to shape us, and we can see the result in both individual- and population-level biocultural diversity.

4

EMERGENCE OF CULTURE AND PEOPLE LIKE US

This chapter prepares you to:

- Define culture with reference to its key features and adaptive function
- Define and discuss cultural relativism (in contrast to ethnocentrism), using examples
- Describe culture's evolutionary emergence with reference to the synergistic interaction of our evolving capacities for cooking, cooperation, and symbolic (e.g., ritual) communication
- Delimit key archaeological indicators of the Cultural or Paleolithic Revolution

I have been talking about adaptation as a bio-genetic process. In this chapter, I want to discuss another option humans have for adaptation: culture. Culture helps us to adapt to various geographically distinct environments by providing us with a means to invent and then pass on customs pertaining to survival-related needs such as for food, clothing, and shelter. Moreover, culture helps communities to stick together or cohere. It gives life meaning.

In this chapter we first define culture and then review how humans evolved into culture-bearing beings. This was no unilineal progression but rather the result of multiple, manifold, intertwining lines of positive feedback. Key in the process was the emergence of our ability to conceptualize and act on our existence as 'we' and, subsequently, of our ability to sustain this feeling, and our commitment to it, meaningfully.

Culture's Adaptive Function

Humanity's original food-getting or subsistence strategies as well our sheltering habits were geographically determined. We had only the resources of our habitats to exploit. The environmental conditions these habitats entailed determined our survival needs. Once we had evolved the capacity for culture (a process we'll get to later), geography also influenced our immediate cultural aims. Where fish lived, we might develop techniques for fishing. Where it rained, shelters would need waterproofing. Where arctic winds blew, we would need warm clothes, and so on. The ability to adapt using culture helped support our spread from Africa, where the human species was likely born and our capacity for culture likely evolved, to populate most of the earth's land mass (see also Chapter 5).

Culture's Features

Commonly, when we characterize culture in introductory anthropology courses, Edward B. Tylor's 1871 definition is invoked: "Culture … is that complex whole which includes knowledge, belief, art, morals, law, custom, and any other capabilities and habits acquired by man as a member of society" (for example, as cited by Langness 2005, 25). This definition, among the first, still works because it highlights three key features of culture: it is learned or acquired, shared socially, and sticky or habitual. Indeed, learned, shared culture becomes so deeply internalized as habit that we often forget it is there, and in fact we may never 'see' it until we knock up against someone with another culture—someone with learned, shared habits that differ from our group's (see Supplement Box 4).

This brings up a fourth point implicit in a part of Tylor's definition not shown above: culture is (4) self-centered. Our deep internalization of our culture leads to 'ethnocentrism,' defined in Chapter 1 as using our own cultural standards to judge other peoples' practices.

Tylor in fact referred to "civilization" in part of the passage that I have not included, positioning his own culture at the top of a progressive evolutionary ladder. This was quite common in Tylor's day, as cultures were thought to all pass through a particular progression of evolutionary stages (for example, from 'savage' to 'barbarian') until they reached the pinnacle of enlightened development—normally and ethnocentrically benchmarked by the anthropologist's culture.

SUPPLEMENT BOX 4: MAKING CULTURE VISIBLE

Culture's habitual nature means that we often forget we have it until someone raised otherwise says or does something to highlight that fact—such as taking the proverbial Martian's (or an anthropologist's) perspective. This occurs in the following excerpt from Horace Miner's infamous 1956 essay on Nacirema body ritual:[1]

> Nacirema culture is characterized by a highly developed market economy which has evolved in a rich natural habitat. While much of the people's time is devoted to economic pursuits, a large part of the fruits of these labors and a considerable portion of the day are spent in ritual activity. The focus of this activity is the human body, the appearance and health of which loom as a dominant concern in the ethos of the people. While such a concern is certainly not unusual, its ceremonial aspects and associated philosophy are unique.
>
> The fundamental belief underlying the whole system appears to be that the human body is ugly and that its natural tendency is to debility and disease. Incarcerated in such a body, man's only hope is to avert these characteristics through the use of ritual and ceremony. Every household has one or more shrines devoted to this purpose. The more powerful individuals in the society have several shrines in their houses and, in fact, the opulence of a house is often referred to in terms of the number of such ritual centers it possesses. Most houses are of wattle and daub construction, but the shrine rooms of the more wealthy are walled with stone. Poorer families imitate the rich by applying pottery plaques to their shrine walls.
>
> While each family has at least one such shrine, the rituals associated with it are not family ceremonies but are private and secret. The rites are normally only discussed with children, and then only during the period when they are being initiated into these mysteries… .
>
> The focal point of the shrine is a box or chest which is built into the wall. In this chest are kept the many charms and magical potions without which no native believes he could live. These preparations are secured from a variety of specialized practitioners. The most

powerful of these are the medicine men, whose assistance must be rewarded with substantial gifts. However, the medicine men do not provide the curative potions for their clients, but decide what the ingredients should be and then write them down in an ancient and secret language. This writing is understood only by the medicine men and by the herbalists who, for another gift, provide the required charm.

The charm is not disposed of after it has served its purpose but is placed in the charm-box of the household shrine. As these magical materials are specific for certain ills, and the real or imagined maladies of the people are many, the charm-box is usually full to overflowing. The magical packets are so numerous that people forget what their purposes were and fear to use them again. While the natives are very vague on this point, we can only assume that the idea in retaining all the old magical materials is that their presence in the charm-box, before which the body rituals are conducted, will in some way protect the worshiper.

As anthropology took hold in the United States, anthropologists there began to see that different societies had different histories, and that their cultures were therefore unique and particular. The technical term for this viewpoint is **historical particularism** and it is most associated with Franz Boas, one of America's anthropological pioneers and, not coincidentally, an immigrant marked by his own ethnic differences from the mainstream US Anglo elite.

Boas was appalled by the misuse of science to support racism, which was essentially what most of the evolutionary ladders ranking other peoples' cultures were promoting by the end of the nineteenth century. Boas counterattacked, arguing (and rightly so, with loads of evidence) that environment influenced bodily form, that there was no such thing as a 'pure' race, and that 'racial mixing' was not harmful. His cultural research directed attention away from **unilineal** (one-way) evolutionary thinking that placed the ruling culture at the top, toward a paradigm that allowed for many lines of cultural evolution. Overall, Boas's work highlighted the need for **relativism**, which evaluates the ideas and practices enacted by members of a given culture by that culture's own standards. They are understood neither as they rank on a baseless developmental scale nor from the ethnocentric viewpoint of the foreigner–theorist, but rather in their own context.

The paradigm shift in anthropology promoted by Boas's work is clearly seen in the changes museum curators made in response to his insights. Prior to Boas, artifacts were generally arranged, no matter where from, by category. Categorically similar objects were grouped together in a bogus unlinear sequence running from 'savage' to 'civilized.' Such an arrangement showcased different approaches that humans might have to the same type of adaptive challenge (meat procurement, say, which might take place using bows and arrows, or high-velocity computer-assisted rifles). Yet organizing artifacts out of cultural context detracted from museum-goers' (let alone scholars') capacity to develop a holistic understanding of a given artifact's place in the cultural world from which it was wrested.

An Inuit basket, for instance, could be displayed with baskets from around the world. It could be placed in between those that are more and less complexly or intensively crafted, for instance; and those could be arranged between baskets likewise, resulting in a continuum. Curators could suggest that over time each culture's baskets might evolve toward one end of the line implied in the way the baskets were organized.

However, displayed in the context of other Inuit artifacts instead, such as those used likewise in food-gathering and preparation, the Inuit basket—and the artifacts now around it—become culturally meaningful. Museum goers now know more about the basket's purpose; they can better imagine what it might be like to use it. When shown on its own or with other baskets

FIGURE 4.1 Franz Boas, modeling for an exhibit as "Hamats'a Coming Out of Secret Room" in a Kwakiutl Indian ceremony. Photo courtesy of National Anthropological Archives, Smithsonian Institution, Negative MNH 8304.

only, that remains uncertain. Plus, the future of the basket is foreclosed; a one-way (and false) evolutionary trajectory is asserted.

A more open-ended, historically sensitive vision of cultural evolution coupled with context-sensitive relativism was what Boas demanded. Regarding the latter, he even modeled himself how certain artifacts were to be used or worn by display dummies so that they might be seen by museum-goers not only in context but also as if in use or action (see Figure 4.1); he knew what that would look like from his ethnographic fieldwork. Such museum display tactics allow the viewer better entry, however virtual, into other cultural worlds. They allow us to appreciate the historically particular nature of each culture, or the indeterminate evolutionary lines along which the shape of material objects—and ideas—might move.

The Emergence of Our Species

We have talked about the adaptive nature of culture, its key features, and the importance of context. We still need to talk about when and how culture came into being. Culture is an evolved capacity—one that became possible only after many interrelated genetically underwritten changes had built up. These changes made our ancestors better able, in brain and body, to make things, communicate using language, cooperate in reaching intersubjectively shared goals and, eventually, identify as a member of a group.

Before we talk about how these four capacities and their entailments underwrote culture's emergence, we should quickly review the emergence of the broader animal family—or actually, 'genus'—into which we modern humans fit. All living things can be grouped into higher and larger classifications until the point at which they have been subsumed by the highest, largest, most coarse-grained classification of all: technically, the 'empire' or 'domain.' Within the empire or domain we have 'kingdoms'; below those we have a series of nested subgroups including the 'phylum,' 'class,' 'order,' 'family,' 'genus,' and 'species.' The subgroups denote finer and finer distinctions. For example, we humans belong to the mammalian subset of the animal kingdom, which includes all animals with milk-producing or mammary glands and hair or fur. Within this classification, mammals with five-digit grasping hands and/or feet—lemurs, monkeys, and ape—fit into the order Primates. The Hominidae family of this order includes humans but also gorillas, chimpanzees, and orangutans (i.e., all great apes—and their ancestors).

Such inclusiveness is a relatively new thing: formerly, Hominidae comprised only humans—gorillas and such were assigned to their own family, Pongidae. However, advances in evolutionary biology led to the collapse of the distinction. This paradigmatic shift in the scientific perspective had cultural parallels: human self-importance was giving way to a more environmentally friendly view, prioritizing connectivity (recall also the rise of One Health, described in Chapter 1).

Nonetheless, at some point focus must be narrowed, and the Hominin subgroup includes only humans and their direct ancestors. The Hominin subgroup includes the genus *Homo*—literally, 'man' (a **genus** is a set of species sharing particular biological features that are even more narrowly defined than in higher-order classifications such as family, order, or kingdom). *Homo* is the broadest grouping with which we shall here be concerned.

Intertwining Lines

Homo first appears in the fossil record just over 2 million years ago. This genus's defining features are notably larger brains and fully opposable thumbs, making a precision grip possible. Beyond that, *Homo* varied, for instance in facial format. As Susan Antón puts it, "This leaves the impression that the emergence of our genus occurred during a time of environmental opportunity, resulting in a kind of species experimentation, the emergence of any one lineage of which might relate as much to luck as to fitness" (2018, 2).

Accordingly, the evolutionary path taken was not unilinear but multistranded. Further, it had various resplices as well as splits, with the latter occurring at least frequently enough to put the brake on species divergence via **gene flow** (see Chapter 2). Also through gene flow, adaptive traits could spread to groups without the receiving groups having had to wait on natural selection. In this sense gene flow can provide an evolutionary short cut. Of course neutral traits and disadvantageous ones could spread through gene flow as well. Either way, groups would branch off, evolve apart, and then their descendants would reunify with other descendants of their forebearers. This is seen, for instance, in the recently discovered comingling between modern humans and other varieties of *Homo*. Not least here were the Neanderthals, a European branch, often caricatured wrongly and unfairly as dumb cave-dwellers (Mooallem 2017).

The discovery of evidence that interbreeding happened shocked many, not least because Neanderthals had been seen by most as a distinct species, *Homo neanderthalensis* (in scientific writing, a species name is listed after the genus). The discovery that Neanderthal DNA persists in modern humans raised questions regarding that distinction—and the correlate truism that species absolutely cannot interbreed.

Although it may seem frustrating that even the category 'species' is fuzzy at its edges, the lessons here are clear. As with the addition of gorillas to the Hominidae family, this change in understanding reminds us that, however useful, classificatory boundary lines are impositions on the complex continuum of life. More concretely, the kind of genetic divergence necessary to completely bar interbreeding—and gene flow—between two subpopulations is generally a major, messy, gradual, and complex process. Science is a complex process too—one in which past answers can be refined and even overturned paradigmatically when enough new empirical evidence amasses.

Tools, Talking, Walking

If you will keep in mind the ideas shared earlier regarding gene flow and fuzzy species boundaries, yet allow me to speak categorically for a bit, I can tell the story of modern humanity's emergence fairly quickly. *Homo's* lineage likely began with ***Homo habilis,*** the first archaic human species known to systematically make and use stone tools. Although *Homo habilis* translates parochially to handy man these tools were just rock hand-axes for smashing or crushing or scraping. They were tools nonetheless. Now this is not to say there were never tools before; however, as Antón has observed, at this point tool sites become notably more abundant, and raw materials now being used were brought in from notably farther away (10 km at some sites; 2018, 4).

Homo habilis certainly made less invasive tool-related modifications to its environment than we do today. Nevertheless, use of tools intensified the species' niche construction (niche shaping) activities. It also supported a trend that would slowly gain speed: brain expansion. *Homo habilis* had a larger brain—relatively speaking—than any similar species at the time or prior. This provided a selective advantage: even a slight enhancement in brain power would come in handy in fending off predators, one of the major environmental pressures faced by *Homo*—and indeed all primates—at that time (Fuentes 2009).

Some brain enhancements selected for in the time of *Homo habilis* may have supported better communication. The capacity for language leaves little archaeological trace, but an experimental research study by Thomas Morgan (see Balter 2015) suggests that it emerged roughly (circa, or c.) 2 million years ago. Those in the study who were taught how to make stone tools with gestures had double and those taught with words had quadruple the success of those only allowed to look at samples.

As Walter Goldschmidt (2006) puts it, our capacity for making tools and our capacity for talking likely evolved hand in hand. Both involve classifying things, connecting them (concretely or mentally), and dexterity. To make a tool, you must have in mind ideas about what types of materials might be best, and how to modify them to gain the most effective tool action. More complicated tools may require you to prepare materials in some way before tool construction. For a multipart tool, you must consider how parts can be connected.

Further, making and using tools can require fine motor coordination and dexterous fingers and hands. Recent discoveries regarding the brain suggest that the manual dexterity necessary may be intimately linked not only with the development of gesture language but also (although it took longer, evolutionarily speaking) with the development of the parts of the brain implicated in the manipulations of lips, teeth, breath, and tongue that are necessary for speech. I'll circle back later to discuss other aspects of communication's evolution; for now let's just note that the 'connectionist' vision of the brain works well not only with the theory that making tools and communicating evolved in tandem but also with the larger picture of interconnected feedback loops painted by current research into most features of our world.

The next notable addition to the *Homo* genus came 50 or so million years later, when ***Homo erectus/ergaster*** entered the scene (*Homo ergaster* used to be thought distinct but now is considered a key type of *Homo erectus*). This species really optimized its bipedalism. Earlier *Homo* species could walk; even nonhuman primates are known to stand on two feet sometimes—for instance, when gathering food. However, this new species walked all the time, habitually, and more effectively. Altered body proportions reflect this. *Homo erectus/ergaster* was taller and had longer legs and shorter arms than other *Homo* species. Due no doubt in part to its bipedal efficiency, this species is the first found in locations beyond Africa—a topic we return to in Chapter 5.

In addition to its more truly upright posture, *Homo erectus/ergaster* also had a more protrusive nose. This may have been selected for in support of the kind of breathing that more active bipedal locomotion, or running, requires. Running in turn may have provided an advantage in hunting, allowing individuals to chase after prey that might otherwise escape. Also, it could be an advantage in the race to avoid being chased down and eaten oneself (Fuentes 2009).

As the threat from predation declined, another selective pressure was on the rise: babies became a bigger burden. *Homo erectus/ergaster* had an even bigger brain—and so a bigger skull top or cranium—than *Homo habilis*. Babies now had to be born less mature or their heads would get stuck, killing them and their mothers. This meant that infants, once born, took somewhat longer to become independent enough to survive on their own.

The energy costs reproductively mature females bore in carrying and feeding their now-more-dependent offspring must have been offset by a turn, however slight, toward more food sharing and group food-gathering efforts. This would have entailed more cooperation, and cooperation on a different order than is seen in other primates, but it is important not to project back in time today's human qualities. Cooperation at this stage was still unlikely to entail the sense of mutuality that we experience when cooperating today (Tomasello 2011)—which I'll come to soon.

Fire and Food

Bigger brains and taller bodies cannot happen without more food overall. Diet evolved, and tools became somewhat more sophisticated in concert. Not to belabor the point, but one adaptation did not follow the other in simple linear fashion; numerous connections within and between systems were implicated. The adaptations that contributed to *Homo's* evolution linked up with each other through multiple, multidirectional feedback loops. Locomotion, breathing systems, brain size, tool-making skills, control of fire, the addition of cooked foods to the diet, and so on all changed in self-reinforcing—and sometimes self-limiting—ways.

For example, changes in the diet that supported advantageous changes in brain size may in turn also have supported intestinal shortening and tooth size reduction. These changes may have been largely due to the advent of cooking, which necessitated first learning how to capture or control fire (starting fire came later).

What made expending the effort it took to control fire for cooking advantageous to *Homo erectus/ergaster*? For one thing, the energy trade-off. As Richard Wrangham has argued, prior to cooking our ancestors would have had to devote half their waking hours simply to chewing. Digesting took lots of energy too. Cooking frees up four to five hours for other pursuits, which now could be undertaken while chatting around a shared campfire. Cooking can thus be a very pro-social affair. Moreover, cooking decreases the physical effort and so the metabolic cost of eating. This in turn means more food energy can be used in other ways, such as to fuel the brain or make neurons (the brain takes twenty-two times more energy than skeletal muscle to maintain; Gorman 2008).

Some assume meat was our main go-to for cooking. It's true that raw meat is very difficult to chew let alone digest; and cooking does soften foods and make nutrients more available. But so does chopping or slicing it finely (Zink and Lieberman 2016). Regardless, a diet of more than one-third protein would produce ammonia toxicity. This problem as well as existing ethnographic and archaeological evidence leads Wrangham and others to argue that cooked tubers—ancestral yams for instance—were likely the centerpiece of emerging *Homo's* campfire cuisine (Wong 2013; and see Wrangham 2009).

Prior to cooking (and perhaps fine chopping), individuals with smaller teeth or shorter intestines would have died off due to an inability to get the nutrition they needed from the then-available, rather tough foodstuffs. Now, however, in a niche constructed to include among the first prepared food in the diet, small-toothed, short-intestined individuals could survive. Genes with the potential for expression as smaller teeth or shorter innards would therefore be passed along into the next generation's gene pool at a higher rate than before. Concurrently, people with larger teeth and longer guts would no longer have a selective advantage. Moreover, a skull less pressed to grow forward or maintain robustness to support super-strong chewing muscles would have more freedom to grow back and up and with thinner walls; that is, a more expansive brain case would not have been a disadvantage.

Interconnected changes like these, and others, accrued as time passed. Various new *Homo* species emerged, including archaic forms of bigger-brained ***Homo sapiens***, whose earliest or most archaic fossils date back to about 300,000 years ago. The subspecies ***Homo sapiens sapiens*** (modern humans, the group we fit into) becomes prevalent starting maybe 100,000 years ago. The line between us and others in the *Homo sapiens* species is fuzzy but, anatomically, *Homo sapiens sapiens* can be distinguished by our highly domed skulls and vertical foreheads. Our noses project more than those of other *Homo sapiens,* and we have bony chins. Also, we have the smallest teeth and jaws, and the lightest skeletons.

Sapiens is Latin for wise or knowing. From its emergence, this new, bigger-brained species of human was smart in a way that others in the genus *Homo* hadn't been. *Homo sapiens sapiens* was even smarter still: the brains of this species were—are—different from those of other *Homo sapiens*, not so much in terms of size, but in terms of what they could and can do. *Homo sapiens sapiens*—modern *Homo sapiens*—were primed, through a rewiring or reorganization of the brain rather than simple growth in its size, to learn. They were primed to carry and enact culture. They were ready to imbue the world with meaning and to talk about it; they were ready to experience, subjectively, a state of being 'we.' When this happened, we went from being simply anatomically modern to being fully or behaviorally modern.

Part of being behaviorally modern entails taking risks and seeking novelty. The relative intensity of these traits is to some degree brain-based. We know this because of research into the constellation of behaviors classed now as attention deficit disorder (ADD). Impetuousness may cause problems in school classrooms today but it also can lead people to important discoveries, and this would have been a boon to our ancestors as our kind emerged. Without anyone fearless or impulsive enough to grab the unlit end of a burning stick or to jump on a floating log to see what might happen we might not have gained control of fire or invented rafts and then ships, for instance. Sure, then as today, some individuals might have died or suffered due to risk-taking that did not pan out. However, just as the group benefits from risks successfully navigated by the minority, the group also benefits from the information that unsuccessful risk-taking produces. In this way, some of the traits associated with ADD—a penchant for novelty seeking, a high capacity for acting on impulse—could have been naturally selected for (Williams and Taylor 2006). Some of the traits associated with autism may similarly have provided a selective advantage of some kind. We are only now realizing the potential benefits to the species of neurodiversity; further research in this area promises to be illuminating.

In summary, rather than a single switch effecting the ancestral changeover into 'modern' humans, many small changes working together synergistically accrued. With this to be kept always in mind, Table 4.1 provides a summary of important steps in the process.

Communication Becomes Necessary

As *Homo sapiens* began springing up in the fossil record, the archaeological record also changed in several ways. By this point in our evolutionary history, some groups may have cared for the aged and infirm, and even buried their dead (or some of them) with intent. That they did so

TABLE 4.1 Becoming human: timeline for culture's emergence[2]

Species name	*Date species appears in fossil record (and new, distinguishing skeletal features)*	*Behavioral developments evidenced in fossil record along with this species*	*Correlate practices or capacities hypothesized to have emerged during this species' time*
Homo habilis (handy human)	c. 2 million years ago (Notably larger brains)	• Systematic tool fabrication and use	• Use of simple language • Rudimentary resource-sharing
Homo erectus/ergaster (upright human)	c. 1.5 million years ago (Full bipedalism, with concordantly shorter arms, more protrusive noses, longer legs, and modified pelvis [babies born facing backward]; smaller teeth and jaws)	• Controlled use of fire for cooking • Social gathering at campfires (hearths)	• Empathy (intensified allocare)
Archaic *Homo sapiens* (Archaic wise or knowing human)	c. 300,000 years ago ('Human-looking,' anatomically: lighter skeleton, reduced brow ridge, higher, more rounded brain case)	• Burial of (some of) the dead • Use of pigment • Long-distance resource exchange	• Skilled in 'engineer' language; some social use of language also • True cooperation (supported by Theory of Mind, Shared Intentionality)
Modern *Homo sapiens* (Modern wise or knowing human)	c. 100,000 years ago; full capacity for became evident c.75,000 years ago (bonier chin, sharper nose; globe-shaped brain case)	• Symbolic art (cave paintings, portable figurines) • Multipart tools (e.g., bow and arrow) • Clothing and self-adornment	• Culture (supported through systematic use of 'ritual' language)

is significant: the capacity for social bonding that such burials reflect is necessary for culture to work. As noted earlier, culture provides humans with a way to adapt to various geographic habitats without waiting for genetic adaptation to kick in; but it also makes meaningful social life and real teamwork possible. In the titular words of Walter Goldschmidt, culture's emergence provided our "bridge to humanity" (2006). Person-to-person or intergroup connections helped our evolving ancestors across.

Language provides one way to connect. Through language, humans can inform other humans of things that they think might be helpful. In other words, humans can share information for the other person's benefit. Such information sharing would have been rudimentary in our early *Homo* ancestors; it is virtually absent in nonhuman primates today.

Great apes, for instance, have trouble understanding that when a human points a finger at something the act is meant to convey information to them (such as "There! There is some food that has been hidden!"). As Michael Tomasello has shown, they do not have the ability to identify with let alone to trust the pointing troop member (Tomasello 2011). This may seem surprising given that the majority of primates live in groups. Remember, however, that among our ancestral species group life was pragmatic (it remains so among nonhuman primates today). Member behaviors were not cooperative in today's sense; they were individualistic and even exploitative (Tomasello 2011, 5). Early *Homo* groups did travel and eat together. Nevertheless, as remains mostly true for today's nonhuman primates, this was basically just to help each other avoid being taken down by predators: there is safety in numbers.

Modern humans are different. Our ability to take in, and desire to give, altruistically helpful communication is keen. It has to be: recall that our big brains mean babies are born much more immaturely than ever before. While nonhuman primates (and other creatures) move from infantile dependency to the so-called 'juvenile' stage without a pause, modern humans have childhood. Childhood per se is unique to humans, whose brains have so much to accomplish in becoming mature. This includes more than just physiological growth: we can no longer maximize our environment without being told or shown how to do it. As part and parcel of having childhood, we are primed to teach and learn.

Inter-Subjective Mental Connections

Sometimes, when we observe someone doing something that we would imitate or that we can see ourselves doing—say, picking up a particular shell and rubbing it just so, to create a hole through which to thread it—our own brains respond as if we already are engaged in practicing that action, too, although we only are watching. That is, some of the neurons or brain cells that control specific actions can fire if we simply watch or even just think about that action. Once discovered, these cells were aptly named 'mirror neurons.'

We need to guard against accepting a facile 'just so' story in which mirror neurons explain all. Nevertheless, neurons with this mirroring capacity seem to be part of one key mechanism for nongenetic information passage (Goldschmidt 2006, 31; see also Ridley 2003, 209–222).

Certainly, nonhuman primates have mirror neurons: these were in fact first described through brain studies done on monkeys that were watching other monkeys, and some humans, bring peanuts to their mouths in Giacomo Rizzolatti's laboratory (Blakeslee 2006). That nonhuman primates can mirror or imitate others overtly, too, is well known (see Figure 4.2). However, human mirroring entails qualitatively and measurably greater sophistication—the kind it takes to enable and support culture.

The emergence in *Homo sapiens* of the capacity for the kind of learning that culture depends on may be related, at least in part, to a boost in the quantity and quality of mirror neurons hundreds of thousands of years ago—a boost that was itself of course the result of feedback coming

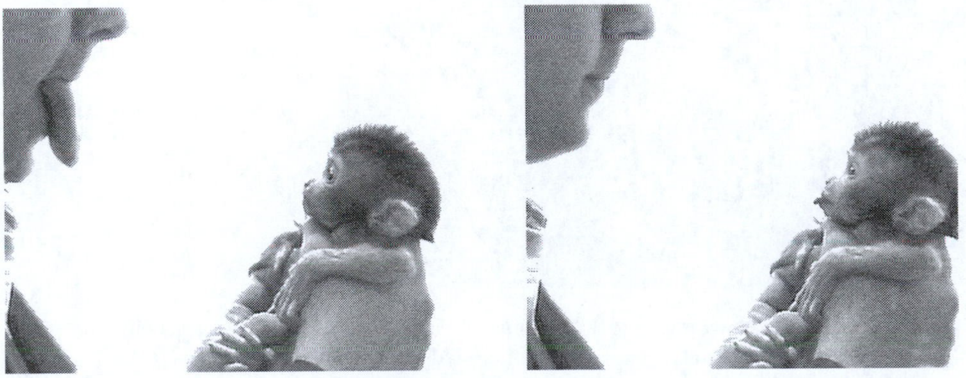

FIGURE 4.2 A newborn macaque imitates tongue protrusion. L. Gross, 2006. "Evolution of Neonatal Imitation." *PLoS Biol* 4(9): e311. doi:10.1371/journal.pbio.0040311.

from a variety of directions related to how our ancestors then lived. No doubt other factors beyond mirror neurons came into play also.

The mechanism notwithstanding, mirroring happens when we learn how to make or throw a spear, or soothe a baby, or weave a basket by watching another member of our group do so. It happens, too, when we watch a film or play that moves us, or go to a favorite team's sporting event. It can happen when we see someone struggle with and nearly drop an armload of packages, or trip while walking across campus. In other words, mirroring is implicated not only in motor learning but also in physical acts of empathy—of placing our own feet in someone else's shoes or of feeling a connection (see Figure 4.3).

Mirroring effectively transfers emotions, or feelings, from person to person. This kind of empathic sharing can help to create and maintain community bonding. We must not underestimate the importance of such bonding: without it we cannot cooperate.

Today humans know that 'two heads are better than one' and 'many hands make work light.' Rather than evolving two brains, however, or two sets of hands, *Homo sapiens* evolved the ability to cooperate very, very well. For example, when gathering food, we may divvy up and even rotate through roles (trail cutter, child carrier, resource locator, food extractor, and so on). On a hunt, some people may climb trees to flush out prey while others act as shooters. Before heading out for plants or animals, groups agree on joint goals. Their work is coordinated and synchronized.

Now is a good time to mention that human eyes today have big, bright white areas (sclerae) surrounding the pupil and iris; nonhuman primate eyes do not. Eyes with big, bright whites—human eyes—are much easier to follow than the eyes of any other organism. We do not know when our ancestors got their big whites, but we do know that, given how natural selection works, highly visible whites could be selected for only if they did not disadvantage the bearer. That might happen if an animal's big whites allowed others to spot food it was eyeing and to grab that food before the first animal has a chance. At a high enough frequency, such exploitation could lead the first animal (the one with the visible sclerae) to starve and die prior to

FIGURE 4.3 Marion Koopman, right, and US Navy sailor Samuel E. Adolt mirror each other; they and Margaret Williams, center, are on board the Fletcher-class destroyer USS Uhlmann (DD-687) at Terminal Island, California. Photo: Los Angeles Times, 1950; UCLA digital collections photo ID: uclalat_1429_b104_66045-2.

reproducing and passing the genetic basis for bright whites along to the next generation. We might have seen such selfish exploitation in early *Homo*. In *Homo sapiens*, however, knowledge of where another is looking must have been used also, and more so, to coordinate interaction in a way that increased the fertility of all whiter eyed parties (*Economist* 2006).

Having some sense of what another is thinking, or what psychologists term **theory of mind**, also provides a potential advantage. The ability to read minds, if you will—to attribute beliefs, intentions, desires, emotional states, etcetera to others; to grasp that these may differ from our own beliefs, intentions, etcetera; and to use our theories regarding what others' minds hold in making predictions regarding how they may act in the future—seems key to developing a sophisticated model of and for cooperation.

More than this, and even beyond empathy and mutuality, successful cooperation depends upon **shared intentionality**: the ability—and motivation—to collaborate with others as a 'we' holding a joint objective in mind. It takes a certain kind of intelligence to be able to relate to let alone enact 'we-ness,' as people do when dancing a tango or bringing in the harvest. One must be able to understand the actions of others as tied to certain intents or goals, form with others shared objectives or conventions, locate and identify what others are attending to, direct the attention of another somewhere else, and so on (Tomasello 2011).

Nonhuman primates (and by extension, our ancient ancestors) do not do these things let alone collaborate creatively, as we do when we put our heads together to come up with better solutions conjointly than we would on our own (as Fuentes argues, "the initial condition of any creative act is collaboration"; 2017, 2). With some modest exceptions non-human primates' work only counts as group work because individuals share proximity or are spatially located near each other as they forage, eat, and so on. Human groups whose cooperation went beyond this minimal level simply had a better chance of survival in light of evolving selective pressures. Those who could cooperate well and, moreover, creatively—as what Fuentes calls "supercooperators" (2017, 4)—had a selective advantage.

The Role of Cooperative Childcare

One push for cooperation was infant dependency, which had been on the rise ever since our brains began expanding our skulls. Earlier, I mentioned that food sharing and group food gathering would have had to increase. Other changes, related to bipedalism, caused infants to be born basically rear-facing instead of facing up or forward, as nonhuman primate infants are today. This change, says Wenda Trevathan (1999), made unassisted childbirth a bit harder, for instance making babies' necks more vulnerable to damage as mothers reached down to guide them up and out. Birthing alone is not impossible; but having someone there to help doesn't hurt. Accordingly, while nonhuman primates seek solitude upon first feeling labor contractions, humans seek the company of others. This propensity was selected for: females who sought assistance as a response to labor contractions would have had more offspring survive to themselves reproduce.

At first, the urge for companionship during labor may have just reflected a variation in how individual organisms responded to pain. Later, our ancestors would have evolved the capacity to realize, intellectually, that the birthing process could entail challenges. They would have evolved the capacity to pass information about these challenges on to others, too, as well as the capacity to learn from shared information, so that personal experience was not the only teacher. The anxiety that knowledge of childbirth's pitfalls must have generated would have motivated expectant mothers to ask for help, and other mothers to provide it (Trevathan 1999). Selected for, the predisposition toward getting and giving support during birth surely had a hand in fostering further growth in then-emerging cooperative and information acquisition skills.

Increasing rates of birth attendance also would have been coupled with, and in turn helped ratchet up, increasing rates of the behavioral strategy called **allocare**. This entails providing childcare services to children who are not one's own biological offspring (*allo* means 'other,' so 'allocare' means, literally, 'other care,' as opposed, for example, to being cared for by your birth mother). Allocare can entail anything from occasional babysitting to the longer-term arrangement between a nanny and his or her charge, or a foster family and foster children. The extended family often provides allocare. Another example of allocare not infrequently seen in the ethnographic record entails breastfeeding an infant who did not come out of one's own womb.

Allocare provides the biggest selective advantage in communities where food is scarce. It happens elsewhere, too, but, when getting enough food to feed the family is a hard thing to do, having other people help care for the children so that able-bodied adults can go out and collect more food is all the more advantageous. Individuals who may not be able to walk as far (for example, grandparents) can make an important contribution to group survival by pitching in with allocare. Moreover, older siblings who help take care of children learn what we would term 'parenting skills' in this fashion (see Figure 4.4).

Allocare seems to have mutually reinforced the natural selection of other capacities that support productive cooperative interaction. Think of the advantage gained by infants and children who are better at getting people to meet their needs. By making certain faces or gestures, young ones provoke allo-maternal or allo-paternal responses in people charged with taking care of them. Those with a better capacity for engaging in something akin to a theory of mind—those

FIGURE 4.4 Middle-childhood Aka girl cares for young child in forest in central Africa. Photo courtesy of Barry Hewlett.

better able to discern what potential caregivers are thinking or feeling—can target such requests for food, hugs, injury care, or whatever it is that they need toward those individuals most likely to be responsive. Children who are better at such things and who therefore are better at getting others to take care of them—and to teach them things—simply have a better chance for survival—and therefore for passing along any genes supportive of such skills.

Similar to birth attendance, then, the early practice of allocare was a likely contributor to the natural selection of perceptive and expressive capacities necessary for productive group life. Like tool making and talking (and all that these entail), allocare had humanizing effects, in that it reinforced the emerging brain structures responsible for enabling and promoting human connectivity. However, something more had to enter our behavioral tool-kit before the real transition to humanity—and the emergence of people with full-blown culture—would occur.

Ritual and the Cultural Revolution

Given the evolved capacity of humans to connect, verbal language is an excellent tool for transmitting the learned, shared information that comprises much of human culture. For instance, we can tell each other how to make regionally relevant houses or clothing. Moreover, we can gossip, which conveys essential information about norms and values—indeed, 70 percent of all talk is actually social gossip (Dunbar 1996).

However, as Goldschmidt explains, there are some "problems with the way language handles the complexity of the world" (2006, 33). Yes, language is excellent for conveying technical information, such as how to make a particular type of blade from a particular type of stone, or that you need lunch, or regarding how to—or not to—behave at the company picnic or in class or at an uncle's wedding. In and of itself, however, verbal language is not always so good for conveying or transferring sentiments or feelings from one person to another. For this, day-to-day verbal language needs embellishing, such as through lowering your voice or clasping the hand of the person you are speaking to. Better than these techniques, there is the language of **ritual**—of action sets performed for their culturally relevant symbolic value.

From an outsider's point of view, ritual action is that which seems unnecessary, unmotivated, or illogical For instance, there is no practical reason to lug a cut tree into one's living room, where it will drop sap and needles, let alone to festoon it with lights so that it poses a fire hazard during the Christmas holiday. But to many Christians, it wouldn't be Christmas without that—and group membership is demonstrated by following through (see again Supplement Box 4).

Rituals include not only religious practices but various calendrical or cyclical celebrations or commemorations, rites of passage, purification- and atonement-related practices, inductions and funerals, sporting events, courtships and graduations. A wedding is a ritual; so is a presidential inauguration. Not all rituals are all-encompassing: scholarly conferences, doctor visits, class meetings, greetings and partings—even phone calls home: all entail ritual aspects or symbolically important communicative elements without which something would be missing.

Ritual communication includes the more kinesthetic or bodily communicative mediums of song, dance, and prayer as well as systematized gestures or actions such as eating together, shaking hands, kissing cheeks, doing the wave, and so forth. These communicative actions are key to the vital function ritual served in creating for our ancestors the human state of feeling connected. Sometimes parts of a ritual, such as drumming, dancing, or drug use are deployed to intensify that function by increasing receptiveness in participants. Through its communicative dimension, ritual allows us "to create the emotional cohesion that makes for effective and

FIGURE 4.5 Ritual event: "The Thanksgiving Dinner," Corner of Grand & Allen Streets in Ridley, New York, 1870. Artist unknown.

continued collaboration and to reduce tensions between potentially hostile persons and groups" (Goldschmidt 2006, 40). As Goldschmidt further explains,

> [Rituals are] multimedia events that make everybody involved feel the same way about whatever the group needs to feel about. If you want to psych up youths to go to battle, you get out the drums and gesticulate in unison.... If you want to solidify a group you eat together and sing and dance in unison.... Ritual communication stands in contrast to the engineer's logic of speech, supplementing it with what I call an 'aesthetic appreciation' that transmits feelings from one person to another—not information about feeling, but the feeling itself.... [Rituals are the] cultural invention that made culture work.... Tools give the cultural mode of life its survival advantage, and language is the device that made culture possible, but ritual was needed to hold the social fabric together.
>
> (40–41)

Rather than simply instructing, as day-to-day survival-related or "engineer's" language does, the language of ritual motivates. It entices our social and cultural commitment. Ritual coerces through its ability to mobilize emotions and transfer them between members of a group (see Figure 4.5).

With ritual communication, it is not the explicit messages—for example, "Turtles are our clan's totem" or "This is the American flag"—that are of the most import, but rather the viscerally felt confirmation that "Our clan (or society) is a wonderful group to which we should be loyal." Feelings are paramount, and they reinforce group identification. Paradoxically, as Goldschmidt notes, imagination and sentiment are essential social catalysts for us humans, although we often focus instead on our rational intelligence as the thing that makes us special (42; and see Fuentes 20117).

The Full Bloom of Culture

Although we may have been capable of true language with the emergence of *Homo sapiens*, notable amounts of tangible evidence that humans are really attending to social connections only begin to be seen in the archaeological record starting about 75,000 years ago. By this time, full-fledged culture—the kind supported by ritual—had emerged. In other words, we became truly human: we began to express ourselves symbolically, such as through art and personal adornments (see Figures 4.6 and 4.7). The oldest evidence of this has so far been found in Africa (see again Figure 4.7). Therefore, most scholars agree that Africa was where behavioral modernity or full-fledged cultural life was born; and it was born to our particular subspecies: *Homo sapiens sapiens*. With our new-found self-awareness, we threw open a door, about 75,000 years ago (but of course earlier in some places, later in others), into what many scholars call the Late Stone Age or Upper Paleolithic.

The archaeological record of culture includes things like small statuettes, often depicting female figures. Less portable but no less impressive were the polychromatic cave paintings, for example of horses and bison, that also began to appear. Color was derived from natural earth or mineral or pigments such as sienna, umber, and ochre, which had previously been found used for body decoration. For that, in addition, we now see string, thread, and sewn clothing as well as shell and tooth beads. More complicated tools arose, as did burial artifacts apparently meant to "support, defend, and amuse the deceased" (Ehrlich 2006, 170)—sure signs that symbolic life was in high bloom.

The change in the archaeological record described indicates, too, that human beings had by now developed the capacity for spirituality and religious beliefs. Some say that without cultural belief systems to justify or at least make sense of the tragedies and trials that inevitably occur within human life cycles, the anxiety that must have accompanied our ancestors' increasing intelligence and self-awareness would have been too much to bear. In addition to providing a mechanism for achieving social cohesion, then, the meanings conveyed through ritual also provided a means to mitigate existential angst.

FIGURE 4.6 Front and back views of a female figurine in ivory from Kostenki, in Russia. About 25,000 years old. Courtesy Olga Soffer.

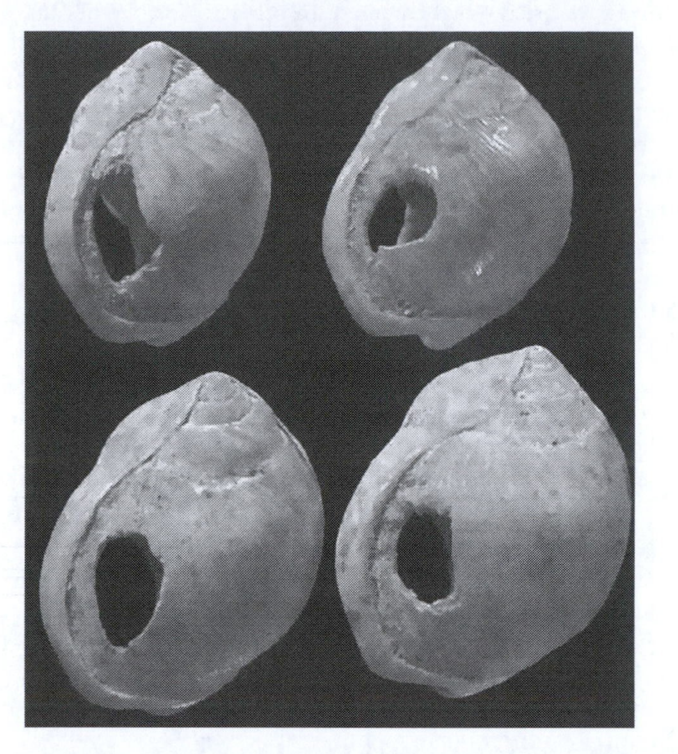

FIGURE 4.7 Shell beads from the 75,000-year-old levels at Blombos Cave, South Africa. Photo by Chris Henshilwood and Francesco d'Errico, CC-BY-2.5.

The transition to culture appears to have been made relatively quickly; many scholars speak of it as the **Cultural Revolution** or (to avoid confusion with historical events in China) the **Upper Paleolithic Revolution**. Of course, this revolution became possible only after many interrelated genetic changes had built up, leaving our ancestors better able, physically, to make things, cooperate in reaching intersubjectively shared goals, identify as a member of the group, and communicate with language. In other words, it was really more "like a slow burn that slowly built into the raging fire of modern creativity" (Langley 2018), when a threshold was surpassed. However, even if we agree with Goldschmidt (2006) that the emergence of ritual practice pushed us over that threshold, beyond a tipping point, to engage in full-fledged cultural life, we still must ask why such a long gap existed between the evolution of our capacity for language and so forth as *Homo sapiens* and our engagement in a fully cultural existence.

Why the lag? (See again Table 4.1.) Some argue for the triggering role of population density, as relatively low as it still must have been back then. The increase would have been important in the lead-up to culture because the more members a group has, the more likely it is that one of the members will contribute an innovation that others see as useful. Further, "you need sufficient human tinder for those sparks of culture to catch" (Mooallom 2017). For himself, Goldschmidt points to the lag time common between any technological advance and a cultural solution to the problems or changes that it brings about. Think about social media and smart phones, and how we are still adjusting to the social (including political) changes these inventions have wrought. Early innovations—including those entailed in using language for teaching and learning—could be optimized only if we could maintain social cohesion, and it simply took us awhile to figure out how to do that. Once we could really leverage our existence in cooperative social groups, which we did through the communicative practice of ritual, we crossed the "bridge to humanity" described by Goldschmidt, and culture took off.

Notes

1 Reproduced by permission of the American Anthropological Association from *American Anthropologist*, Volume 58, Issue 3, pp. 503–507, 1956. Not for sale or further reproduction.
2 See also "What Does It Mean to Be Human?" hosted by the Smithsonian at http://humanorigins.si.edu/, particularly the Human Characteristics subpage (National Museum of Natural History, n.d.).

5

GLOBAL EXPANSION, HUMAN VARIATION, AND THE INVENTION OF RACE

This chapter prepares you to:

- Outline the immediate evolutionary history of *Homo sapiens sapiens* and explain (in relation to the concept of race) how humans populated our geographically diverse earth
- Recount the history of the race concept as applied to human subgroups and explain the difference between biological race, racialism, and racism
- Define geographic cline and explain the value of making geographic ancestry-based distinctions between subpopulations in certain limited circumstances
- Explain the genetic argument against race, with reference to genetic variation within and between diverse groups

The balance of evidence today indicates that the baseline capacity for culture discussed in Chapter 4 emerged originally when our ancestors all lived in Africa. Since that time, we've spread out around the globe. Archaeological, human skeletal, and genetic evidence provides insight into how we spread, and into how migrating groups responded when they moved into regions that were colder or hotter, or higher or lower, or drier or wetter; had differing food resources; or varied on other dimensions in relation to where they had done most of their previous evolving. To survive, humans adapted. For behaviorally modern humans, most adaptations were cultural. But our bodies sometimes adapted too.

The creep of humanity across the earth underwrote the emergence of a good deal of diversity in how geographically distinct groups look physically, as well as in bodily structures and processes that occur unseen inside of us. As interesting and important as those internal differences may be, most discourse today about diversity focuses on superficial variation—hair texture and skin color, for instance—the kind of variation now broadly sorted into a handful of racial categories. That's too bad, because differences in eye shape, nose form, skin shade, and so forth represent only the tip of the diversity iceberg. Moreover, the scientific basis for grouping such features together to separate us into so-called racial categories has been greatly misconstrued.

This chapter supports a more sophisticated understanding of human diversity by examining how the diversity of environments we met with when *Homo sapiens* populated the world affected various subgroups of our species differently—but not so much that separate biological races really were created. For one thing, many adaptive physical variations evolved independently of each other. Skin color, for instance, need not—and often does not—correspond to height, eye shape, nose width or length, hair texture, blood type, red blood cell shape, or any other trait in a given population. Most traits associated with the so-called races do not clump together in predictable, stable patterns—which they would do if humans really could be subdivided into biological races. Moreover, patterns that *do* exist correlate with geography, and generally with regions that are fairly limited in size, so that knowing the location of one's particular ancestral homeland, and its evolutionary pressures, has much more predictive value in regard to such traits than does one's so-called race. Biologically speaking, humans are of one race, not many.

The Spread of Humanity

It is by now well-established that *Homo* emerged in Africa and that, by about 75,000 years ago, and likely even earlier, behaviorally modern humans—that is, humans expressing their full capacity for culture—had come into being. However, the fine details of exactly when and how humans like us took over the world remain the subject of much debate. While results of these debates will not change the substantive content of this book, it still is worth describing some of the issues contested, both for the practice it gives us in weighing theoretical options and the insight it begins to bring regarding the construct 'race.'

In the classic 'out of Africa' model all humans can be tracked back through one line to the cradle of civilization. According to this model—classically a 'single origin' model—members of an expanding *Homo sapiens* population walked first into various parts of Africa and then some ventured out into other continents, largely via what is now the Middle East (see Figure 5.1). While roughly correct, recent evidence complicates this story greatly, as we soon see.

Location-Linked Diversity: A Small Variation on the Basic Human Theme

Each of the new environments that we entered posed slightly different adaptive challenges. Our basic human nature—our brain-based intelligence and creativity, capacity to empathize and form social bonds, and aptitude for linguistic and other symbolic expression—was born in Africa. Through migration, however, we augmented our basic humanity with new and relatively superficial, region-based biological diversities. We adapted culturally, too, of course; and most diversity is biocultural rather than simply biological or simply cultural. Here, however, we will focus on geographically based explanations for location-linked bodily adaptations.

For example, by the time humans had spread across the earth, rather than everyone having darker skin, as was the norm in sub-Saharan Africa, some populations had evolved skin that was lighter in color—eventually including 'white.' Likewise, rather than everyone making lactase (by which we digest milk) only in their first few years, peoples whose new environmental adaptations included reliance on milk-based foodstuffs—which were neither necessary nor available in our earliest days in Africa—evolved the ability to make lactase well into adulthood. Such traits increased a population's fitness or chances for survival in the regions where natural selection encouraged their prevalence.

Importantly, non-Africans have been non-African only for a short amount of time relative to the time-depth of humanity's existence in Africa; and those who branched off represented only a small part of the total African gene pool. Therefore, the degree of genetic variation in non-African populations is much lower than the degree of variation found on the African continent. Moreover, most genetic variation found outside of Africa represents a subset of variation also found within it. This explains why we may find more genetic similarity between people native to, say, China and Ireland, than we may find between people native to, say, the neighboring African countries of Zambia and Zimbabwe.

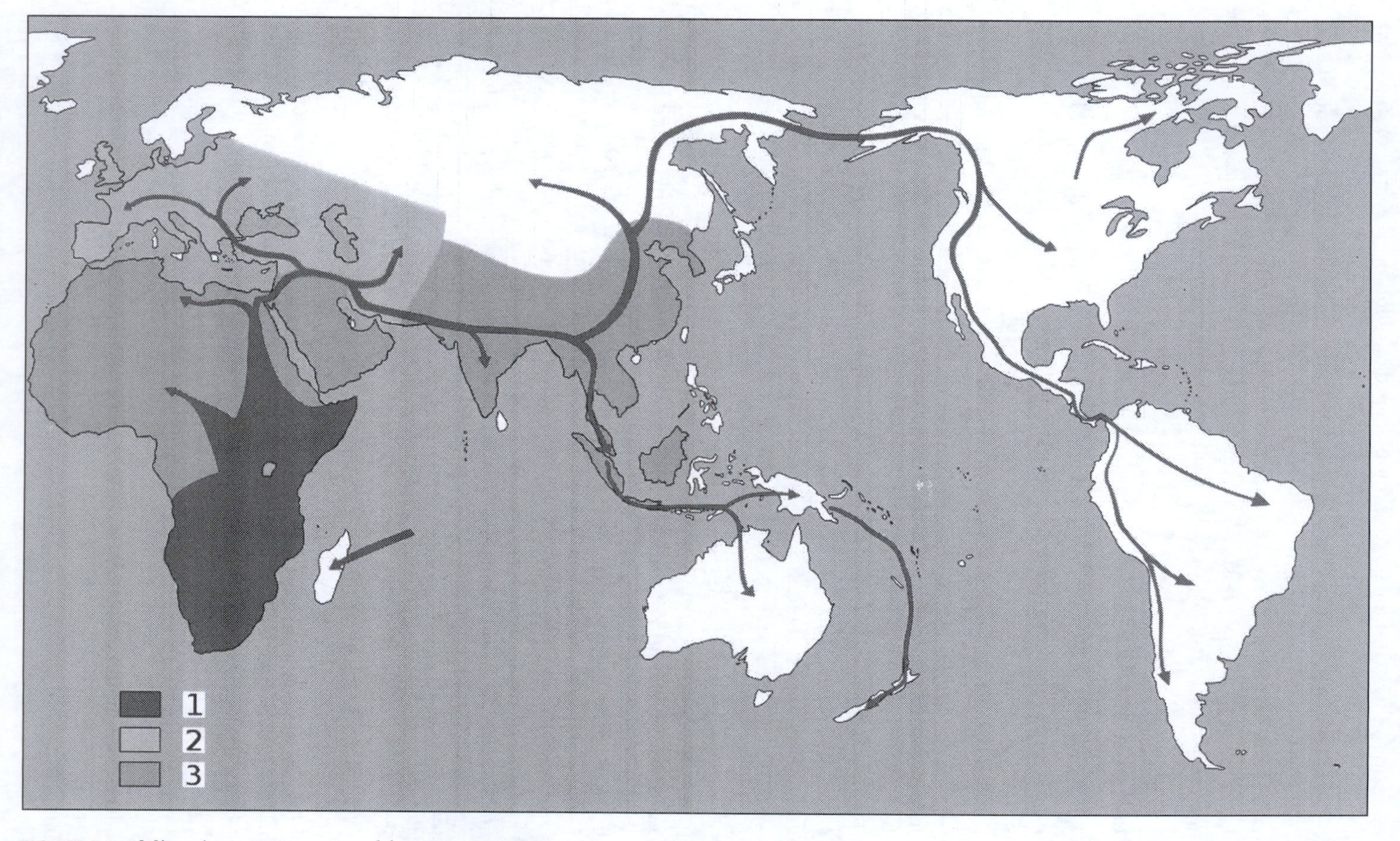

FIGURE 5.1 Migration routes ancestral humans may have taken in populating the continents. 1 = *Homo sapiens*; 2 = Neanderthals in particular; 3 = earlier species. Adapted from *Spreading Homo Sapiens by Magasjukur2, CC-BY-2.5*.

Migration Waves

Complicating the single origin, out of Africa model is the fact that migration did not occur in a single radiating move. Many theorists believe, based on skeletal and genetic data, that small groups of humans left Africa in several waves. Sequential outmigration waves were likely spurred by challenging changes in climate or in the ecosystems that the migrating groups had been part of.

Moreover, not all emigrants were behaviorally or even just anatomically modern. Remains found in Asia and Europe suggest that some members of the species *Homo erectus*, an archaic and extinct branch of the human lineage, wandered out of Africa well before *Homo sapiens* even existed. Nonetheless, by far the largest waves of migration—those of anatomically and behaviorally modern humans or *Homo sapiens sapiens*—came 60–50,000 years ago.

When the newly outbound populations met up with groups that had emigrated earlier and survived, what happened? Until recently there were two main schools of thought. Both were fairly linear. One held that those who left first were decimated by diseases newer population waves brought in, or driven out or otherwise replaced by the newcomers. This idea, in which a single line ascends while other versions of *Homo* dead-end, is known as the **replacement hypothesis**. The other school of thought held that earlier emigrant groups (the ones that survived, anyhow) were absorbed by newer populations. This idea is known as the **assimilation hypothesis**. Either way, the newer populations ruled.

However, some theorists believe that emigrating *Homo* populations—including even the first, most archaic (non-*sapiens*) groups to journey out of the African continent and survive—continued to evolve concurrently, without replacement or assimilation being necessary. This model is perhaps most commonly called the 'multiple origins' model.

In a minority, linear, simplified version of this model, ancestral lines evolved in parallel, in the same direction without comingling, to produce humans of our kind. A visual depiction of this version of the model would look like a candelabra. In the dominant version of the multiple origins model—the version best fitted to the data we have and one that appreciates complexity—gene flow connected ancestral lines. Indeed, a trellis provides a better visual; and to differentiate this trellised (or webbed, or networked) version of our 'multiple origins' we can term it the **multiregional metapopulation model**. A **metapopulation** is a group of separately located populations of the same species that interact in some way and thereby maintain their identification as one species. Accordingly, regional populations periodically intertwined versus remaining wholly independent. Genes flowed between various *Homo* populations whenever they met up with one another as they traversed various regions of the earth.

With continued gene flow, regional populations maintained distant kinship; some may have been assimilated also. Regional populations never diverged enough genotypically from the metapopulation for speciation to occur. Or, as Hannah Devlin puts it, summing up current thinking, rather than one ascendant human group winning the earth's dominion in simple, linear fashion, "reality is messier… with populations diverging and then interbreeding again. Rather than the tree of life it's more like a dense, thorny bush" (2018). Our single branch is turning out to be a thicket.

The Neanderthal Situation

Evidence supporting the idea of a human metapopulation comes, most recently, from Africa, suggesting that even the emergence of *Homo* as a genus was multiregional. Yet because we have been talking about *Homo sapiens* as an established species, I'll focus here on evidence from research on Neanderthal peoples.

Neanderthals occupied Europe and some of western Asia (including what is now Israel) from about 130,000 to about 30,000 years ago. Physically, in relation to us, they had protruding midfacial areas, heads that bowed out in the back, large brow ridges, and a very heavy and robust bone structure, all to support big muscles. One likely explanation for this is that Neanderthals' forebearers wandered out of Africa early, so that their line diverged from that of their African *Homo* cousins just prior to the emergence of *Homo sapiens* (and then *Homo sapiens sapiens*). Thus, without any real cultural capacity, Neanderthals would only have been able to evolve physically to cope with the European environment: cultural adaptation was not, at least originally for them, an option.

Common knowledge used to hold that Neanderthals were a dead-end offshoot of our ancestral line. According to the replacement-oriented single-origin model, when humans fully capable of culture arrived, the Neanderthal version died off. These 'cavemen' just couldn't compete, or so it was said; and modern *Homo sapiens* replaced them. Scholars holding this view labeled Neanderthals *Homo neanderthalensis,* denying them our species' appellation (*sapiens*), casting them as what lay people might term nonhuman.

Archaeological evidence tells a different story, however. For instance, some Neanderthal groups buried objects such as flowers with their dead. They also made jewelry and deployed bird feathers; at one site in Gibraltar they took only dark ones, likely for ceremonial or aesthetic reasons. They also used toothpicks, made glue from birch bark, and made and used pigments, maybe even painting their bodies or faces with them. In short, material evidence suggests that Neanderthals were indeed, at least toward the end and to some degree, meaning-making beings (see Mooallem 2017). They may have branched off after the capacity for culture emerged, or they may have developed these behaviors independently—or learned them from other groups in the human metapopulation. The fact that there are so many geographically separate Neanderthal sites demonstrating some form of culture gives support to the latter idea.

Genetic evidence does show that modern human and Neanderthal peoples interbred, and to do so they would have had to run into one another at least now and then. While no contemporary humans native to sub-Saharan Africa have Neanderthal genes, bits of the Neanderthal genome can be found in all other living populations. Collectively, about 20 percent of the Neanderthal genome remains in circulation: mostly genes relating to cold adaptations such as thicker hair and skin (Sample 2014).

Some hold-outs interpret the genetic evidence as indicating only a common ancestor (i.e., one from prior to when our lineages split), or simple assimilation; but because Neanderthal genes are so widespread, and given the archaeological evidence also, the majority see it as supporting the multiregional metapopulation model, with its ebb and flow of lineal splitting and splicing. In other words, Neanderthal peoples seem to have been one of many threads in a long, intertwining line of "shifting clusters" making up the *Homo* metapopulation (Mooallem 2017). When clusters bumped into one another, they sometimes interbred and the continued gene flow this entailed kept them reproductively compatible as well as providing some groups an evolutionary short cut toward attaining certain traits. Sometimes too, perhaps particularly in more recent cases of comingling, one group learned behaviors or techniques from the other.

Scholars now mostly classify Neanderthals as *Homo sapiens neanderthalensis,* representing them as part of our own species but at the same time opposing them to us (*Homo sapiens sapiens*) and us to them as two different biological races or subspecies. A biological **race** is a clearly differentiated species subgroup with features that regularly co-occur. In Neanderthals, as compared to modern *Homo sapiens*, the skeletal features mentioned earlier (notably robust bones, larger brow ridges, etc.) and some others co-occurred with regularity: they exhibited **concordance**. With the end of the Neanderthal line the last real (biological) human racial divide was extinguished.

The Invasive Human Species

In summary, once humans could walk, some groups left Africa by foot, and this happened several times over, most intensively with groups in the *Homo sapiens* line. Although some groups died off, genes still flowed now and then between the continually increasing number of groups that survived to comprise a human metapopulation. Individual lines separated and intertwined as circumstances allowed it, until eventually, maybe 30,000 years ago, only *Homo sapiens sapiens* were left. By about 10,000 years ago, a single human species with no racial subsets (biologically speaking) populated nearly every habitable area of the globe.

This notion—that we are race-free—may seem counterintuitive. To understand why we think race exists biologically, and why we create it socio-culturally, we must ask how human variation was explained prior to the emergence of modern science.

Early Ideas about Human Variation

In prior times, as now, people around the world told origin stories to explain where they came from. In many early origin myths, people knew themselves as the only humans. For instance, many indigenous group names translate simply to 'People' or 'Human Beings.' Other groups they may have encountered (for instance when a party from outside passed through their territory) were not in this number. They might be cast as another kind of being (think of the various types represented in Tolkien's Middle-earth: Hobbits, Elves, Dwarves, and so on); they might more simply be seen as animals, or gods or monsters.

What explained variation? Each group had its stories. Among Europeans, one theory held that certain descendants of Adam and Eve devolved or degenerated while others progressed to a more perfect state (naturally, storytellers' groups did the latter). Some theorized that all humans began in a so-called 'savage,' animal state, progressed to what was termed 'barbarism' (at which point they could at least talk), and that eventually everyone would reach the most advanced state of so-called 'civilization' (see Langness 2005).

Other versions of Europe's savage–barbarism–civilization continuum were generally static; they did not include the option of change: human group-to-group variation was part of the world's order. Similarly, the Mbuti from Congo explained that the supreme god made different races from differently colored clay. And in another twist on this theme, one kind of clay was used: color variation among groups represented how long a given group's clay forbearer was left in the fire.

Notwithstanding, 'race' did not have much currency until relatively recently. Contact between very different groups of humans did not happen very frequently until the invention of transportation modes that supported it. Certainly, groups sometimes met in ancient days: had they not, there would be no metapopulation. Sometimes groups met when searching for food or a good place to camp; trade networks also brought people together. For instance, merchant caravans passed through many regions on foot, by camel or horse, or by wagon, and merchants did see variation first hand. However, the physical variation they encountered turned up gradually as they passed over the land, with, for instance, hair texture grading slowly from curlier to straighter on average. Sure, some differences would be noticed. However, which were deemed of interest and what was made of them depended on who was doing the looking and interpreting, and why.

Classifying Human Subsets

One early way differences between human groups were understood was by fitting them into pre-existing conceptual models of personality or character. This is exactly what Carl Linnaeus did when he penned his original taxonomy, *Systema Naturae* (1758). As Stephen Jay Gould

(1994) reminds us when describing the evolution of racial taxonomy that I will here summarize, Linnaeus was the Swedish botanist, zoologist, and physician infamous for classifying the plants and animals of the known world as part of the Scientific Revolution's effort at knowledge expansion. When it came time to classify humankind, Linnaeus decided to break us down into four groups, in keeping with the legacy of Greek medicine, also known as the humoral system.

The ancient Greek or Galenic humoral system teaches that the body contains four liquids, or humors (blood, phlegm, yellow bile, and black bile), and that each is associated with one of four "complexions" or temperaments: sanguine (cheerful), phlegmatic (lethargic or unemotional), bilious or choleric (quick-tempered), and melancholic (sad). Humors also tracked to the basic elements—wind or air, water, fire, and earth—as well as the four seasons. As long as the four humors are in balance, a person is healthy. Imbalance can be treated by removing excesses, such as by purging or vomiting, or by correcting for deficiencies, as through special diets (Sobo and Loustaunau 2010, 84).

As Gould tells it, Linnaeus extended the scheme further than before, mapping the human race upon the cardinal directions (north, south, east, and west) and then linking each with a humor and temperament (see Figure 5.2). In dividing by four, Linnaeus came up with races that we might today term American, European, Asian, and African, matching each with the colors dictated by the four-fold Galenic scheme: red, white, yellow, and black.

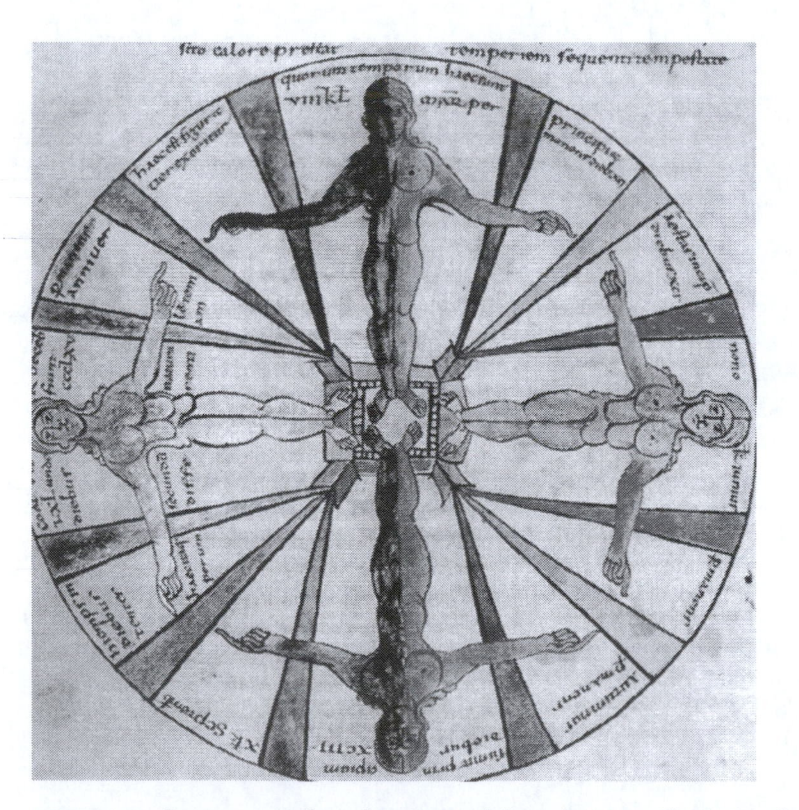

FIGURE 5.2 Four seasons, four elements, and four characters illustration from twelfth-century version of Isidore of Seville's sixth-century *De Natura Rerum* (On the Nature of Things). Linnaeus built upon this legacy of four-fold thinking in his race typology.

Despite the value placed on divisions by four, at the time Linnaeus wrote there was widespread belief among his contemporaries in the unity of the human species and an understanding that each individual contained all humors, temperaments, elements, and so on. Linnaeus believed the differences he charted graded gently into one another, and his writing does not include racial rankings.

Like the priority he placed on the number four, the focus on temperament in Linnaeus's scheme may strike us as strange, conditioned as we are to look to skin color as the key to race. Cross-culturally, however, it is not bizarre at all. Many groups used non-skin-based heuristics. The Lakota people of what are now the Dakotas, for instance, are said to have called white people *wasicu*. This means, literally, 'fat takers,' or 'eaters of fat'; it also means 'non-native.' It refers to the non-native's disregard for preserving natural resources for future generations; it refers to greed. In using the term, the Lakota distinguished whites not by the color of their skin but by their behavior.

Back in Europe, Linnaeus's fourfold scheme was widely accepted; after all, it fit nicely into existing ways of understanding and classifying the world. However, it was not long before someone challenged him, ostensibly on the basis of—as all good science is meant to be—observation. The challenger was Johann Friedrich Blumenbach, who in 1795 introduced the five-race schematic that will be more familiar to readers today.

Turning away from temperament, Blumenbach focused on what people looked like. He added the color brown to the four-fold white, red, yellow, and black framework. He made another change, too. Before I describe it, I must note that Blumenbach believed in the unity of humankind; he ascribed differences between human groups to climate and habitat. That is, they were impermanent. Nonetheless, he proclaimed people from Mount Caucasus in Georgia (a nation located at the junction of Europe and Asia, on the Black Sea) to be closest in appearance to the created ideal or original Adam and Eve, and thus justified naming his central race after them (that is, Caucasian). While geography still was important, Blumenbach's scheme made physical appearance fundamental.

Biological Race

Its origins had been laid, but a full-fledged ideology of race, complete with rankings and a biological (rather than environmental) foundation, did not gain in popularity among Europeans until the age of discovery segued into the age of conquest, and more Europeans wished to enslave or colonize more non-Europeans—particularly (but not only) those living in Africa and the Americas. At this time in Europe and then too in some of Europe's colonies, the idea that people were born into fixed social stations and that some naturally could dominate others was being displaced through a series of social and political revolutions. In contexts now favoring liberty and equality, those in support of colonization or slavery needed a strong justification—and they found it in the hierarchical construction of race that was Blumenbach's legacy.

What better way to justify unequal social arrangements than to claim that it is natural for a biologically more intelligent and civilized human group to enslave, colonize, or otherwise oppress another that is biologically dumber and less civilized? The popular construction of race helps people do this because of the role biology was given. People longing to justify certain socio-political arrangements came to assume that there were categorical, biologically (today, genetically) based differences between broad subgroups of humanity—and that subgroups could be divided based on visually identifiable traits such as skin color.

By contrast, as a scientific rather than popular term, 'race' is akin to 'subspecies.' Although even in biology its use is now questioned it commonly refers to a subgroup of a species in which certain defining traits always co-occur. The term is used in regard to many kinds of animals—but not modern humans, because, from the perspective of biologists, no human subspecies or races presently exist.

Racialization

Visually assessing people and fitting them into subgroups based on how their bodies look (and perhaps move, sound, and so on) is called **racialization**. The simple classification of people in this way is distinct from **racism**, which includes an evaluative component. In racism, not only are people classed by so-called race, but also the races are rated and ranked (see Fluehr-Lobban 2006, 2–4).

Everyone today who has grown up with the idea that human races exist—everyone enculturated into the belief that there are separate races, and that which race one 'belongs to' matters—will practice racialization. They will see race in people, and they will use race to organize and explain aspects of their world. With racialization, race becomes a key descriptor of every human being. This is why doctors, police officers, and others who must quickly describe patients, suspects, or clients do so by mentioning race—often prior to other descriptors ("The suspect is a white female of average height"; "You know Don—he's the black fellow who sits in the middle"). Today there is really no such thing as what some liberal thinkers naively term 'color blindness.' Holding to the idea that one does not racialize, particularly when touting as well the related belief that the United States is a meritocracy, entails denying not only the value of our diversity but also the systematic ways in which racism is maintained and actualized—an issue we'll return to (and see DiAngelo 2018).

Can racialization exist without racism? Yes. We can acknowledge and appreciate difference without ranking it. This book characterizes human biocultural diversity and pays tribute to human variation. Without variation, a species may be more prone to extinction or decline, a fact demonstrated in examples throughout the book.

However, even the most neutrally deployed racialization still has a fault: the construct of race, on which racialization is built, has no biological reality. That is, when we look at the biology of the so-called races, and when we examine their genetics, *we see more difference within groups than between them*. Biological races simply do not exist.

Nonconcordance

One major challenge to the race construct comes from genetic science, which, again, has repeatedly found more genetic variation within the so-called races than between them. Whereas race has been selectively defined based on a very few of the most visible physical characteristics, human variation actually consists of all kinds of visible and invisible traits and related genes: we may or may not have the capacity to digest lactose as adults; we may or may not have lots of the intestinal transmembrane conductance regulators associated with vulnerability to cholera (see Chapter 8); we may or may not be genetically programmed for male pattern baldness; our earlobes may be long or short; our breasts small or large; chests hairy or not; cheeks dimpled or not; ear wax crumbly or moist; fingerprints loopy, arched, or whorly. We vary by myriad traits whose distribution across humanity does not overlap with race as the lay public knows it (see Diamond 1994). You may well be less genetically like a person you consider as belonging to your own 'race' than you are to someone in another 'race' entirely. Indeed, as Richard Lewontin discovered in the 1970s, 80–85 percent of human variation occurs within so-called races; only 10–15 percent exists between these groups (Lewontin 1972, 2).

These numbers still astonish us because of our culture's deeply held belief in race and our reliance on visual cues to mark it. That is, Lewontin's statistics describe genetic differences, not just easily seen ones: many of our genes store information used by our bodies in making proteins expressed under the skin and internally. As many communities have asserted in support of our common humanity, we all have red blood. We also all have lungs, bones, a liver, and so on. In addition, features that make the human species unique—our brain-based capacities for logical and creative thought, and for empathy and the ability to cooperate and form communities—these are longstanding and account for much more of our humanity than racialized visible features that are evolutionarily newer and geographically varied.

Moreover, the genes containing information used in building the visible traits we most often associate with race are not actually distributed in race-like clumps. They neither occur only in certain so-called racial groups, nor do they co-occur, or occur together, in any predictable way. Rather than **concordance** (when one gene set, gene, or allele is a predictor for or index to the fact that another particular gene set, gene, or allele will be present), so-called racial traits exhibit **nonconcordance**.

Another way of grasping this is through the "jaggedness principle," developed by Todd Rose (2015) after considering Gilbert Daniels' observations regarding 4,063 pilots whose bodies the US Air Force measured along ten dimensions. Cockpits had been built to fit the 'average' pilot. This makes intuitive sense, at least in our culture. But averages are statistical myths: they are mathematical creations. Indeed, Daniels determined that fewer than 3.5 percent of the pilots had measurements equal to the average on any three dimensions. Not one was average on all ten. The implications of this were huge: designing cockpits for an average pilot meant designing them for no one. What to do? The Air Force introduced adjustable seats, controls, and so on, which greatly reduced crashes and other aviation mishaps.

Where is jaggedness in all this? Well, a person may have a smaller-than-average head circumference, a bigger-than-average chest, longer-than-average legs, and so on. When plotted against population averages, the line made by connecting the points representing these measurements on any given individual looks quite jagged. The points do not line up, one on top of each other, in a neat stack. In other words, they are nonconcordant. Just because you have short legs does not mean you have a small waist, and so on. We all vary, and we do so in varying fashion. The jaggedness principle describes this kind of nonconcordance—and with that, we are back in the realm of race. Like our body measurements, our racialized traits also form a jagged line when mapped to an idealized norm or a so-called 'racial type.'

This may seem counterintuitive. We experience the world through a racialized lens. Yet our belief that races are out there is similar to the pre-Copernican belief that the world is flat. It looks flat from my window even today when in truth planet Earth is marble-shaped. Also, in truth, racialized traits do not necessarily co-occur. Many are inherited independent of or nonconcordant to other racialized traits.

Traits that do co-occur usually do so in subsets of people within a larger 'racial' category—subsets having to do with specific ancestry. So, for example, we do find certain patterns of concordance among people whose ancestors come from a particular village in England, or a particular village in Equatorial Guinea, or a particular village in China. However, these concordant traits are not found concordantly in all people from England, or Equatorial Guinea, or China; and they definitely are not seen among all whites, or blacks, or Asians.

Moreover, so-called index traits like skin color are not really distributed in a way that overlaps entirely with the racial groups we have invented. While it is the case that many people living in sub-Saharan Africa have darkly pigmented skin, so do some people indigenous to other areas: certain Indians, Melanesians such as the people of Fiji and Papua New Guinea, and Australian aboriginals have very dark skin. Likewise, some Africans have eyes in the so-called Asian shape, with epicanthic folds, although they are not 'Asian.'

Another way to think about nonconcordance is to consider the differences between, say, Haitians, Sudanese, and Ugandans. All may be racialized as 'black' but these groups are very different, historically as well as physically (Relethford 2000; Bruner and Manzi 2004). Build varies, as does height: African peoples can be tall, such as among the Maasai of Kenya and northern Tanzania, or they can be short, such as among the Mbuti who live in the Congo region. Likewise, facial structure varies: it is longer and narrower to the north, and shorter, broader, and flatter to the south. In terms of skin color, which is taken so often as the ultimate marker of race, average color aside, there is more pigmentary variation between people living within sub-Saharan Africa than in any other geographic region in the world.

The logic of grouping all African people together starts to seem nonsensical once we consider the wide variation of traits within Africa. Likewise, Peruvians, Costa Ricans, Mexicans living in Chiapas, and Mexicans living in San Diego are very different in average height, facial structure, skin pigmentation, and so forth, putting the lie to any idealized lumping of them into a broad, racialized group. Or for another example, there are the long-term English, French, Irish, Polish, and Italian populations. All may be classed as white today, but there are differences among them large enough to have at one time supported belief that there were various 'races of Europe' (see Figure 5.3) including some labeled as 'nonwhite' (see Fluehr-Lobban 2006, 180–183).

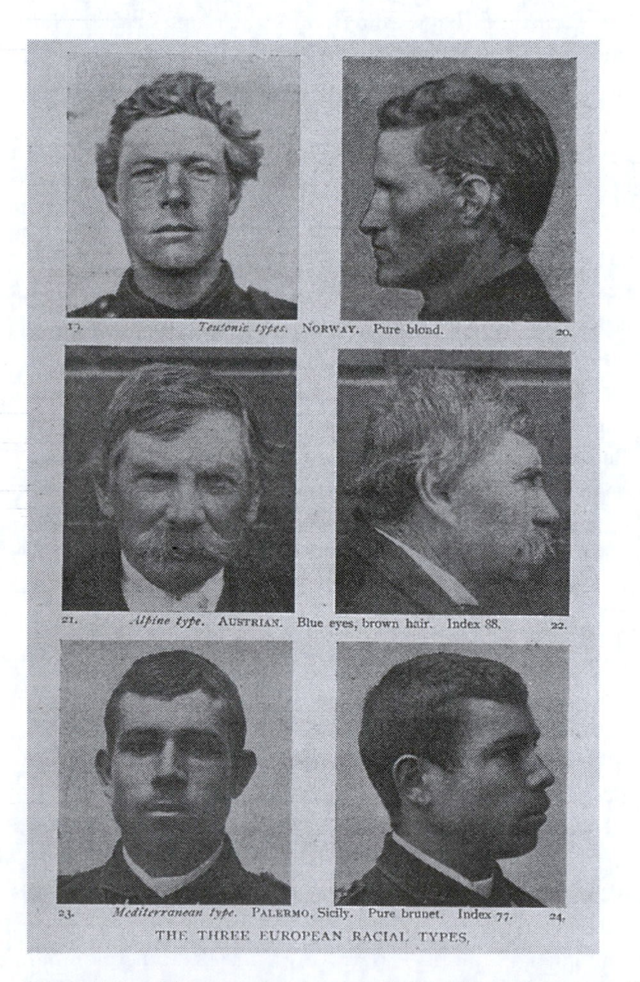

FIGURE 5.3 Photographs showing "The Three European Racial Types" from *The Races of Europe: A Sociological Study (Lowell Institute lectures)* by William Zebina Ripley, between pages 120–121. Courtesy of the Wellcome Library, CC BY 4.0.

Belief in 'Race' Fosters Clinical Mistakes

Some subsets of populations do bear being defined biologically, including genetically, not least because of the clinical implications that ancestry may have. People whose ancestors are from region X may be more likely to suffer from condition Y or be more susceptible to disease Z. However, in making these distinctions we must not confuse ancestry with race. It is crucial not to wash out difference by throwing people of varied ancestry (for example, Basques and Catalans, both of whom live in what is now Spain; or people from Spain in general and people from Italy, Albania, and the United Kingdom) into one racialized bucket. Such over-generalizations have little real utility in physical terms and can be detrimental medically if particular propensities are overlooked because they don't fit a patient's 'racial profile.'

Richard Garcia (2003) has spoken out against racial profiling by clinicians. Race is a constant descriptor in medical records, as if it can tell us something important about a patient. Instead, it seems to cause otherwise smart people to fall back into stereotypic thought patterns. For instance, Garcia tells of a black teenager with a cyst in his head so large that it had "squashed his brain against his skull." Emergency room staff initially wrote off this youth as simply an intoxicated gang member. In another example, Garcia's childhood friend, Lela, suffered from cystic fibrosis (CF) for much of her early life without a proper diagnosis because of an assumption that people in her racial category ('black') did not get this disease. As Garcia tells it, "Only when she was 8 did a radiologist, who had never seen her face to face, notice her chest X-ray and ask, 'Who's the kid with CF?'" (2003, B15; for more on CF, see Chapter 8).

Things are further complicated by the high frequency of 'mixed race' children, such as Garcia's own blonde, fair-skinned daughter, whose mother is 'black.' Any doctor who simply eyeballs his child may inadvertently and erroneously cross off of the possible diagnoses list any disease 'known' to predominate in 'nonwhite races.'

Forensic Anthropology and 'Race'

While most scholars and scientists have long discarded racial categorizing in favor of finer-grained inquiries into geographic ancestry, 'race' remains in use among forensic anthropologists who do skeletal identification for law-enforcement agencies, even though those whose work concerns blood groups tend to reject it. Why?

As forensic anthropologist George W. Gill explains, while race may not be useful for analyzing blood, it is for analyzing bones (2000). Certain skeletal traits, says Gill, do prove reliable in "assessing ancestry," as forensic anthropologists term it. Specifically, a number of individual methods, such as those entailing midfacial measurements or femur (thigh bone) traits, have an accuracy rate of over 80 percent on their own; used in combination, these methods can predict race with near-certainty. This is a fact.

However, there are two problems here. One is that living people generally use skin color as the primary or key index of race; while they can see some effects of skeletal structure (for instance on the face), they do not look first at a person's skeleton—or even his or her facial structure—to decide what race box to plop that person into. A person with a 'white skeleton' may actually have been very dark-skinned, and so on. Another problem is that the racial classification scheme is, biologically, bogus. As Gill explains:

> The "reality of race" ... depends more on the definition of reality than on the definition of race. If we choose to accept the system of racial taxonomy that physical anthropologists have traditionally established—major races: black, white, etc.—then one can classify

> human skeletons within it just as well as one can living humans.... I have been able to prove to myself over the years, in actual legal cases, that I am more accurate at assessing race from skeletal remains than from looking at living people standing before me. So those of us in forensic anthropology know that the skeleton reflects race, whether "real" or not.
>
> *(Gill 2000)*

In effect, then, 'racing' skeletal remains can work. We can determine, at a coarse level, whether a person was black, white, Asian, and so on. However, such classing is based on convention, not biological fact. Because the expectation for it has been institutionalized, job security for forensic anthropologists can hinge on fulfilling requests to 'race' remains. That begs the question: does it provide added value to a forensic report?

Some critics chide that racing skeletal remains is like determining whether the creature encountered in a forest glade was an elf or a hobbit or a gnome when none of those creatures really exist. Others note that 'racing' bones diverts attention from facts about a person that are likely to be more crucial to his or her identity, identification, and perhaps cause of death. For instance, someone whose remains were identified as 'Asian' may have been a business traveler who self-identified first and foremost as Norwegian. So, while it is possible to do or provide racial classification in forensics, like racial classification in clinical medicine, it can pose a problematic diversion. Just because we can do it does not mean that we always should. This is particularly so in arenas linked to the biological sciences, which squarely reject the concept as applicable in human beings.

Census-Type Data

I do not contend that all racial data are useless simply because race is not a biological entity. Race exists as a cultural construct, and it has real ramifications in terms of life chances and experiences. Therefore, we do need to collect such data sometimes.

Race data can be used for tracking whether we are meeting certain goals within particular institutions, for instance. If we did not track people by race, we would not know, for example, about particular problems in educational access or achievement, or inequities in healthcare. I myself have collected race data in the context of pediatric quality improvement research. I generally used self-reports, although the 'eyeball method' also has merit in that context, if only because it can reflect how clinical professionals are likely to categorize an individual, and so it may better predict the reaction or treatment that individuals are likely to get (as Dr. Garcia has already warned us).

Ethnicity

It does seem inappropriate, however, to use the term 'race' when what we are measuring is really not biological, and so there has been a slow shift toward use of the term 'ethnicity.' A full-fledged discussion of ethnicity could take us far from our path in this chapter, so here let's just note that, technically speaking, **ethnicity** is tied to notions of shared national or regional origins and shared culture—but not necessarily shared biological heritage.

In the United States today, ethnicity is reflected only in government census classification distinguishing 'Hispanic or Latino' people from those 'Not Hispanic or Latino.' These options are used to supplement the question of whether a person is 'American Indian or Alaskan Native,' 'Asian,' 'Black or African American,' 'Native Hawaiian or Other Pacific Islander,' or 'White' in terms of 'racial origin.' In this scheme, US Hispanics who have more long-term ancestors from

Europe (for example, Spain) than from, say, Peru, may class themselves as 'White Hispanics'. Those with more long-term ancestors from the Americas might class themselves as 'non-White Hispanics,' as may those whose long-term ancestors were mostly brought to the Americas (for example, Panama) from Africa as slaves.

The US census is just one of many census systems grappling with the problem of how to subdivide a population. Because each nation has different ideas about what kind of background data are relevant in describing the population, each nation's census forms have different options (census categories differ over time as well as space). Today, a 'black' person in England could be counted by the census there as Caribbean or African; but in Guam, as in the United States, there is only one category for black individuals. The English also subdivide 'Asians,' while in South Africa a person from India and a person from China would be counted in the same category (American Anthropological Association 2007a).

In many nations racial and ethnic designations used by the lay population differ from those used officially. In the United States the general public holds that there are plenty more ethnic groups than just 'Hispanic or Latino' (and its converse, 'Not'). As well, many feel that the categories are themselves much too broad, and so speak of Mexican Americans versus Cuban Americans and so on. We also speak of German Americans, Polish Americans, Indian Americans, African Americans, Japanese Americans, and the like. Note that all of these titles make an overt reference to peoples' shared ancestral origins and broad cultural heritage (including linguistic, religious, and so on). In effect, and although on their own they also can lead to dangerous stereotyping, ethnic classifications come much closer to the fine-grained groupings related to geographic ancestry that doctors and others would find useful.

Of course, ethnic categories can themselves lead to problematic generalizations. Some ethnic categories are so broad as to reveal insufficient detail about ancestry. Also, ethnicity has a very strong cultural component—ethnicities can be learned over time, or forgotten. White middle-class American ethnicity, for example, is attributed now to people whose ancestors came from all over Europe—even though important subgroup distinctions were once made among Europeans. Today, members of this ethnic group do not generally think much about their particular familial European history. Nevertheless, they are communally identified through commonly preferred television shows, modes of dress, and viewpoints, including views on who qualifies as a member of their group.

That said, while the US census category 'White' has long been unitary, by the time you read this even white people may face the question 'Where are you *really* from?' in terms of national origins. Although this question could fuel regressive white–white divisiveness, a more optimistic hope is that it will bolster interracial empathy, simultaneously dislodging whiteness's official status as a neutral racial category or as the default for those not forced by the state to think about their origins (Emba 2018).

Geographic Clines

As Blumenbach knew so many years ago, we know today, too, that the boundaries between so-called racial groups are porous. More than that, rather than hard dividing lines, there are blurry areas between them. Earlier we learned how traits once thought to be confined to one group actually are found (sometimes at quite high rates) among another. Pigmentation in the skin provides a good example. Many groups of people living in India are much darker than the group of people classed in America as 'black.'

Rather than being linked to so-called race, human variation is linked to geography and to the adaptive challenges posed by particular regions. Such challenges generally do not appear and disappear as if controlled by an on–off switch; they don't simply exist or not exist. Rather, they grade in and grade out, or arise incrementally and dissipate similarly, just as vegetation and air quality change when one goes up into and then descends down out of the mountains. The incrementally changing distributions of traits over geographic regions, which are related to incrementally changing challenges, are known by biologists as **geographic clines**.

The Example of Skin Pigmentation

Because they are geographic, clines are best demonstrated on maps. Maps of skin color distribution, for instance, such as in Figure 5.4, show that pigmentation or darkening of the skin with melanin is heavier in areas where people are more exposed to a certain form of ultraviolet radiation (UVB, or B-range ultraviolet radiation), and that it gets lighter farther away from those regions. Why? As Nina Jablonski and George Chaplin explain it (American Anthropological Association 2007b; see also Lenkeit 2009), melanin serves as a barrier, keeping UVB from penetrating through to the bottom layers of our skin. Still, why would blocking UVB give a population a selective advantage? The answer seems to lie within the **melanin–folate–vitamin D triangle** or, more specifically, how well this trio is balanced in relation to the body's needs given where a group is living.

Some UVB is necessary for the creation, in the human body, of vitamin D, which the skin itself creates in response to exposure. Vitamin D helps the body absorb calcium, which helps build strong bones and teeth. Sunlight-generated vitamin D works better toward calcium absorption than the dietary variety which, in any case, is not available in all environments (fish oil is a great source). However, too much vitamin D can poison us, and this is more likely to happen to those who live in sunny climates but without much melanin in their skin. In other words, darker skin in such regions serves an adaptive function. So does body-covering clothing, which of course is a cultural adaptation that didn't exist prior to the emergence of culture.

In addition to leading the skin to overproduce vitamin D, too much UVB also can lead to the destruction of folate or folic acid. Here, too, melanin is helpful: folate is essential for successful reproduction, and so those who were protected against losing it were likely to have a larger number of healthy offspring, increasing the frequency of the genes that code for more melanin production in the next generation's gene pool. Too much of this protection, however, means that not enough vitamin D is produced. Keeping the melanin–folate–vitamin D triangle balanced in relation to environmental UVB enhances group fitness, underwriting selection for lighter or darker skin depending on exposure levels.

In other words, skin pigmentation patterns around the globe—at least prior to mass migrations and jet travel—reflect the importance in attaining just enough melanin to enable ample calcium absorption via vitamin D production while still protecting folate stores from degradation given the degree of UVB exposure. Lighter skin, with less pigmentation, was naturally selected for in populations that ended up living in areas with low UVB exposure, such as Scandinavia. Conversely, populations living in certain parts of Africa and South Asia have very dark skin. Populations graded from darkness to lightness in geographically clinal patterns, just as UVB penetration of the atmosphere appears as a gradient when mapped.

An apparent exception concerns people living in regions of Canada and Alaska, where sunlight is very scarce at certain times of the year—but who are relatively dark in color. The explanation for this has to do with their diet, which is high in vitamin-D rich foods, including seal and walrus; the high degree of reflected sunlight they are subjected to when it bounces off ice and snow; and their relatively recent arrival in that region (evolution typically takes a long time).

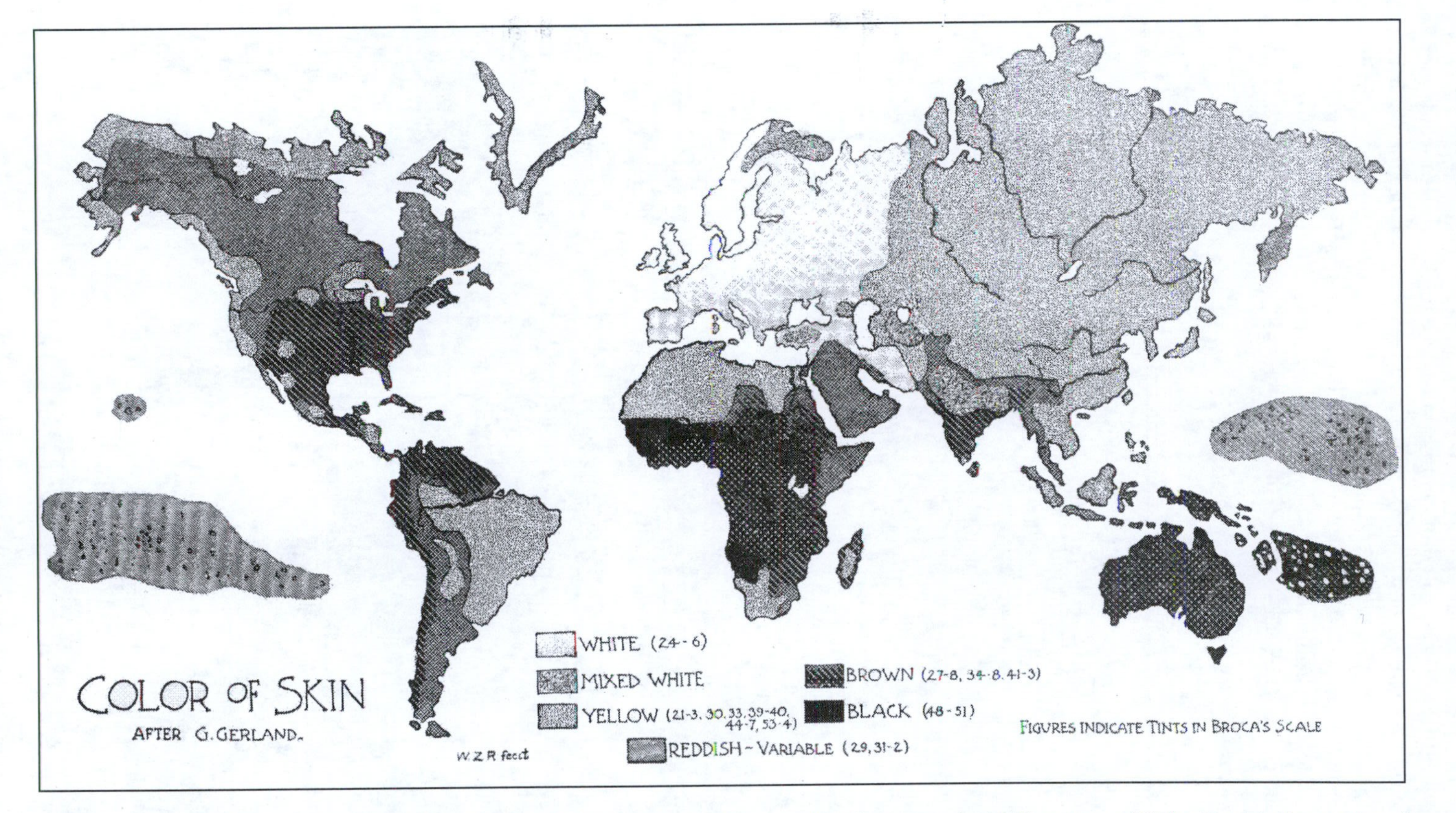

FIGURE 5.4 Map demonstrating clinal distribution of skin color published in *Popular Science* in 1897 as part of "The Racial Geography of Europe: Blondes and Brunettes III" by William Zebina Ripley.

Other Good Examples

Other traits that are good candidates for demonstrating geographic clines include hair color (see Figure 5.5), head size, ABO blood group, and whether or not one carries the sickle cell allele or can digest milk sugar as an adult. Recall that sickle cell disease is found in geographic regions where there is malaria, and lactose tolerance is found where there are sheep or other dairy animals. Or take blood type, distributions of which may be linked to past exposures to particular epidemic diseases: people with A or B types appear to be protected from particular diarrheal

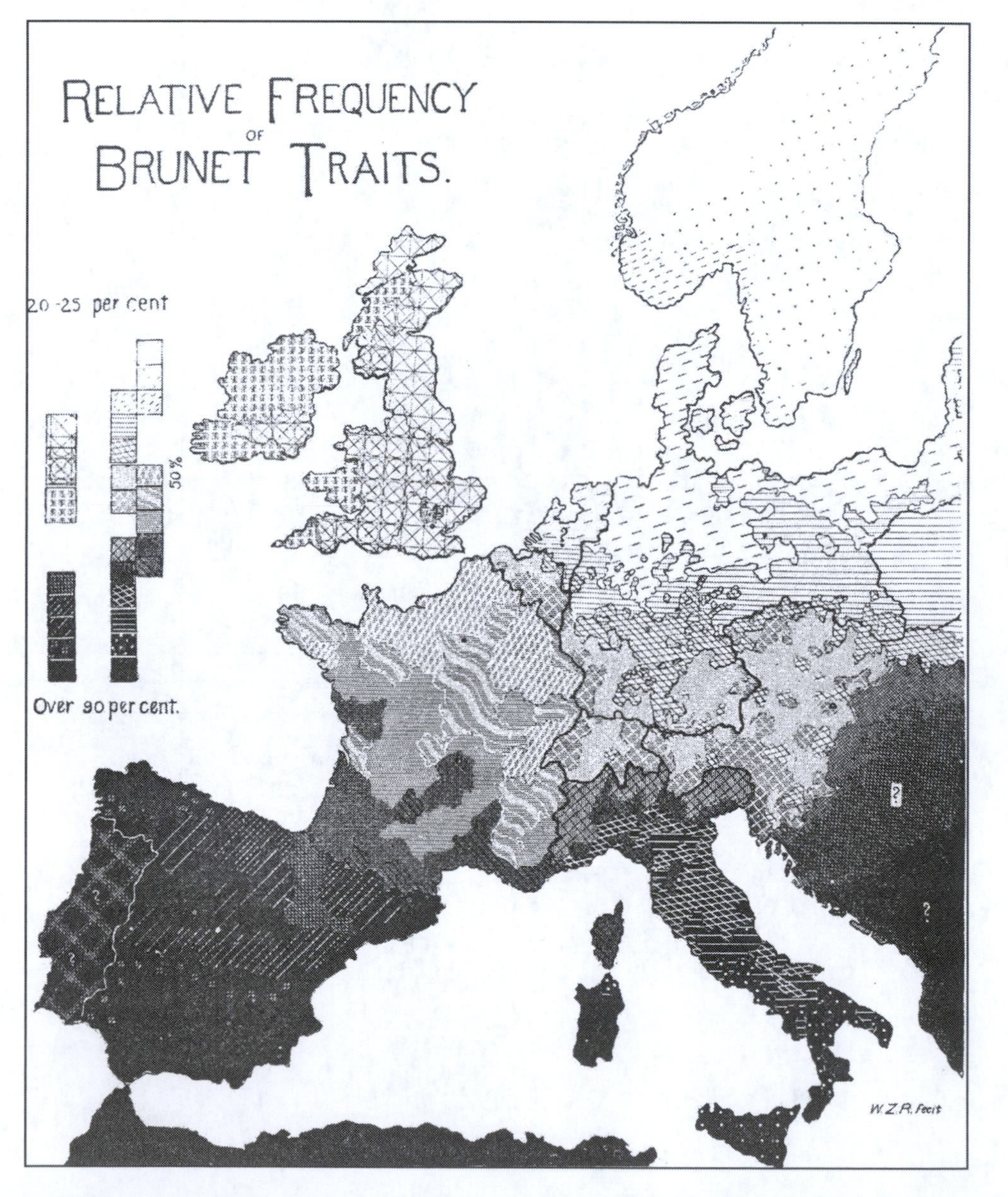

FIGURE 5.5 Map demonstrating clinal distribution of brunet hair published in *Popular Science* in 1897 as part of "The Racial Geography of Europe: Blondes and Brunettes III" by William Zebina Ripley.

diseases (and the AB type more so), many of which swept through the areas in which such blood types are most common. Type O seems to confer some advantage over malaria, and so is more commonly found in areas where that disease is historically endemic (Ridley 1999, 140–144).

Even body build is clinally distributed. For instance, a tall, slender build is of greater advantage in regions that are very hot, such as in northeastern Africa. There, having more body area allows for more sweating and thus better thermoregulation. Stocky builds, which help to conserve body heat, are better for colder areas like the Arctic—and not surprisingly are found there (McElroy and Townsend 2004, 81–82).

Bad Examples

We should not make the error of assuming, based on this logic, that all traits are clinal. Nor should we assume that contemporary pockets of particular traits always resulted from clinal pressures. Not only does widespread human migration muddy the waters; so, too, does a fact that we must always come back to: despite above-discussed exceptions, most traits are not purely biological, let alone genetic.

Take, for example, Jamaicans, who at the 2008 Olympics won eleven medals in track and field for their tiny island nation, including six gold. In 2012, they won twelve (four gold, four silver, four bronze). This was expected based on their 2008 results—but in 2008, their success took the world by surprise.

Racialist thinking sprung into action as people tried to explain their outstanding success (and racist thinking, too, as 'really hard work' and 'dedicated, improvement-directed practice' were overlooked as possible reasons). It is true that the geography of Jamaica and other environmental factors might have led to selection, in the island, for some kind of biological advantage. As David Shenk notes, the media had found one "within hours" of the 2008 successes, calling it Jamaica's "secret weapon": the protein alpha-actinin 3, which powers muscles to contract forcefully and fast. The protein, it turned out, was linked to the *ACTN3* allele—and 98 percent of all Jamaicans had it (Shenk 2011, 100–101). Yet, as Shenk reports, "no one stopped to do the math" (p. 101). It turns out that in the United States 80 percent of all people also have a copy of the allele.

Nobody stopped to do the geographic investigations either: it turns out that Jamaica is an island with numerous microclimates and varied human habitats. Moreover, Jamaicans themselves are extremely heterogeneous genetically (let alone otherwise). Like African Americans, whose heritage is much like their own, Jamaicans do not represent a genetic island; they generally have a mix of West African, Indigenous, and European ancestry, but proportions range widely in individuals, and many have Asian heritage, too. Partly because Jamaica served as a crossroads for the slave, sugar, and rum trades, the nation became very cosmopolitan. Athletic clusters, such as are found in Jamaica, "are not genetic, but systemic" (Shenk 2011, 101)—they thrive in the interaction of biology and culture.

In Jamaica, people have long revered track and field—a fact highlighted by field and track athletes themselves when interviewed. In the past, great athletes left the country for others where they might better develop their careers. However, in the late 1960s, returning athlete Dennis Johnson helped the small, poor nation begin to develop an elite training program (Shenk 2011, 109–111). This plus the intense love of running and can-do mindset fostered by Jamaican culture (who can forget the Jamaican bobsled team's emergence? [see Figure 5.6]) much better accounts for Jamaican Olympic success than does trying to explain it as clinal, let alone 'racial.' Particularly in a context where poverty limits opportunities, so that sports achievement becomes a golden ticket worthy of self-sacrifice and devotion, Jamaica's Olympic success makes sense.

FIGURE 5.6 Jamaican athletes Ricky Mcinstosh, Nelson Stokes, Dudley Stokes (*driver*), and Michael White at the start of the four-man bobsled race at the La Plagne Winter Olympics, France, 1992. Photo by Leo Campbell, courtesy of N. Christian Stokes.

Race's Social Significance and Self-Perpetuation

The information in this chapter is not meant to suggest that race does not influence our lives. It surely does, just as our behavior influences it: for example, people who seek to marry people more like themselves may be aiding the sexual selection of particular racialized traits—traits that members of a particular group find more beautiful. Indeed, some argue that this kind of selection explains the population-wide emergence and persistence of racialized traits for which we now have no satisfying adaptive explanation (for example, larger buttocks in certain populations, less body and facial hair others; see Diamond 1994).

On a more sinister level, racialization often gives way to racism—the prejudicial and biologically determinist belief that some races are better than or superior to others—a belief often grounded in reductionist thought. Such prejudice can have and has had terrible effects on populations deemed inferior as well as, some would argue, on the moral health of those who are racist. Here I should note that a lack of race literacy leads many to imagine that racism occurs only in evil or ignorant individuals. Likewise, it individualizes our views on racial firsts, such as Hiram Leong Fong or Jackie Robinson, the first Asian senator and first black major league ball player, respectively. Voicing their achievements as self-achieved firsts aptly lauds their talents. Yet it diverts attention from the fact that they were, in truth, the first people in their groups that whites *allowed* to do the things they did. The implicit message regarding the imputed overall inferiority of people of color diverts our attention from systemic or institutional factors (such as sporting league rules) that kept others like them from such professions to begin with.

Similarly, casting racism as an individual flaw blinds us to its structural embeddedness: even nice people can be racist because racism is part of the social structure as it now stands (DiAngelo 2018)

SUPPLEMENT BOX 5: ARCHAEOLOGY CAN PUT RACIST ASSUMPTIONS TO THE TEST

The first documented African American homesteader in San Diego County, Nathan "Nate" Harrison, settled atop Palomar Mountain in the later nineteenth century and lived there until his death in 1920. Reputedly reclusive as well as destitute and lazy, Harrison left a legacy of artifacts behind. Eighty years later, Seth Mallios and his students excavated the remains of the house he built and maintained on his 160-acre claim (Figure 5.7). In the process they not only unearthed, preserved, and cataloged more than 30,000 artifacts: they also helped bust some of the racist myths that, over the years, had made their way into the historical record regarding Harrison's life (Mallios 2019).

For example, as Mallios reports (2019), legend had Harrison living a lonely life, with few possessions to his name. He supposedly subsisted on handouts, as would be necessary given his reputed idleness. Some claimed that Harrison "never did a solid day's work"; indeed, he was said to be "absolutely allergic to labor of any kind."

However, the archaeological findings paint a different picture. For one thing, they suggest that Harrison often had guests, including women and children, who left behind things like cosmetics and marbles. He was not antisocial: he did not live like a hermit.

Moreover, he was industrious. The archaeological assemblage contained a wide variety of horseshoes, horseshoe nails, spurs, spur disks or rowels, and saddle buckles, as might be found

FIGURE 5.7 Aerial photograph of San Diego State University student archaeologists excavating the foundation of the stone cabin in the summer of 2006. Courtesy of Seth Mallios.

on any working ranch. Student archaeologists also uncovered multiple sheep shears and hundreds of sheep bones, dozens of which had ample butchering marks—from hatchets, knives, and saws: it was clear that these animals did not die of natural causes: they were slaughtered for consumption. Furthermore, detailed laboratory analysis revealed that nearly all of the bones came from animals that were over 3 1/2 years old at the time of death. In short, the archaeological evidence revealed that Harrison deliberately and strategically delayed slaughtering any given sheep until it could no longer produce high-grade wool.

Harrison must have done well as a sheep farmer: Mallios's team also unearthed an array of high-status objects including fancy suspenders and garters, various coins, silver-plated cutlery, and other "ornate goods," some of which had been made overseas, for instance in Germany and England. These luxuries suggest that a large proportion of Harrison's income was disposable (i.e., available for spending based on want rather than need).

Why does evidence regarding Harrison's life contradict the stories people circulated about him? In short, because those stories reflected racist assumptions. Mallios explains: "Archaeology is especially well-suited at busting historical myths. Whereas written accounts are carefully crafted by authors who are often all too aware of their audience, archaeological artifacts are originally deposited in the ground with far less agenda, bias, and purpose. It is for this reason, that they reflect a more democratic history."

Institutionalized racism is racism expressed through our social and political institutions: our criminal justice system, our banks and mortgage lenders, our hospitals, our city planning boards, and so on. It entails differential access to society's goods, services, and opportunities, such as to own land, get an education, hold a certain kind of job, or buy a house. Institutional racism makes it hard for members of disenfranchised groups to achieve their full potential. It also can lead to a group's direct physical persecution, for example through lynching, enslavement, or even genocide.

Although sometimes dramatically active, institutionalized racism often is expressed through inaction. Take (the mostly black population of) Flint, Michigan's lead-filled water supply, brought to light in 2016, or the US government's slow response to hurricanes that hit (again, mostly black) New Orleans in 2005 and (mostly Hispanic) Puerto Rico in 2017. Longer-term neglect, such as that which leaves schools serving people of color in ill-repair and ill-supplied, also evidences institutionalized racism. If we are to alter the institutional policies underwriting these kinds of inequity, racism's systemic dimensions must be acknowledged.

The effects of racial thinking can reach not just structurally upward into our institutions but also physically inward, into the genome. For example, in promoting sexual selection it can lead to small population differences that were not there to start with. For instance, Tay–Sachs is a lethal metabolic disease found more frequently in Jewish people of specifically Eastern European or Ashkenazic heritage. The gene sequence entailed in Tay–Sachs may have provided a heterozygous advantage against tuberculosis (TB), which would have been a grave threat to survival of ghettoized Jews, such as those living in Nazi-occupied Poland (McElroy and Townsend 2004, 90). In this, Tay–Sachs carriers are much like those who carry the gene for sickle cell disease: while the latter have higher survival odds in the face of malaria, the former have higher survival odds in the case of TB—the risk for which goes up when forced by racism to live in close, impoverished quarters.

One Race, Not Many

We will return to the contemporary ill effects of a belief in race later, after we build our understanding of the basis for present-day social structures. This chapter has set us up to do that successfully because older, unsubstantiated ideas about race should no longer divert us.

We have looked under the surface of the race construct to see what it attempts to summarize about human biology. We have learned how poorly fitted it is to its categorizing task due to the many variations in human ancestral habitats—including some human-made—and the resulting broad range of diverse and variously overlapping adaptations that different gene pools carry. These pools—we humans—are not so easily or clearly divisible into a handful of broad racial categories. Rather, diverse human traits are best understood as distributed along continuums, many of which have very little to do with one another. Further, along a given continuum, there are so many slight gradations in a trait that deciding where one racialized type ends and another begins can be a futile business indeed. A more productive approach to human diversity is one that uses a holistic lens.

PART II

Socio-Political and Economic Factors

In Part I, we adopted a systems point of view toward human diversity, taking into account the multi-directional interactions between various system components—interactions that sometimes led to the emergence of something—or some trait—entirely new. We learned how advantageous variation within a population can be to group survival. From a systems perspective, many of the variations now existing between human populations are the legacy of human–environment dynamics in which particular traits, under particular circumstances, provided an adaptive advantage.

Although we did not ignore humanly created environmental factors in Part I, in Part II they come to the fore. Part II focuses on the consequences for human well-being of how we subsist or make a living, and how we group and organize ourselves socially and politically. Ideas introduced in Part I about synergy take on increased importance as we explore the truly interactional, co-creational nature of our biologies and our cultures.

Cultures include pragmatic knowledge, such as how to raise certain crops, and agreements on how groups of people should organize themselves, for instance for exchanging goods and services. Part II focuses on how the socio-political and economic structures humans create to such ends, along with habitat features, relate to the kind of human diversity already described. As an added feature, Part II highlights health as an index of how well—or how poorly—a group is adapted to the circumstances in which it lives, because as goes health so goes fitness.

Chapter by Chapter Overview

Part II begins with Chapter 6's examination of health-related correlates of the forager (hunter-gatherer) way of life. It begins by exploring the ramifications of foraging for nutritional status and other aspects of human health, including that of our microbiomes (in keeping with our systems focus). Chapter 6 then describes the selective advantages of breastfeeding and of co-sleeping with infants and children, and explains how and why these evolutionarily appropriate patterns might be bypassed in contemporary circumstances. In this way, we begin to ask whether some 'modern' practices may in fact make little if any evolutionary sense.

Foraging is the most basic human way of making a living. A more intense method is settled agriculture. After characterizing the subsistence continuum and intensification process implied here, Chapter 7 describes the impact that a shift from foraging to agriculture can have for socio-political and economic organization. We ask how and why domestication happened where and when it did. We then investigate the ways that, through complex feedback loops, sedentism, surplus accumulation, specialization, and social stratification evolved, and how they led (as they continue to lead) to poorer health for large portions of agricultural populations. Agriculture-related health effects seen in the archaeological record, including the skeletal record, are examined; and to provide a point of contrast, we also explore the biocultural effects of industrialized agriculture—the kind we practice today.

In the first two chapters of this section, then, we move from our initial explorations of the adaptations that fostered humanity's emergence, via natural selection, to those adaptations made by humanity in response to the very ways of life that behaviorally modern humans had created. We examine how subsistence strategies affect not only health but also social structure and how then, via social structure, they affect health again. This kind of multiplicative feedback can and does happen because human systems work much like complex adaptive systems, in which everything is connected.

With Chapter 8, we begin to bring in a few new theoretical frames to further our quest to comprehend the bioculturally interactive nature of human variation. After instruction in the basic language of epidemiology, and a review of the role of the human immune system in the body–environment relationship, readers learn to use the disease ecology perspective to explain how the agricultural lifestyle contributed to the spread of infectious diseases. This lesson is extended to show how humanly created changes in our environment today have fostered drug resistance (or increased staying power) in certain pathogens. As well, our own bodies have adapted, in certain ways, in certain populations, to some pathogens, so that some populations are more (or less) prone to certain disease than others.

A second framework emphasized in Part II focuses on what is termed the 'political economy' (put loosely, contemporary global capitalism and its antecedents). The biocultural diversity fostered by, for instance, the class system is certainly humanly created, and not all for the good. In Chapter 9, readers learn to use the political economy perspective to gain insight into the unequal impact of social stratification on different human groups, particularly as seen in health indicators and outcomes. But Part II does not end on a negative note; readers learn about the social justice approach as one way to help combat the humanly induced biocultural variation reflected in health inequity statistics.

Central Lessons of Part II

By the end of Part II of the book, readers will have sharpened their understanding of the relationship between biology and culture entailed in the assertion that we are biocultural beings. They will be able to identify and explain how our social structures—which humans, by dint of their evolved cultural capacities, have created—can themselves foster and maintain population-level biocultural differences. These differences range from naturally selected-for genetically based resistance to particular diseases in some populations, to developmentally derived cognitive deficits related to toxic exposures or nutrient shortages in others, to context-dependent enhancements to athletic abilities in still other groups.

As well as being able to identify and explain such differences, readers will have gained practice in seeing connections. Readers will have grown more skilled in seeing how the ways that a society engages with the environment (through their subsistence mode) will have cascading, ripple-out effects, and in understanding how populations that seem to be living in wholly separate worlds may in fact be intimately implicated in each other's life chances and challenges.

6

FORAGING

A Human Baseline

This chapter prepares you to:

- Characterize the typical foraging or hunter-gatherer subsistence strategy
- Explain the ramifications of foraging for nutritional status and other aspects of human health
- Describe the selective advantages of breastfeeding and explain how and why the evolutionarily appropriate pattern might be bypassed
- Describe the selective advantages of co-sleeping and explain how and why the evolutionarily appropriate pattern might be bypassed

In previous chapters, we learned about human adaptation, including the kinds of adaptation that led to the evolution of behaviorally modern human beings—people like us. Our original **subsistence strategy**—our approach to the task of extracting food—was to hunt and gather, also known as **foraging**. We collected all kinds of plants, insects, and animals to eat. With the emergence of our capacity for culture, we spread out both within and beyond Africa.

Because different regions have different kinds of natural resources, one of our first major tasks on moving to any new environment entailed figuring out how to adapt our foraging skills so that we might extract ample amounts of food and water locally. In some geographic locations, exploiting resources that differed quite substantially from those we had previously evolved to depend on also meant we had to develop new tastes. There were various fruits, leaves, herbs, grains, legumes, fish, crustaceans, insects, reptiles, birds, and animals wherever we went. Yet, no matter how our diets differed from place to place, we remained foragers for thousands and thousands of years. In this chapter we explore what our forager legacy really means in terms of human biocultural variation.

Nutrition Basics

When we eat, we capture and repurpose energy originally sent to Earth by our sun. The sun shines down on the plants, the plants grow with its energy, and we eat some of the plants as well as some of the animals that eat them first. We are aided in the process by microbial helpers that live inside of our digestive systems and assist us in extracting nourishment as well as in other ways, soon described (see also Chapter 1). This nourishment involves three fundamental substances: protein, carbohydrate, and fat.

Protein is a body's building block. It is in turn created from amino acids, found in diverse combinations in various foods items. Complete protein sources such as meat do exist, but we also can get protein from juxtaposing foods with complementary arrays of amino acids. Some examples are corn and beans, whole grain cereal and milk, rice and peas, and peanut butter and whole grain bread.

While protein provides matter, energy—the energy to build body parts from that matter, or to use it, for instance, in running or picking onions—comes from carbohydrates (sugars, starches). Fat serves as an energy store, and for insulation; it also is essential for the proper functioning of the nervous system.

The three key classes of nutrients—protein, carbohydrate, and fat—are called **macronutrients**. They are pulled from various foods through the digestive process and then put back together in different ways to structure our bodies or serve as fuel when needed. To correctly process the macronutrients we also need trace amounts of what are therefore termed **micronutrients**: vitamins and minerals without which the biochemical processes entailed in keeping us going just cannot happen. Likewise, water is essential to human survival.

So is fiber, an indigestible carbohydrate. Fiber not only keeps the contents of our digestive tract moving along; it feeds key inhabitants of our gut's microbiome. Intestinal bacteria get their fuel from fermenting the fiber we eat. In the process, they produce enzymes that help us break down foods, as well as certain vitamins and other compounds vital to our body's functions. Some help keep the gut lining healthy, for instance by lowering inflammation; others regulate appetite, immune function, mood, or **metabolism** (the biochemical process of breaking down and repurposing food components for bodily use). Some fight pathogens, often through raising the acidity of the intestinal environment (Schueller 2014).

Prior to the days of the supermarket and trans- and intercontinental shipping (and even still today in many regions of the world) macronutrient and micronutrient sources varied depending on what was geographically available. For people who lived near rivers or bodies of water, fish and crustaceans could provide many nutrients; not so for those in the desert. Likewise, certain fruits that were abundant in central Africa simply were not available to those living in the north.

Further, many of the food items we think of today as essential to a good diet were unknown in older times. Bread as we know it, for instance, did not exist until well after the invention of agriculture. Milk, too, was not standard fare. Indeed, to digest such things properly, populations exposed to them adapted genetically, over time. Given that, what was our diet really like prior to farming, and how did we keep ourselves fed?

The Foraged Diet

Our human ancestors, like most foragers today, were relatively tall and lean. We know this from skeletal data and by making ethnographic extrapolations. We also know that their lifestyle demanded plenty of physical activity. Therefore, as Stanley Boyd Eaton and colleagues argue, they "must have existed within a high energy throughput metabolic environment characterized by both greater caloric output and greater caloric intake than is now the rule" (Eaton, Eaton, and Konnor 1997, 208). In other words, they ate but also used (metabolized) lots of food.

FIGURE 6.1 Gathering food in the Okavango Delta, Botswana, 2004.

What kind of food did they eat? One common misconception about our early diet is that it was heavy on meat; this is reflected in the misnomer 'hunter-gatherer,' and it is why some scholars prefer the term 'forager.' In reality, most food likely came from plants (see Figure 6.1). When I say 'plants,' I do not mean cereal crops such as oats, rice, maize (corn), barley, or wheat—those are more commonly cultivated than foraged. Rather, I am talking about roots, fruits, nuts, legumes, leaves, and other noncereal plant items. These make up about two-thirds of the average forager subsistence base.

Moreover, we often foraged for insects and arachnids (in common parlance, 'bugs'); this would have been quite common in the past, and even today many people gather and eat them. Indeed, bugs make up a substantial portion of the protein that many people get in their diet. They are much more efficient at converting what they eat into 'meat' than are cattle, and their 'meat' is highly nutritious—with grasshopper, for instance, providing three times the protein of beef, ounce for ounce. Plus, bugs have a significantly smaller footprint on the environment than livestock; they are "natural recyclers" and actually thrive under low-cost conditions. Bugs can be quite tasty, too (Goodyear 2011).

If the thought of eating grasshoppers and their ilk elicits an ethnocentric 'ugh,' even in light of sustainability issues, consider the following. Many Americans and Europeans, for example, happily eat what to others is putrid, moldy, and rotten: cheese, which is indeed made by decaying milk, sometimes with added bacteria, too. Furthermore, processed foods contain bugs. According to the US Food and Drug Administration, peanut butter can have up to 30 insect fragments per 100 grams before it is subject to seizure or grounds for a citation. We'll talk more about the industrial food system in the next chapter.

What of the quality of the forager diet? Again, we turn to Eaton and colleagues, who have studied the question extensively. They conclude that "our preagricultural ancestors would have had an intake of most vitamins and minerals much in excess of currently recommended dietary allowances" (1997, 208). Indeed, their diet was "nutrient rich" (212)—except with proportionally lower carbohydrate and fat intake. This is important, because of the links between health problems such as diabetes and heart disease and an overabundance in the diet of carbohydrates and fat (especially saturated fat, which is much less abundant in the typical forager diet). Further, any excess protein ingested by foragers generally would have been offset by higher fiber and potassium intake and lower levels of sodium, as well as more physical activity.

Also, higher fiber intake would have boosted the diversity of the gut microbiome, which is important because, as mentioned earlier, various microbes serve various vital functions for the human body. To best grasp this, let's look to recent advances in comparative research, such as Jeff Leach and Rob Knight have been doing (see Schueller 2014). The Hadza of Tanzania, the only remaining true foragers on the earth today, are central here. They get 100-plus grams of fiber a day, on average, mostly from baobab fruit and wild tubers. Compare that to the US average of 15.

Largely because of their high-fiber diet, the Hadza have the most diverse microbiome yet seen. This seems to protect them from some cancers, diabetes, heart disease, and obesity, as well as from depression and anxiety (Schueller 2014). In evolutionary terms, keeping their human hosts healthy and happy, or able and wanting to socialize, seems a good survival strategy for organisms that only get fed when we eat, and spread when we interact (that is, through human–human contact; Kohn 2015). Likewise, getting us to eat not just any foods but those that they (our microbial friends) find nourishing offers the microbiome a selective advantage. This explains why some human food choices or cravings are microbially instigated (Sheikh 2017). Not all of those cravings are to our benefit, but here we can learn from the Hadza: keeping our diet nutritionally sound—not feeding and thereby propagating unhelpful microbes—helps ensure that our microbially induced food choices serve us well.

So much for solids. What about liquids? Water, in pure form, is not always in constant or easy supply. People who lived in geographic regions without ample potable (clean, drinkable) water sources had to discover or invent ways to get good water or they would not have survived: we are ill-fitted to a water-free environment. We can get water from food, of course, such as juicy melons or tuberous roots. People who raise milk-giving animals can get water from fluid milk, once they evolve the adaptive ability to digest it. Cleanly prepared beverages that contain water, such as beer or wine, also provide fluids.

Most foragers do not need to make alcoholic beverages—but in communities in which large quantities of beer or wine are regularly drunk, often because potable water is lacking, people who are less vulnerable to over-intoxication or drunkenness have a survival advantage. Who is less vulnerable? Group members able to increase production of the enzyme that breaks down alcohol. In beer or wine drinking cultures of old, these group members had increased fitness; more of their genes were passed along into the next generation. Further, cultures with alcoholic beverages central to the diet have rules regarding things like when, with whom, where, and how much to drink. In comparison, those whose ancestors hail from regions with plenty of good, clean water may be quite vulnerable to alcohol's ill effects (Ridley 1999, 191).

Drinks aside, the data clearly support the hypothesis that our early human ancestors' forager diet was different than ours in terms of the ratio of plant to animal items, and the balance of protein, carbohydrate, and fat. Geographic variation affected the specifics of what they took into their bodies for nourishment, and how they metabolized foodstuffs. It had secondary effects on other bodily systems, such as cardiac health—and on culture, too.

It is less patently obvious that the nutritional value of the food items themselves was different. Yet, based on current agricultural science and ethnographic data from contemporary foragers, we can assuredly say that it was.

First, most foraged food items are consumed within hours of being collected and with minimal processing. That is, they are eaten fresher and rawer than most food today. If I pick an apple from a tree and eat it directly or even one week later, I get a better deal nutritionally speaking than I get when I buy an apple (or potato, or whatnot) from the grocery store—one that's traveled on a truck across the country and had been in cold storage for some time even prior to making its trip. As time passes it loses some freshness and nutrients degrade or seep away at least somewhat. A lack of attention to storage conditions, such as temperature and humidity, can make matters worse, as can infestation with pests, including molds and fungi. When nutrients break down or degrade their **bioavailability** to the eater—the degree to which the eater can extract and make use of them biologically—is negatively affected.

Cooking also can break down components in a food item. This increases the bioavailability of certain nutrients by freeing them up, so to speak (see Chapter 4). However, cooking also can leach nutrients from food or decrease their nutritional value. When we boil certain vegetables, for instance, some of their nutrients are left behind in the pot water; and the longer we boil them the more nutrients are broken down or drawn out. Some forms of cooking add unhealthful components, such as when frying, or grilling on certain charcoals. Toxic compounds also can be introduced to food on contact with cookware or cups and dishes, for instance in the form of lead used in glazes. Processing foods with added chemicals further changes their nutritional profile. Hydrogenating the fat in peanut butter so that the paste and oil do not separate may be a boon to those who hate to stir the stuff, but the process creates a form of fat that definitely is not good for us.

Another factor is what goes into the food to begin with, such as from the soil that the food is grown in. Soil has (or does not have) nutrients. These nutrients are plants' food inputs. To some degree, whatever is in the soil goes into the plant. Fertile, nutrient-rich soil produces more nutritious carrots than depleted soil does. Soil aside, plants grown in the wild contain certain components or chemical compounds—**phytochemicals**—that can provide ingrown or natural protection from predators and disease. Artificially selected plant species often have had these protective phytochemicals bred out, not purposively but as a secondary effect because chemical pesticides allow plants without them to survive. Yet many phytochemicals found in wild species have antimicrobial or antioxidant properties (and provide more robust colors and flavors). These phytochemicals often are healthful for human beings to eat, but typical industrial supermarket produce is lacking in them. (see Pollan 2006).

For animals, something similar happens, but in this case in addition to being linked simply to inputs (think of grass-fed versus feed-lot beef), outputs become important as well: how much the animal exercises affects the fat content of its meat. Wild game is much leaner than most supermarket meat, coming as it does from animals that not only romp and play more than feed-lot or factory-grown 'animal product sources' but that also run from predators. Indeed, the fat wild game does contain is proportionally more 'good' than 'bad' (saturated).

When we compare and contrast present-day human populations eating a modern diet to early humans eating as foragers would have, the difference to health is palpable. Biocultural diversity in rates of heart disease, diabetes, digestive disorders, and so forth reflect diversity in human subsistence patterns.

The Foraging Life

Foraging is not simply a dietary strategy; it is a lifestyle. I provided a snapshot from the Hadza earlier, but to better envision this way of life, let's consider the !Kung, whose lifestyle is loosely representative of the lifestyle that must have been experienced and enacted by our ancestors. The following description is drawn from the work of Richard Lee (2000 [1968]).

!Kung Foragers

Lee worked with !Kung people to get data regarding work time allocation, nutrition, calories taken in, and so forth (note that the exclamation point printed prior to 'Kung' represents a clicking sound that we do not make in English). The !Kung live in the Kalahari Desert, which spans the middle-west of the southern-most part of Africa (see Figure 6.2). The group of !Kung that Lee and colleagues worked with lived in the Dobe area, which had eight permanent water holes and, at the time of the research, fourteen independent and self-sufficient camp groups.

The research was conducted more than fifty years ago, and things have changed drastically for the !Kung as a result of incursions into their territory and an onslaught of modernization efforts. When they hosted Richard Lee, however, they subsisted nearly totally via foraging. As such, the old data are a vital source of information. In keeping with ethnographic tradition, I speak of the findings in the 'ethnographic present,' describing the snapshot they present as if current, to enliven the description.

Each !Kung camp fends for itself. Only rarely are foodstuffs traded between camps. However, people move often, whether between camps or as camps themselves are moved to new locations in the region. Moving is a constant feature of !Kung life; most camps at least change locations with the seasons. In summer, when the rains come, there are many temporary water holes to camp around. In the dry season, camps cluster around the permanent water holes. Lee's fieldwork spanned nearly one and one-half years, which is important because of the impact of seasons on !Kung foraging life. Short studies, in which researchers buzz in and out of a community quickly, lack a comparable comprehensive perspective.

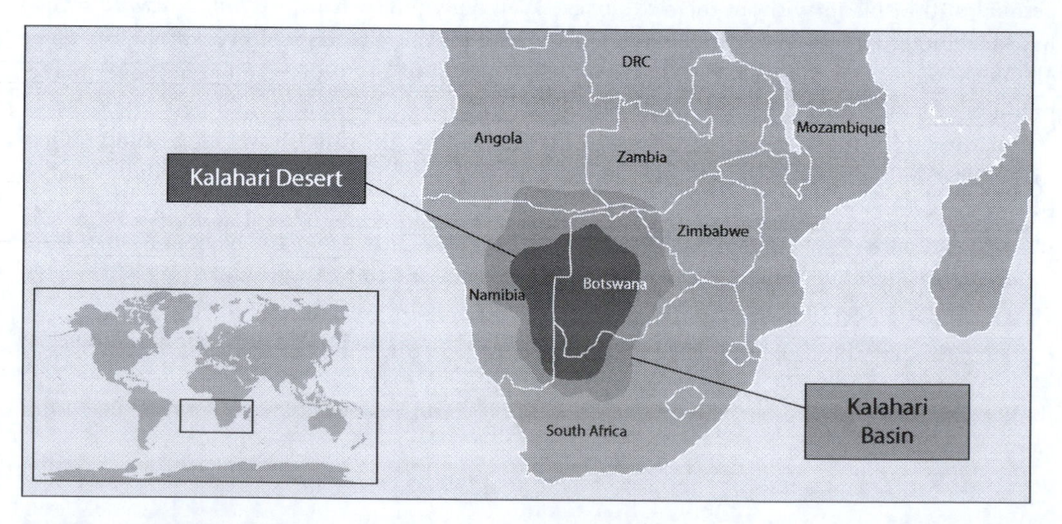

FIGURE 6.2 The Kalahari Desert (shown in dark gray) and the surrounding Kalahari Basin (in lighter gray). Redrawn from original by Cay Leytham-Powell, CC BY-SA 4.0.

SUPPLEMENT BOX 6: NO ROOM FOR AN ARROGANT FORAGER

In an oft-quoted story regarding his time with the !Kung (Lee 1969), Richard Lee contributes an ox to a wintertime feast as a thank you to his forager hosts. In return, Lee receives a barrage of insults: people derided the ox as too thin, too stringy, too old, a "bag of bones." He endures ten days of such harassment in sadness and confusion, looking all the while for a solution.

When the time comes to slaughter and share out the ox, Lee is relieved to find that his hosts had been joking. The ox provided enough meat for two days and nights of fasting and dancing.

Lee understands that he'd been being teased—but he cannot tell why. So he asks a friend about the incident. The friend explains, smiling, "It is our way":

> Say there is a Bushman who has been hunting. He must not come home and announce like a braggard, "I have killed a big one in the bush!" He must first sit down in silence until I or someone else comes up to his fire and asks, "What did you see today?" He replies quietly, "Ah, I'm no good for hunting. I saw nothing at all [pause] just a little tiny one." Then I smile to myself … because I know he has killed something big.
>
> In the morning we make up a party of four or five people to cut up and carry the meat back to the camp. When we arrive at the kill we examine it and cry out, "You mean to say you have dragged us all the way out here in order to make us cart home your pile of bones? Oh, if I had known it was this thin I wouldn't have come." Another one pipes up, "People, to think I gave up a nice day in the shade for this. At home we may be hungry but at least we have nice cool water to drink." If the horns are big, someone says, "Did you think that somehow you were going to boil down the horns for soup?" To all this you must respond in kind. "I agree," you say, "this one is not worth the effort; let's just cook the liver for strength and leave the rest for the hyenas …." Then you set to work nevertheless; butcher the animal, carry the meat back to the camp and everyone eats.

Later, another !Kung friend corroborated this lesson on "obligatory insults," but Lee remained confused as to why someone who would be sharing out meat should be treated so disdainfully. "Arrogance" was his friend's answer:

> Yes, when a young man kills much meat he comes to think of himself as a chief or a big man, and he thinks of the rest of us as his servants or inferiors. We can't accept this. We refuse one who boasts, for someday his pride will make him kill somebody. So we always speak of his meat as worthless. This way we cool his heart and make him gentle.

The message was clear: there is no place in forager society for self-superiority. Those who tend toward arrogance must be pulled down from their self-built pedestals before they begin to threaten the egalitarian social order, for instance by not sharing freely, bossiness, or condescension.

Diet

From any !Kung water hole, food is within walking distance, although the distance walked does increase to up to ten to fifteen miles in the dry season. In any case, people generally go to collect food every three or four days; a supply of no more than a few days' worth of food is usually on hand, and whatever has been hunted or gathered is always shared with the camp, and so quickly dissipated. Even if people did want to store food, it would not necessarily last well. The !Kung have no refrigerators or pest-proof storage lockers.

Food kept on hand, however shortly, consists mostly of plant-derived items. The main staple item in the !Kung diet is the mongongo nut. A **dietary staple** is something that a group eats a lot of (bread, tortillas, or rice, for example), and most staples are plant-based. Grown on trees, mongongo nuts are wrapped in a thin layer of edible fruit flesh. Lee reports that the nut is so abundant that more pounds of it rot on the ground than get eaten—even though mongongo nuts account for half of the vegetable diet of the !Kung, weight-wise. Average daily consumption is 300 nuts per person, and they are eaten throughout the year.

The !Kung also have access to eighty-four other species of edible plants, most of which are roots and bulbs or fruits, berries, and melons. Small mammals, birds, reptiles, and insects make up the remaining proportion of their diet (see Figure 6.3).

No matter what else is eaten, mongongo nuts are central. On average, the nuts contain five times the calories and ten times the proteins of cooked cereal crops. The average daily portion yields about 56 grams of protein and 1,260 calories. When we include other food sources, the average !Kung daily intake is 2,140 calories. Fiber data were not collected but we do know that the !Kung average 93.1 grams of protein each day. By weight, meat accounts for 37 percent of this. Mongongo nuts account for 33 percent and other vegetable foods for 30 percent, putting the total vegetal contribution to the !Kung diet—pound for pound—at least at 63 percent; note, however, that plant food generally weighs less per cubic inch, therefore accounting for more dietary volume, than meat.

What does this all mean? Besides providing ample evidence for the argument that foragers subsist largely on plants, it also demonstrates that the !Kung are well-nourished, at least insofar as what has been measured. Today's United States Department of Agriculture (USDA) recommended daily allowance (RDA) of protein for adults is 46–56 grams. Recommended caloric intake for active women and men of average build and nineteen to fifty years old is 2,200–2,400 and 2,800–3,000, respectively (it's 1,800–2,000 and 2,200–2,400 for sedentary women and men). !Kung adults are very physically active—but they also are generally of smaller stature than

FIGURE 6.3 San people in Deception Valley, Botswana, start a fire by rubbing sticks together, 2005. Photo by Isewell via Wikimedia Commons, CC-BY-SA-2.5.

the average American. According to Lee the RDA for them was estimated at 1,975 calories and 60 grams of protein per person per day.

Hours Worked

So much for !Kung diet and nutrition status. Now we must ask (as did Lee) about the time and effort it takes to feed oneself and one's family as a forager. While men and women both work a roughly equivalent number of hours per week, women (who focus more on gathering than hunting) bring home two to three times as much food as do men. The number of hours worked per week specifically gathering or hunting for food is what we would call 'part time.' In fact, it is twelve to nineteen hours—not even 'half time' by our standards. That is all it takes for !Kung people to get the food they need to eat. With the living so easy, children are not expected to contribute substantially to foraging efforts; this will not happen until they are married (between the ages of fifteen and twenty here). The very old and infirm, also, are well-provisioned by able-bodied adults in the camps.

Lee's foraging figures do not include time spent making and fixing tools. Gross, based on Lee's 1979 report on the same, put the figure for that at 6.3 hours per week, complementing his average of 17.1 hours per week in food-getting labor (1992, 265). Lee did not include housework (for example, food preparation, firewood collection), which Gross puts at 18.9 hours per week. Neither Lee nor Gross includes child-care.

Hourly workweek statistics for the United States and other workforces also ignore the time it takes at home to feed, wash, and otherwise rejuvenate the workforce for another day on the job. Home-based maintenance of human resources is essential for production, but as it is not paid for directly by employers it is left out of the waged labor ledger books. Moreover, and in any case, even with the labor necessary to re-create or maintain the means of production included in weekly work-hour totals, the !Kung, and those with comparable forager lifestyles, have quite a lot of time left over for leisure pursuits, including spending 'quality time' with extended family and friends (see Figure 6.4).

FIGURE 6.4 Hazda group at leisure in Tanzania. Photo by Rachel Naomi David.

Food Options

At this point you might ask if the !Kung work only part-time getting food because there is not much food to get. Maybe they spend their off-hours lying around too weak to enjoy themselves? Not at all.

It turns out that the !Kung, and likely too our forager ancestors, gather only a small proportion of all the edible plants and animals known to them. Ninety percent of the !Kung's plant-based diet is drawn from 23 Dobe-area species when there are 85 edible species known to be available. Animal wise, the local area has 223 species known to the !Kung, who classify 54 of these as edible—and regularly hunt only 17. Why? They are being picky, eating only what best appeals. They can afford to be picky because the environment amply provides. As per the nutrition data recounted earlier, the !Kung get not only plenty to eat but good nutrition also via the foodstuffs that taste best to them.

Seasonal Stress

I do not want to paint too rosy a picture here. Dobe seasonality does take a toll. For most Western people, at least in the middle classes and above, there are no seasons. If we want strawberries in December, we can get them. Many of us have freezers, and some know how to can vegetables and preserve fruits. If something perishable (fish, fruit) is on sale or in good supply we can buy a bunch of it, freeze or otherwise preserve it, and eat it six or more months from now. We also have access, should we desire it, to nonlocal food, which extends quite substantially our menu of available eating.

If we lived as traditional foragers, our diets would shift with the seasons. Yet **seasonality** means more than this. Not only would what we eat change: how much we eat would also be affected. For many of the few foraging groups left in the world, the difficult season is much more difficult than it was for the !Kung when Lee stayed with them. This is largely due to the marginal lands they have been pushed onto and incursions into their traditional foraging grounds. Be that as it may, even for some pristine foraging traditions there are certain times of the year in which certain kinds of food—and in dire cases all foods—become very scarce. Giving the body a cyclical break from constantly having to process certain kinds of nutrients can be a very good thing; but without food for long enough, people do suffer and can even die.

Exceptions to the Rule

Anthropologists are famous for taking exception. When people make generalizations about humankind, someone always manages an anthropological veto: "Not among the [insert group name here]!"

This does not mean generalizations have no value for enhancing understanding. It is helpful to know that some aspects of human life are more or less universal. However, we must be careful that generalizations remain a starting point for discussion and investigation rather than a stopping point. Otherwise, we are working with stereotypes, not generalizations (Galanti 2008, 7). Furthermore, the uncontextualized assertion that *all* foragers everywhere and throughout time have eaten a well-balanced, nutritious, mostly vegetal diet (and without too much work having gone into it) would be plain wrong. In general, it has been the case—but not everywhere and not always.

Exploring an exception here is thus a necessary step in our learning. It will be illustrative not only of biocultural diversity but also of one of the key principles behind the initial evolution of such diversity: geographic variation in resources. People adapted to a myriad of environments when they spread out around the world, and although many ended up like the !Kung in fairly hospitable areas, others had to invent new ways of hunting and gathering better fitted to their new environments.

The Inuit Are Hunters

The Inuit, who live in the Arctic areas of lands now called Canada, Alaska, and Greenland, are one such group. As described by Ann McElroy and Patricia Townsend (2004, 15–28), the Inuit survive (traditionally) mainly by hunting. Besides the challenge of finding enough to eat in order to stay well-nourished in a frigid region that supports almost no edible plants and very few animal species, the Inuit also must keep from freezing to death.

Inuit bodies differ from !Kung bodies. As we know from Chapter 5, these differences are not 'racial.' They have to do with the unique geographic regions in which each group lives. The Inuit are quite stocky (although not fat), and they have somewhat shorter limbs. Their bodies have evolved to have less surface area than bodies adapted to hotter climates. They are, therefore, less subject to what in the Arctic habitat would be excessive cooling through too much body heat loss. There are internal differences as well, for instance in terms of the kind of fat Inuit have—they are able to keep making 'brown' or extra-body-warming fat into adulthood whereas other humans stop making it in childhood. Plus, Inuit blood vessels can very quickly dilate; quick dilation is helpful to reheat cold fingers and toes (McElroy and Townsend 2004, 81–82).

Bodily adaptations are helpful to be sure. Without a cultural boost, however, via adaptive clothing, shelter, and food-related practices, Inuit survival in the Arctic would be impossible. Yet survive they have done, for thousands of years, and very well indeed (until industrialized nations began to invade). Although Inuit culture as a whole provides much to learn from—take for example the bone sunglasses or goggles that the Inuit make to protect against snow blindness, their social structure and childrearing practices, or their cosmological relationship with polar bears—here we will focus on how they make a living, traditionally. We will ask how they have subsisted in an environment where most of us, without Inuit knowledge and skills, would perish quickly (see Figure 6.5).

The Inuit do eat some plants. Berries grow abundantly in August, for instance. Inuit also gather and eat kelp or seaweed, sour grass, and sorrel. Other plants ingested by the Inuit are those dried for use as medicinal herbs and teas.

FIGURE 6.5 Inuit in Canada killing salmon with spears. Canadian Geological Survey, Frank and Frances Carpenter Collection, Library of Congress, LC-USZ62-112765.

As a result of their minimal dietary dependence on plants, Inuit carbohydrate intake is relatively low: about one-fifth less than average American consumption, and maybe one-tenth of the average consumption of carbohydrates in poorer nations. Rather than subsisting mainly on plants, the Inuit subsist mainly on animals. They are hunters.

The energy costs of Inuit subsistence are large. About 160 days a year are spent in high-intensity activities such as hunting, trapping, fishing, and traveling by dogsled (which entails lots of running, pushing, and pulling). When in transit, a new snow house must be built or tent erected every night. Still, the Inuit spend more than half of their time relaxing, visiting each other, trading, having feasts, or building and repairing their equipment.

With such great reliance on animal sources for food, the Inuit protein intake is relatively high: about 200 grams per day. Fat, too, is eaten in what to the !Kung—and even to today's Americans—would be huge quantities.

Nevertheless, the Inuit are neither fat nor vastly unhealthy. For one thing, in part due to the work they do to survive, the Inuit overall are lean and muscular. Men expend between about 2,700 calories a day (3,100 in peak hunting periods). Women do not hunt, but they can burn nearly as many calories, particularly when carrying children on their backs, digging for clams, or breastfeeding. Remember that sedentary US women and men—that is, most US adults—use only 1,800–2,000 and 2,200–2,400 calories each day, respectively.

The nutritional content of the diet also contributes to Inuit health status, typified by healthfully low blood pressure, low cholesterol levels, and low rates of heart disease. How can that be with a diet so high in flesh foods? For one thing, the meat they eat is significantly lower in saturated ('bad') fats than present-day feed-lot beef, and higher in polyunsaturated ('good') fatty acids. The omega-3 polyunsaturated fatty acids found in fish and blubbery animals, such as seal, whale, and polar bear, boost the healthfulness of the mix. Moreover, while the foodstuffs eaten are low in vitamin C, the Inuit maximize access to the C that is there by eating meat raw; eating the plankton in the stomachs of raw fish, walrus, and caribou; and eating whale skin.

However, calcium intake is less than optimal in the Inuit diet. The ingestion of dried fish and bird bones as well as the soft parts of animal bones to some degree compensates for this (Inuit molars and jaws are relatively strong, which is adaptive when dealing with this kind of hard, crunchy food). Still, some Inuit do suffer from mild deficiencies in calcium, particularly in winter when sunlight is scarce, and so the body makes less vitamin D, for which exposure to sunlight is necessary. With less vitamin D, calcium absorption is less effective. In some populations a lack of sunlight without supplementation translates to rickets among children. Among the Inuit, who nurse their children extensively, this is not a problem (breast milk has calcium). Nursing women, of course, bear the burden of this, and Inuit adults as a whole—particularly those over 40—have an elevated risk of bone loss and fractures (high phosphorus intake contributes to this).

In summary, the traditional Inuit subsistence strategy provides a reminder of the general lesson that there always are exceptions to the rule (in this case, the rule of a two-thirds vegetal diet). While the Inuit are mainly meat eaters, if we had chosen a mainly vegetarian group to illustrate this fact the general point would remain the same. So would the specific point made: what foraging constitutes depends on a group's circumstances. In this case, our focus is geographical circumstance, and what is therefore available. Internal biological adaptations, such as in regard to digestive processes, also come into play.

Real Loss, Naive Nostalgia

Before moving on it is worth observing that the traditional Inuit subsistence strategy is on the wane, due in part to the way the current economy has affected the environment (and see Schell 2012). Distantly generated industrial pollution has made its way into the Arctic food chain.

Climate change also has taken a toll, as Creighton Backpack Journalism students learned when producing *Mother Kuskokwim*, a gripping film about life in the Yukon–Kuskokwim River Delta of Alaska (O'Keefe and Guthrie 2014). The loss or fraying of connections to the land as diet has changed has had a negative impact. In Canada for instance, Inuit today are more likely than others to experience food insecurity. They have a shorter life expectancy than most Canadians, with more high blood pressure and arthritis. They also engage in more heavy drinking, perhaps to try to dull the pain entailed in losing access to their cultural heritage (Wallace 2014).

As the foregoing suggests, the loss of foraging traditions compromises much more than nutritional status. To enhance the latter, however, some people living where industrialized food systems predominate promote what they call 'stone age' eating or the **paleolithic diet**. It is the case that many present-day substances passing as food or drink are unhealthful. The 'clean eating' movement, which prioritizes whole (unrefined) food without preservatives or added sugar and salt, is responsive to this fact. The 'paleo' diet takes things further, however; and people interested in such a diet should proceed with caution, not least because there really was no single 'stone age' or 'paleo' diet. What any group ate depended on geography. Further, many 'paleo' plans rely much too much on animal-based foods in comparison to the small amount of this that most paleolithic foragers ate; apparent food equivalents might not actually be so equivalent due to genetic evolution and to changes in how we grow or raise food today; and the impact of diet is not singular: our activity levels and other factors mediate and moderate diet's effect on health.

Nutrition aside, adopting special diets can cause us to obsess about eating. We may get so caught up in classifying foodstuffs as healthy or not—good or bad, clean or dirty—that our eating itself gets disordered. This, in turn, has health consequences: it can result in food phobias and problems as serious as anorexia nervosa (severe self-restriction of food intake) or bulimia nervosa (cycles of binge eating followed by self-induced expurgation).

Nourishing Babies

Our brief exploration of Inuit foraging considered some costs and benefits of nursing. Most investigations or descriptions of subsistence ignore the fact that infants cannot digest, let alone chew, food like mongongo nuts or seal meat. We simply cannot gestate (grow and carry in the womb) babies until they are mature enough to do these things: giving birth to more mature, big-headed children would kill us. Breast milk provides a way around this conundrum. The bosom's evolutionary success stems from the fact that it enhances survival of the population by providing nourishment for infants who otherwise would not (could not) eat. Yet there is great diversity in how cultures handle breastfeeding and, concurrently, in health outcomes across populations.

Perfect Nourishment

The nourishment provided in breast milk is perfect for infant needs, such as brain growth; plus, it is easy to digest, and this suits the infant's body, which is immature on all counts. Infants cannot digest the same kinds of food that even toddlers do.

More than just proteins, fats, carbohydrates, and water for the infant, breast milk contains a variety of helpful microbes as well as microbe food: oligosaccharides. The microbes provided through the milk begin the process of colonizing the infant's gut to build its microbiome in ways that vary cross-culturally due to differences in mothers' diets but, in short, the microbiome creates compounds used by the baby's body. The oligosaccharides fuel portions of the microbiome. Oligosaccharides also can attract, and thus protect an infant from, certain pathogens, which latch onto the oligosaccharides rather than attacking the infant's intestinal wall (Gura 2014).

While protecting the infant's health in this way helps the infant grow, breast milk also ensures it doesn't grow too much. That is, when we compare children who were breastfed to those served formula, we find lower body mass and less propensity toward obesity among the breastfed group. Observational studies suggest that this is due to practices that surround breast and bottle feeding respectively. Bottle-feeding caregivers mix up the formula, fill up the bottle, and give it to the infant to suck from. When the bottle-fed infant is full, like the breastfed baby, it turns away from the (bottle's) nipple. However, if there is formula left in the bottle, the caregiver tends to try to get the infant to drink a little more: formula is expensive, and many people prefer not to waste it. Formula-fed infants can thereby get overfed and used to feeding past the point of feeling full. This can have an impact that lasts into later years.

Perhaps because breasts are not see-through and the costs of making milk are not so transparent, women tend to leave breastfed infants alone once suckling concludes. Moreover, all things being equal, breastfeeding (nursing) women only tend to make as much milk as the infants in their care want. The body is set up to supply the volume of milk that the infant demands through a feedback loop that connects the infant's sucking (stimulation) to the breastfeeding woman's nursing hormone production apparatus, and thereby her milk making propensity. The nurser and nursee comprise, in effect, one system. The more an infant suckles, the more milk a nursing woman makes. Breastfeeding women automatically create individually tailored serving sizes for infants in their care (see Figure 6.6).

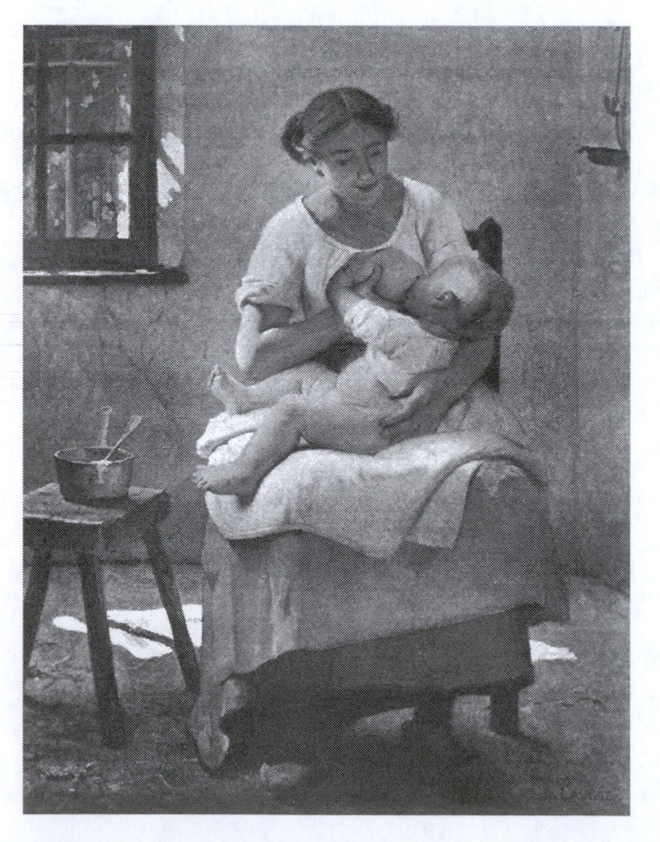

FIGURE 6.6 Depiction of breastfeeding by Désiré François Laugée, 1823–1896.

Other Benefits

In addition to nutrients, and microbes to populate (and fuel to feed) the microbiome, suckling infants also receive antibodies in breast milk. These help the infant to resist certain population-relevant infections and diseases directly. What happens is this: just as the infant's digestive system is immature, so, too, is its immune system. The nursing woman's immune system fills in for the infant's immune system while it is under construction. The **colostrum**—the yellowish fluid that infants can glean from the breast in their first day or so of life, before the real milk comes in (remember, milk production is stimulated through suckling)—is packed with antibodies.

Another benefit of breast milk is its sanitary nature. The milk itself has no time to be contaminated as it moves from the nipple to the infant's mouth. With formula, chances of contamination are rife. One must ask, "Is the water clean? Are the bottles clean? Are those nipples clean?"—questions mostly irrelevant to breastfeeding.

Beyond being good for baby, breastfeeding is good for those who nurse. Although breastfeeding culture and so the distribution of its benefits varies from population to population, breastfeeding can protect nursing women from certain forms of cancer. That includes breast cancer and possibly ovarian cancer. It also provides what one might term a 'natural' form of birth control because, while a woman is breastfeeding, she's often not ovulating. This has to do with the balance of hormones in her body. When ovulation is suppressed, fertility is suppressed. This can be beneficial to both the physical health of the mother and her mental well-being.

Notable here is the interconnection between the breastfeeding woman's body and the immature body of the infant. Parent and child form a system in which, as shown previously, much more than calories flow—and in which the flow is not just one-way. The infant's immaturity, related to the human need to give birth before the child is too big for passage, is offset by adult care. The parent's body can produce for the infant ample antibodies as well as foodstuffs. So, while the first three trimesters of our nine-month gestation are spent within the womb, a fourth is spent just outside of it. The infant freshly delivered still can be understood as if part of the mother's own body. Alternately, we might say that the boundaries between infant and mother are remarkably porous. Of course, an even greater number of porous boundaries can be implicated—particularly where allocare entails milk-sharing.

Milk-Sharing

I have, until this point, used the phrase 'breastfeeding women' rather than mothers because not all women who breastfeed are genetic mothers to the infants breastfed. Some women who breastfeed are what we call 'wet nurses': women who keep their breastfeeding capacity alive by suckling infants who are not their genetic progeny. Others may be biological relatives or even friends simply lending a hand (breast) with childcare.

As discussed in Chapter 4 in relation to allocare, there is evolutionarily adaptive significance to such behavior. In part because of this legacy, milk-sharing happens virtually everywhere—even in contemporary US settings. This may come as a surprise: given our individualistic norms, mainstream US culture stigmatizes milk-sharing, calling it 'unhygienic' at best. However, experts generally see milk-sharing as a safe option, particularly when done directly (via suckling) but also when milk is handled carefully in regard to storage receptacles, temperatures, and durations (Palmquist 2017).

In the United States, about one-third of milk sharers share with family members. The other two-thirds come together via local community-based networks and milk-sharing websites, which number in the hundreds. Parents ask questions and sometimes even exchange baby pictures. Even when mediated by the internet, the process still is generally intimate and manifests an allocare-related community-building function (see Chapter 4). In one study, 12 percent of nursing mothers shared their milk directly, nursing the recipient infants (Palmquist 2017).

Who are US milk donors? Recent research reveals that most are conservative, home-schooling Christians who see the practice as part of a lifestyle independent of government authorities. They tend to have conservative ideas about nature, gender, and 'appropriate' or 'good' mothering, viewing themselves as the voice of tradition (Falls 2017).

On this note, it is certainly true that without breastfeeding, until very recently, infants just would not have survived. If a population's infants did not survive, then that population would have perished. Breastfeeding (whoever the source) was, by necessity, the norm.

Duration of Breastfeeding

As part of our evolutionary legacy, breastfeeding used to extend a lot longer than it generally does today. Data collected by John Whiting and Irvin Child for nonindustrial societies (1953) showed that the average age for weaning (the age that a child gives up breast milk completely) was two-and-one-half years old. By today's mainstream American standards, that may seem amazing. In most cultures, however, for most of time, what we now call "extended" breastfeeding has been the normal way of life. Indeed, of the fifty-two societies in the study, only two practiced weaning prior to a child's first birthday.

In this, the cross-cultural, pre-industrial norm was much more in keeping with current World Health Organization (WHO) recommendations to breastfeed until offspring are two years old and that only breast milk be given until the age of six months. Not even water should pass an infant's lips until then. The WHO also recommends that we initiate breastfeeding within the very first hour of life and then breastfeed on the infant's demand, whenever the infant is hungry.

Why Do Breastfeeding Patterns Vary?

If these are the recommendations, why is there variation? At the individual level, some variation is due to biology. Some women—and some babies—cannot do it. For instance, an infant with a mouth condition such as a cleft lip or palate may have problems latching onto the nipple to suckle, or a woman may have a low lactation response, perhaps due to hormonal or nutritional factors. If a woman is having a hard time with breastfeeding, it is not necessarily a problem for the infant in a traditional culture because of allocare: others who have recently given birth may be more than happy to nurse the baby. Today, of course, baby milk formula is available.

What about variation at the cultural level? In many cultures, breast milk is a valued, life-giving substance, and breastfeeding a valued practice. In others, breast milk may not be seen as healthy. This is perhaps more likely to be true in modernizing, male-dominated societies that disvalue almost anything to do with the female person or body and equate formula with the modern West. In certain situations, buying formula and the bottles needed to feed a baby on formula is considered a good way to show off one's wealth as well as to confirm one's commitment to modernity or 'development' (see Figure 6.7).

Other ideas that vary across cultures, and affect breastfeeding practice, include ideas about who has ownership over a woman's body: the woman or infant, or a husband or boyfriend? Such ideas play a very important role in determining who gets to manipulate whose breasts.

At the same time, some corporations are very interested in seeing our beliefs turn against breast milk. Some of these corporations have gone so far as to disseminate falsehoods about the

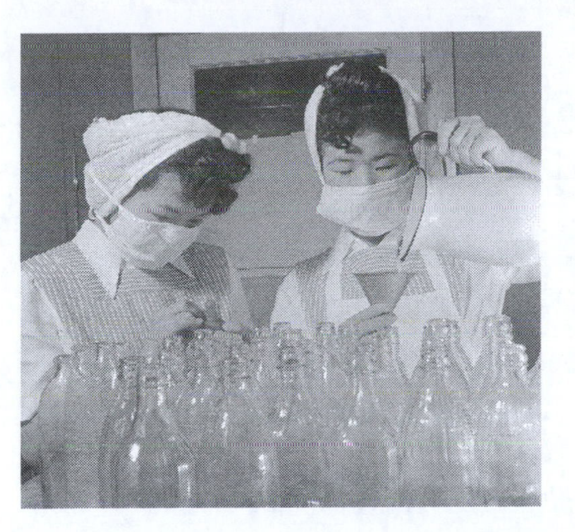

FIGURE 6.7 Ruth Uchida, left, and Haruko Nagahiro, "milk kitchen girls" at the Heart Mountain Relocation Center, Wyoming, prepare formula diet for the center babies using sterilized bottles, 1943. Photo by Hikaru Iwasaki for the Department of the Interior. War Relocation Authority. US National Archives and Records Administration. NARA record: 8464103.

value of formula. Some have been taken to court and been fined and told not to do this. Laws have been passed in order to protect women, infants, and families from this kind of exploitation.

Finally, for some people, breastfeeding simply is a luxury that cannot be indulged. When a mother works outside of the home and needs to keep her job, she may not be able to breastfeed exclusively for six months, let alone continue breastfeeding for over two years. This is despite the passage of laws, in countries like the United States, that support breastfeeding but are hard to enforce. Formula allows many women to stay employed or to gain employment as needed.

Population-Level Ramifications

In addition to the rise in numbers of women working outside of the home, the drop in breastfeeding that formula and the devaluation of breastfeeding support also is correlated with rises in infant malnutrition-related illness and infant death rates. Death is sometimes due to lack of nourishment, such as when poor people stretch formula servings by adding less powder and more water; it also can be due to mixing formula with contaminated water or serving it via contaminated bottles or nipples. Moreover, we see less resistance to illness in infants with no access via breast milk to maternal antibodies.

We also see increased population-level rates of breast and possibly ovarian cancer in woman. Furthermore, we see populations themselves increase. In other words, there are more babies born. Why is that? With less breastfeeding, there is less ovulation suppression. Once ovulation restarts, conception becomes possible again and so birth rates go up when breastfeeding goes down. Bigger group size in itself has many cascading effects, some of which we will explore in later chapters.

For now, a key point is that breastfeeding is an evolved, adaptive behavioral system—one that was selected for among early humans as part of a constellation of adaptations (including earlier birthing) because it enhanced our fitness. It also is one that, because of our capacity for culture, we now can contravene—for better and for worse.

Sleep

Another evolved adaptive behavioral system seen in all foraging societies is co-sleeping. Before exploring co-sleeping a word or two about sleep in general is in order. Without sleep, we die. With a shortage of sleep, our decision-making capabilities, memories, attentions, and visual-motor skills as well as our immune systems, metabolisms, and ultimately our bodies as wholes suffer. We are less productive and have more accidents (Nunn et al. 2016).

While little effort has yet gone into studying sleep cross-culturally we do know that a constant exposure to artificial light throws off the body's biological clock. Blue wavelengths in particular—the kind given off by most screens—are the worst, perhaps because blue light significantly lowers melatonin (a hormone important in sleep regulation). A good way to reset one's biological clock is to go camping—without one's devices. Even just one weekend spent in the bright daytime sunlight, using only the campfire's light at night, can be enough for a reset (Stothard et al. 2017).

Note here that not all human populations sleep in one consolidated nighttime chunk. Some adopt a **polyphasic** sleep pattern, in which sleep is taken in more than one segment. Infant sleep is always polyphasic. In the West today, when we do see partitioned sleeping, it is usually an individually adaptive response to immediate conditions, such as working a split shift. Still, there are a few groups, such as the Pirahã, a foraging group in South America, within which people sleep in multiple bouts as a rule (Nunn et al. 2016); and historical sources do indicate that even Westerners did not always practice consolidated sleep: some European populations used to sleep in two chunks or biphasically. Between what was generally termed 'first sleep' and 'second sleep' people would pray, write, check the animals, stoke the fire, and so on.

What Is Co-Sleeping and Why Do It?

Whether they tend toward consolidated sleep or to sleep in two or more activity-punctuated phases, foragers generally co-sleep. **Co-sleeping** is simply sleeping in very close proximity—around the same fire or in the same room (where rooms are slept in), and sometimes in the same bed (or hammock, or whatever other furnishings, if any, a group uses for normal sleep). In many societies whole groups—not just nuclear families—co-sleep. Note that co-sleeping does not include accidentally sleeping together, such as when a person falls asleep on a couch watching TV while holding a baby. Co-sleeping is purposeful, and uses sleeping arrangements that are culturally recommended for legitimate, full-fledged sleep.

Moreover, co-sleeping benefits the interdependent parent–child system that we have been discussing. For one thing, co-sleeping supports breastfeeding. It's just a lot easier to breastfeed when junior is right close up to your breast than when, depending on your infant's cries to awaken, you must get up, walk down a long hall, walk down the stairs, turn on the lights, fetch the by-now desperate baby, and so on.

Our belief in separate beds for babies is recent. It has a lot to do with an increase in consumption generally, including of houses with numerous dedicated sleeping rooms. It also has to do with culturally constructed understandings about spousal rights and obligations, such as are expressed through the idea of the 'marriage bed.' The mainstream US idea that children should not be in this bed ties into the cultural value placed on independence, according to which we want children to grow up to be able to do things on their own, including getting themselves to sleep. We much prefer this to interdependence expressed by, for example, relying on another person's close proximity for getting oneself relaxed. Mainstream Americans believe very strongly that, even for the littlest infant, sleeping in one's own bed fosters the kind of independent spirit upon which America was built.

A review of the research suggests this is wishful thinking (McKenna and McDade 2005). Further, co-sleeping is common across cultures even now, as it has been throughout history. The basic adaptation of co-sleeping addresses not only breastfeeding but also sleep regulation among infants, which can be handled by fathers or other caregivers in addition to people with breasts.

A Developmental Bridge

Remember, infants are immature; their little bodies are still developing. When born, a baby cannot even hold up its head. It cannot grasp a rattle or hand, let alone a spoon or morsel of food (which of course it cannot bite or chew or digest anyhow). Infants certainly cannot walk or talk. Their immune systems are immature. So it should not come as a surprise that infants' cardio-pulmonary (heart and lung) systems, too, are as yet not fully developed. Even the brain's capacity to monitor how much oxygen is in the blood can still have some maturing to do once a baby has left the womb.

With a fully functional respiratory regulation system, sleepers with low blood oxygen levels can arouse themselves to breathe more deeply. When one's system is not yet ready to enable that, death can result. Indeed, this seems to be a key cause of 'Sudden Infant Death Syndrome' or SIDS, a heartbreaking condition. The baby simply stops breathing, cannot arouse itself, suffocates, and is discovered dead.

As James McKenna and colleagues have shown (see, for example, 2005), the proximity of a co-sleeping individual to an infant serves, in effect, as a developmental bridge, giving the infant's cardio-pulmonary system time to mature. What happens is this: the sound and feel of an adult breathing and of that adult's heartbeat help in turn to regulate an infant's breathing and heart rate. Studies in which people have had video cameras trained on them as they sleep demonstrate that when an infant is co-sleeping with its mother or another mature caretaker, each automatically assumes healthy and safe positions in which to sleep. Both are on their backs. Often the adult takes care of thermoregulation, for instance moving a blanket off an infant to effect a cool down (in addition to respiratory monitoring, thermoregulatory issues may be related to SIDS). With a breastfeeding woman, the infant can breastfeed at will. Neither adult nor infant must fully wake for breastfeeding or thermoregulation to happen.

Population-wide correlation studies have found that where co-sleeping rates are higher, SIDS rates are lower. We see this today in Japan, where sleeping with many people in the same room is quite common and culturally appropriate. In fact, Japan has one of the lowest SIDS rates in the world. We also see this kind of negative or inverse correlation (more co-sleeping, less SIDS) with overcrowding, such as in situations of poverty.

So although mainstream US ideals suggest that co-sleeping is strange and abhorrent, it has been the typical sleeping arrangement around the world across cultures and throughout time—because co-sleeping is an evolutionarily beneficial or adaptive pattern. It is an adaptation whose benefits, like those of breastfeeding, mainstream Americans are canceling out through a cultural emphasis on independence.

Morbidity and Mortality among Foragers

It will be obvious by now that the availability of food and water is not the only thing affecting foragers' health. My focus has been on adaptive behavioral constellations, but I'd like to switch gears here to highlight health threats entailed in the lifestyle that hunting and gathering entails more generally. What are a few of these? Infection is a real problem, to start. Before the invention of pharmaceutical drugs like penicillin, a good proportion of people died from infections gone bad. A tooth infection, for instance, could end up taking a life. Dental problems in general have been a serious challenge for humans ever since we came into being. Infection can also come on with an injury, such as when gangrene sets in, and this can be lethal. There are also parasitic infections—infections with organisms like worms, and so forth.

While infections are now mostly a problem only for people without access to health care, back in time they were everyone's problem. Chronic parasitic or bacterial infections, for example, led to many deaths and much disability and pain among foragers.

Early foragers also died (as foragers still do) from exposure to severe weather or predators, and from traumatic injuries. Today, if one falls out of a tree and breaks one's back or cracks one's skull, or is mauled by a lion, or experiences complications during childbirth, emergency medical services can be called upon. However, this option is a new cultural invention. There was no 911 service for early humans. Like infection, traumatic injury was a major cause of death.

As is true among the poor today, those foragers who survived often lived with the history of their infections or wounds written into their bodies: notable limps and missing fingers, teeth, eyes, and so forth differentiated individuals much more in the days before emergency—and plastic and reconstructive—surgery. On the bright side, most cultures accepted such diversity as part of the human experience and had no expectation of the kind of bodily homogeneity or 'perfection' projected in today's movies, magazines, and social media.

For better or worse, Western culture has penetrated most remaining forager groups. The outlook is not bright, as the current Inuit and broader Arctic situation (discussed earlier) demonstrates. What is happening there has happened elsewhere; only most indigenous groups have not managed to hang onto tradition as long as Arctic peoples have.

Over the years, countless foraging cultures have been destroyed or damaged through intrusion into their native lands, such as for mining, logging, or to claim acreage for agriculture. Many suffered non-native disease epidemics. Some were enslaved. Land grabs and environmental damage pushed many into sharecropping, wage labor, or worse among dominant groups. These receiving communities disrespect and misunderstand the displaced foragers. Adrift from their native lands and practices, often suffering physically with the diseases of poverty, their mental health declines. Yes, some foragers have created new lives with old skills. In India, the Irula formed a cooperative for snake catching. They sell snake venom and use snakes to catch rats for their host communities (Anonymous 1984). Other groups—particularly those with some land rights still intact—have come together to promote cultural renewal (see Chapter 9). Without land, however, extinction becomes the most likely outcome for most foraging cultures. This reduction in the diversity of human lifeways damages our flexibility and resilience as a species.

Nonetheless, foragers who are not harassed by 'civilization' are well-nourished generally, with a well-balanced, highly nutritious, heavily plant-based diet. They also are fit; they have an active lifestyle in which the subsistence work of hunting and gathering is complemented with plenty of leisure time. Most early human foragers probably lived a similar lifestyle to that of the !Kung. We are reasoning backward here: we do not have much beyond some skeletal remains and artifacts to show what life really was like for early humans. Still, it is reasonable to hypothesize that, like the Dobe !Kung, early human foragers probably did not have such a hard day-to-day life after all.

7

AGRICULTURAL REVOLUTION

Another Great Divide

This chapter prepares you to:

- Characterize the subsistence continuum and explain the process of intensification it summarizes
- Explain why the Agricultural Revolution happened and how certain geographic regions supported this revolution better than others
- Describe and explain the bodily consequences or implications for biocultural diversity of the Agricultural Revolution, as seen in the archaeological record
- Discuss the biocultural ramifications of industrialized, monocultural approaches to subsentence
- Characterize the Agricultural Revolution's effects on sociopolitical and economic organization

Foraging is one of the most basic subsistence types that ever existed. The only thing more rudimentary is scavenging. Whereas foraging entails picking an apple off a tree, **scavenging** entails waiting until it drops to the ground. The forager hunts; the scavenger picks up carrion or road kill. Scavenging entails taking what nature has left out on the countertop, so to speak. With foraging, though, one's relationship to the environment is a little bit more manipulative. Compared to the scavenger, the forager has 'intensified'; the farmer, even more so. This chapter explores central consequences of intensification for human biocultural diversity.

Intensification

To begin, let's first have a look at the continuum presented in Figure 7.1. The various types of subsistence are represented. 'Subsistence' refers to how a group makes a living—to how it gets food. **Intensification** is the process of doing more to get food: more intensified strategies manipulate or interfere with (construct) the environment more than less intensive strategies do. The more intensive the strategy, the more tools or technology it uses and the more complicated those technologies will be. More intensive subsistence strategies leave a bigger mark on the environment, changing it from its original form much more notably than less intensive strategies do when extracting food. Through multiple feedback loops, the diverse environmental changes we make as we intensify in turn change us.

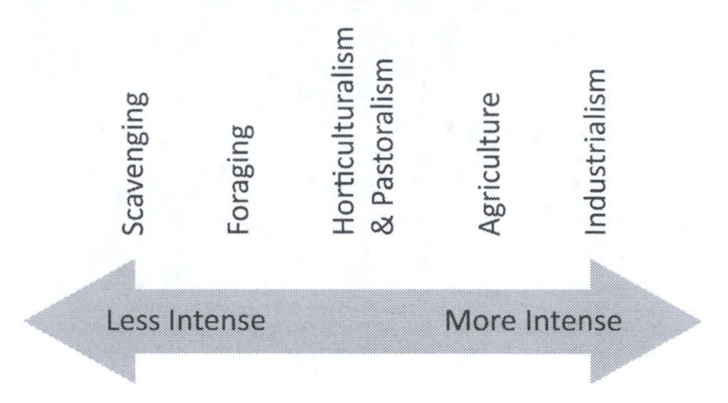

FIGURE 7.1 Intensification continuum. Human manipulation of the environment is lower in subsistence forms to the left and higher in those to the right.

The continuum shown in Figure 7.1 organizes subsistence strategies from less intense to more intense, from left to right respectively. As we move to the right along the continuum, intensification increases. Beyond minimal-impact scavenging, then foraging, lie gardening or **horticulture** and herding or **pastoralism**. Sometimes there will be overlap; for instance, some foragers also may garden a little bit. In any case, horticulturalists and pastoralists are not settled farmers. Like foragers, they generally move around, following the seasons. Any gardening that is done is done without the help of irrigation, animal-powered plows, or fertilizer.

Once a people settle down and begin to manipulate the land with plows, special watering techniques, and fertilizers—once they embark on **agriculture**—a sea change occurs, and much of it stems from the very fact of settlement. Because changes were radical, we often term the start of agriculture, about 10,000 years ago, the **Neolithic Revolution**, as it rang in the new (*neos*) stone (*lithos*) age. Of course, as with other key transition dates provided in this text, the '10,000-year' mark for this revolution is more an *aide-mémoire* than a hard assertion: change takes time. Further, although stone is called out because it took about 5,000 more years for metallurgy to emerge, some scholars prefer the more generic label, **Agricultural Revolution**. Regardless, the lifestyles that we see on the left side of the continuum—for humans, foraging in particular—predate agriculture by a very long time.

Why Intensify?

Let's think of a foraging or pastoralist group that does a little casual gardening—a group that, more than likely, lives a life of fairly good nourishment and relative leisure. Why in the world would they seek change? Why would they settle down?

From the mainstream American point of view, settling down supports a desire to acquire. Americans like having lots of stuff and often assume that others feel the same. We assume that foragers would not be nomadic if they had the option of owning material goods. We assume that agriculture opened this option for such foragers (or pastoralists, or horticulturalists), who happily embraced it. The idea that people 'just want more' thereby comes into play in folk theories regarding the Agricultural Revolution—but not in scientific ones.

Agriculture is harder in many ways than foraging: intensification involves more labor. If we compare the workday of a forager to an agriculturalists' workday, we can easily see that the forager has the lighter workload. The workload goes up for peasant farmers and rises even higher for manual laborers who support industrial agriculture: while the peasant agricultural worker has been shown to work about nine hours a day, the industrial laborer works about eleven (see Schor 1993).

Note here that these figures do not include child labor, which is crucial in agricultural societies and was long depended on in our own until well after the Industrial Revolution. Note too that longer hours do not necessarily get people more of anything except exhaustion.

Necessity, Geography, Curiosity

What really spurred the Agricultural Revolution in places where it happened? In a phrase, climate change. Societies became agriculturalists out of necessity. They had to.

One reason for this could be that human beings over-exploited the flora and fauna where they lived. They pushed the ecosystem beyond its carrying capacity. Over-exploitation happens when, for example, people go fishing and take too many young fish: they deplete the fish population's reproductive potential. For that reason fishing crews today are advised to throw back little ones that 'when bigger' can replenish the fish population.

Over-exploitation probably played a part in supporting the Agricultural Revolution. However, most scientists favor climate change as the crucial driving force. The idea here is that when the climate changed, the **flora** (plants) and the **fauna** (animals) changed in response, and the typical diet constricted. People had to figure out another way to make a living in geographic regions that were affected.

Climate data accord with archaeological data: approximately 10,000 years ago, the climate got warmer and drier in certain areas. People could not make a living through foraging anymore because the species on which they had depended could not survive well enough in the new conditions. People therefore had to figure out some other way to get food on the table. They either had to walk to another part of the world where the environment was more favorable to the foraging lifestyle or, if they were lucky, they found some candidate plants (or animals) for domestication right where they were. For some, simple supplementation via casual gardening—horticulture—or herding was enough. For others, full-out agriculture was the only answer to the necessity of getting fed.

In addition to necessity, geography played a role. Being in the right place—a location with great floral and faunal candidates for domestication—is something that Jared Diamond (1997) calls "**geographic luck**." The area now known as the Fertile Crescent, in Western Asia, near what is often termed the Middle East, had lots of such luck. Specifically, the Fertile Crescent covers Mesopotamia and the Levant (see Figure 7.2). Various continental plates come together there in a way that supports a high level of biodiversity in the environment. Happily, what is true at the species level is true at the ecosystem level: biodiversity has a protective effect in the face of new environmental pressures (climate change included).

The Fertile Crescent was the first place in the world, at least according to current archaeology, that saw domestication of plants and animals. The population there when the climate changed was fairly dense, relatively speaking. Remember, human beings first evolved in Africa, and they first got to the rest of the world on foot via the Middle East. By the time the climate changed in the region, humans had in fact spread around the world, with populations growing sparser the farther away they were from the African hub. Nevertheless, some of the other places humans lived seem to have experienced the great impact of climate change a little bit later. Moreover, they generally did not have the advantage of such relatively easy access to eligible domesticates (we'll discuss exactly what an eligible domesticate is a bit later). For such reasons, the Fertile Crescent stands out.

Beside necessity and geography we must consider a third factor in explaining the Agricultural Revolution: curiosity. People, being people, are curious. Take, for example, my dear childhood friend, who got a toy car wheel stuck up her nose when very young. You likely know someone who did something similar with a pebble or a bean. Why do children stuff things in their

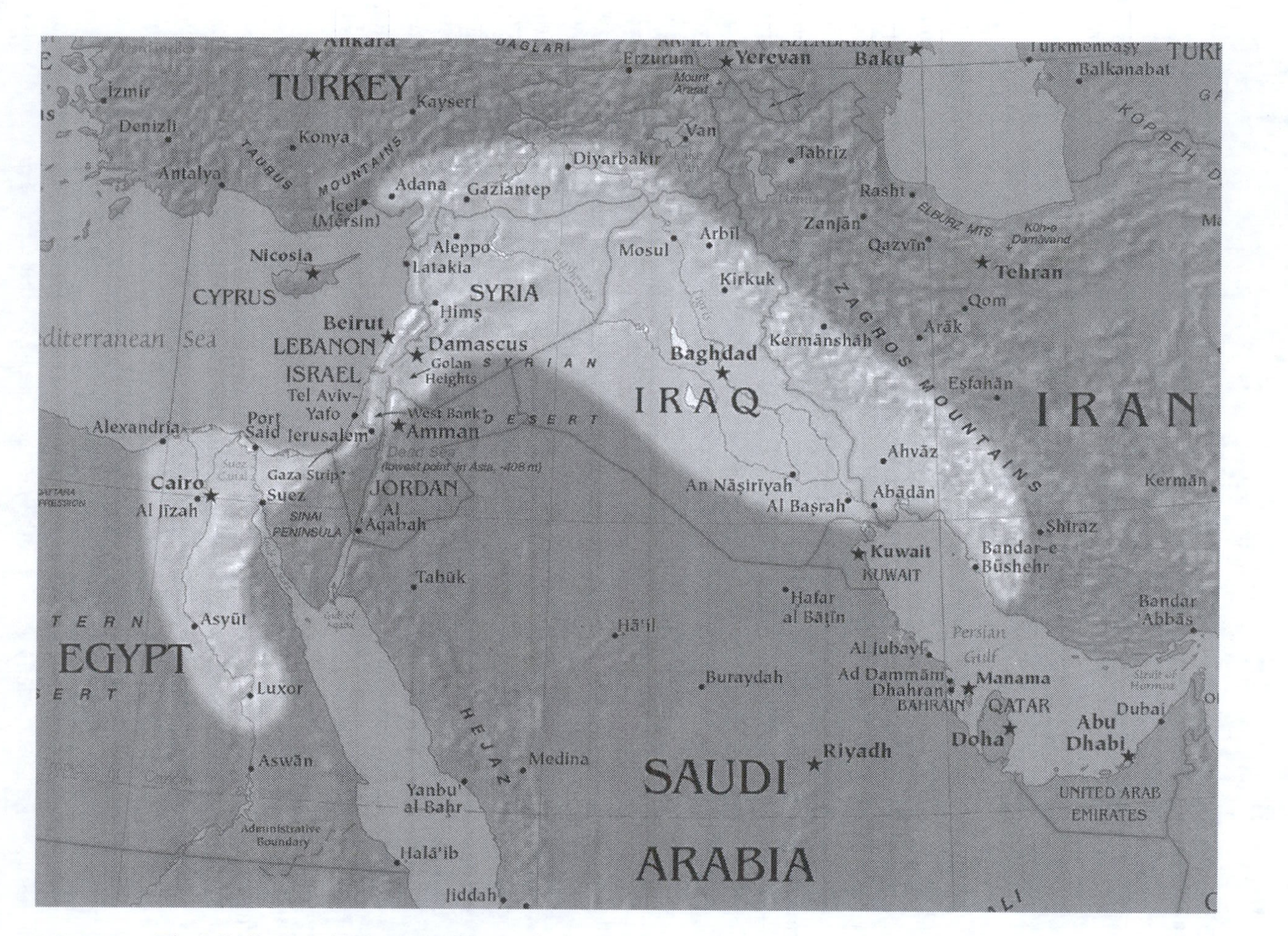

FIGURE 7.2 The Middle East (CIA Map). The Fertile Crescent runs in an arch from the Mediterranean Sea to the Persian Gulf; it is rimmed on the north by Turkey and on the south by the Syrian Desert.

noses? They are curious. Happily, human beings have an evolved capacity to learn from what we observe after applying our curiosity. The child who gets the rock (wheel, bean) stuck up their nose once very rarely does so again.

We use our senses to get to know the world. Our capacity for empirical observation—for noticing things by looking around (feeling, tasting, smelling, hearing, and so forth) has been key to our species' survival. Fortunately for us, we noticed at a certain point that when a seed falls on the ground, it sprouts. Then we noticed that we could control where a seed was planted to sprout. We noticed that if we put it in one kind of soil it grew faster than if we put it elsewhere. We noticed that if we chose the sweetest fruit to eat, and seeds from that fruit got planted, the bushes or trees that grew up from them yielded sweet fruits. Why did we notice? Because we are curious.

So, in addition to necessity and geography, curiosity was a very important part of the equation that supported the Agricultural Revolution. A group might have needed to intensify, and a group might have found itself in a spot where there were lots of eligible domesticates; but if that group—that population—did not have innate curiosity and an associated intellectual capacity to derive meaning from what was observed, agriculture could never have happened.

Those Who Could, Foraged, Until They Could No More

Of course, not everybody was forced into an agricultural lifestyle by climate change 10,000 years ago. In some areas, the need for agriculture did not come into play until more recently. In fact, up until c. 2,000 years ago, about half of the world's population foraged for a living. Today the proportion doing so is nearly nil.

Why did forager numbers keep going down if 10,000 years ago those people who needed to intensify did so? The answer has to do with the power of settled agriculturalists to expand and their evolved desire to do so. In some cases, foragers have basically been pushed out by farmers who expanded into the foragers' territory to take over fertile lands. Nowadays, of course, there are corporate interests at work—logging companies, mining companies, and so on—pushing foragers out of business, either literally or through the pollution they produce, which poisons the foragers' livelihood.

Eligible Domesticates

Returning to the Revolution, what was it about places such as the Fertile Crescent that enabled the adaptive emergence of agriculture? It was proximity to eligible domesticates: people who lived there were close to plants or animals that could be domesticated—plants and animals with certain features.

What were those features? Diamond (1997) provides a good review. Beyond being visually noticeable, he says, the best plants to domesticate are those that can be stored for a long time without rotting. They likewise have seeds that can lie dormant until planted (seeds that do not sprout when in storage)—and that can be easily gathered in the first place.

Wild plant species generally have seeds that are easily dispersed. Wild wheat seeds actually shatter from their shafts in the breeze when ripe. A person walking through a wild wheat field can, by simply brushing a plant's stalk, cause such shattering. Wild wheat seeds (also called grains) are therefore hard to gather—except when the wheat grows in a mutated form, wherein seeds are more firmly attached to the shaft.

Early farmers would have favored and thus promoted this shatter-proof variety, despite its natural selection-related disadvantage. Why? Because it would have been easier for people

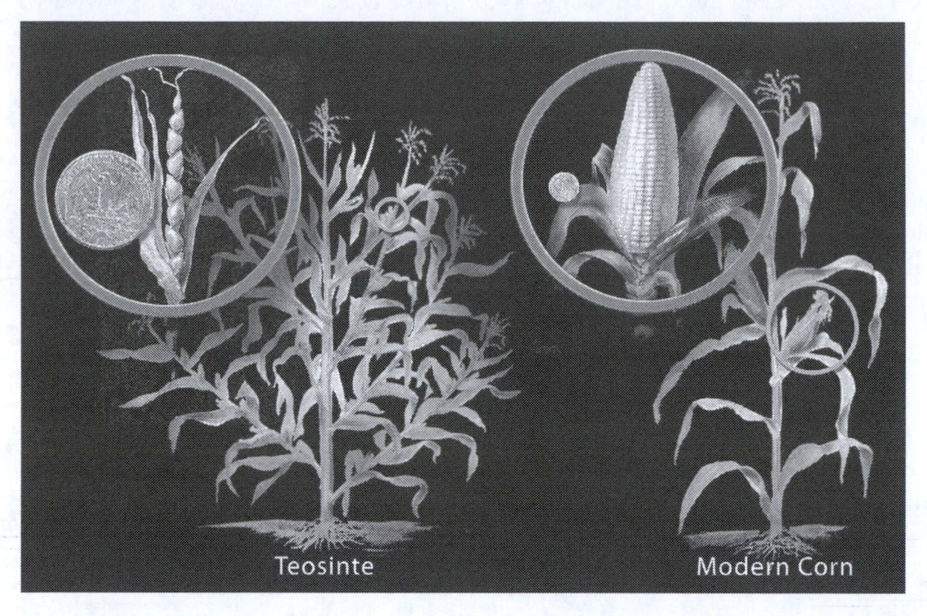

FIGURE 7.3 Ancient and modern corn. Among other things, ancient corn has multiple branching stalks and more, smaller ears with fewer, harder kernels; modern corn has a single, nonbranching stalk and more, softer kernels on fewer, larger ears. Courtesy of Nicolle Rager Fuller, National Science Foundation.

to gather grains to eat from such plants. They also would have selected for softer, plumper grains or seeds rather than the tough-coated seeds that do best for the plant in the wild (see Figure 7.3).

To be further viable as domesticates, crops should be fast-growing and have a good yield. Like wheat, other seed grain crops like rice, barley, peas, maize (corn), and millet fit the bill. For better or for worse, however, these originally were found only in certain geographic regions.

With animals, Diamond notes, similar rules of thumb apply. The animals that reproduce and grow to maturity quickly are more favored than those that like, say the elephant, have only a few offspring that mature slowly. Animals good for domestication also must be people-friendly (they should not, for instance, eat them), and they should be sociable with other animals and docile, too. Animals like pigs, goats, sheep, and cattle are thereby excellent candidates; lions, zebras, and so forth are not. Eligible animals were available, originally, only in certain geographic regions, adding to the geographic luck of those living there (Diamond 1997).

Revolution at What Cost to Human Health?

Agriculture was good. It allowed some populations that might have perished to survive. It eventually helped fuel the Scientific and Industrial Revolutions. However, it was not without costs. The Agricultural Revolution had ramifications for fitness in how it affected health and well-being.

Exposures

The agricultural lifestyle differs vastly from that of foragers. To practice agriculture, a people must settle. They can no longer live as nomads. They must stay in one place. This increases their chances for certain kinds of exposure.

An **exposure**, technically speaking, entails close proximity to a chemical, a **pathogen** (germ), radioactivity, or extremes of weather. One of the costs of settling down to do agriculture is an increase in a group's exposure to environmental toxins. This might, for instance, include lead, which can lurk in water, soil, and food as well as in dishes and utensils. Lead, taken in, can poison various organs and tissues; it interferes with a number of bodily processes. One key symptom of lead poisoning is abdominal distress; others include headaches, irritability, and seizures. Coma and death also are possible. Lead poisoning is especially dangerous to children because of the developmental problems it can cause (see Supplement Box 7).

Indeed, archaeologists have identified sites where lead poisoning is thought to have been common because of the presence of lead in the environment. For example, in locations where clay used for making pottery had high lead levels, people would have suffered.

If foragers had passed through such an area, they might have been exposed to a little bit of lead, but not much, because the foraging lifestyle typically is nomadic. Foragers follow the seasons, moving to different parts of the environment when different animals are going to be there or different fruits come ripe. They move around. They might come back and revisit camps on a seasonal basis, but they do not stay put—and this minimizes chances for toxic-level exposures.

Agriculturalists, on the other hand, do not move. They stay in the same place for a lifetime. The longer people stay in one place, the more likely it is that local environmental toxins are going to penetrate their bodies, and the more likely it is that they will suffer ill effects.

SUPPLEMENT BOX 7: INTENSIFICATION, EXPOSURE, AND THE DEVON COLIC

Devon, England, is famous for, among other things, apples. Turning apples into cider is a good way to store them or extend their shelf life; and so cider—apple juice, hard (fermented) or soft—became Devon's drink of tradition.

Agriculturalists, no matter what the crop, all must figure out modes of storage. One danger here is that storage containers can contaminate food. For instance, cider jugs made from clay that contains lead or from pottery whose glaze was made using lead (lead makes a glaze shiny) could lead to unhealthy levels of lead exposure. And in fact, in Devon, lead poisoning was a problem. However, as Ian Maxted explains, lead poisoning's history in Devon, and its link to the apples, was more complex (2000).

Lead poisoning once was so common in Devon that it was referred to as the Devon or Devonshire Colic. Colic is a kind of stomachache often attributed to babies who cry a lot (it's seen much more frequently in modern Western cultures than in others where babies are held more and breastfed on demand). Devon Colic was a disease mostly of adults, however. The first written description of the colic appears in 1655. The people in Devon at that time connected the colic to drinking cider; they generally explained the connection as due to cider's natural acidity.

Common sense long held, but in the 1760s one Dr. George Baker hypothesized that the lead in the beverage was to blame for the colic. The abdominal distress of the colic also was seen in known cases of lead poisoning. Sure enough, it turned out that lead was used in making cider: it was used in the presses and in the cleaning process. Chemical tests done by Baker confirmed high levels of the contaminant.

The published results were not kindly taken to by cider makers, who did what they could to deny or rebut them (see Figure 7.4). Nevertheless, science triumphed eventually. Cider press makers removed the lead and Devon Colic became a disease of the past.

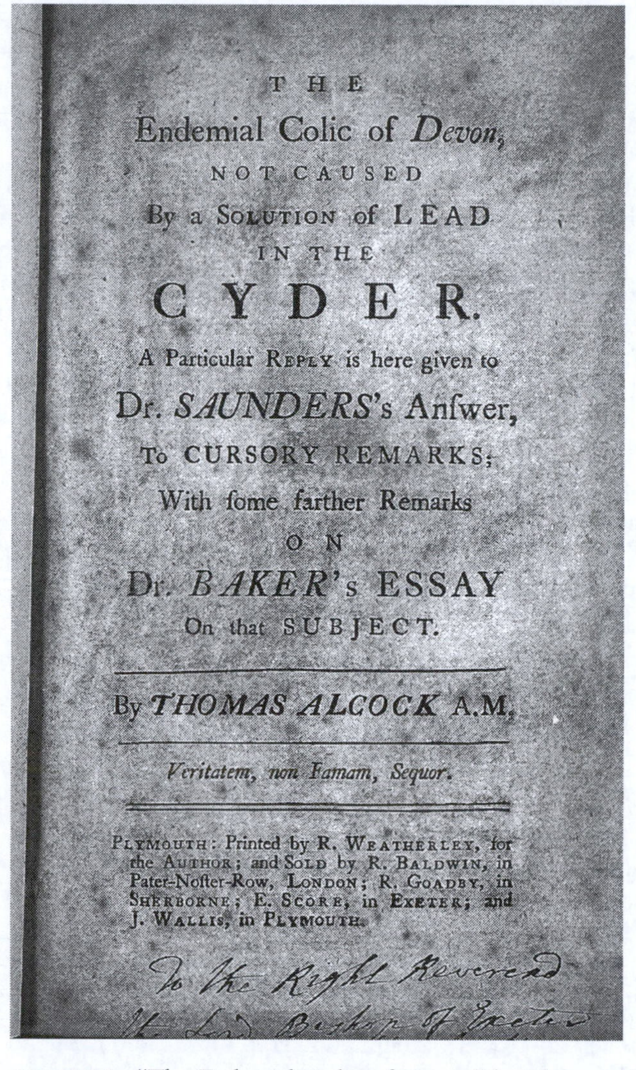

THE
Endemial Colic of *Devon*,
NOT CAUSED
By a SOLUTION of LEAD
IN THE
CYDER.
A Particular REPLY is here given to
Dr. *SAUNDERS*'s Anſwer,
To CURSORY REMARKS;
With ſome farther Remarks
ON
Dr. *BAKER*'s ESSAY
On that SUBJECT.

By *THOMAS ALCOCK* A.M.

Veritatem, non Famam, Sequor.

PLYMOUTH: Printed by R. WEATHERLEY, for the AUTHOR; and SOLD by R. BALDWIN, in Pater-Noſter-Row, LONDON; R. GOADBY, in SHERBORNE; E. SCORE, in EXETER; and J. WALLIS, in PLYMOUTH.

FIGURE 7.4 "The Endemial Colic of Devon," by Thomas Alcock, a cleric and a cider maker. Credit: Wellcome Collection. CC BY 4.0.

Skeletal Health and Growth

Beyond using material artifacts like pottery and cider presses to get a sense of how agriculture affected health, skeletal remains and teeth are also revealing. For instance—and for much of this and the next few sections we will rely on classic works by Clark Spencer Larsen (2006) and Mark Nathan Cohen (1989)—the hard work of agriculture is written into the skeletal joints. Agricultural populations demonstrate higher rates of degenerative joint disease, reflective of more demanding activity patterns. They also have thicker bones—again a reflection of a more physically demanding life (exercise stimulates bone manufacture).

When we look at the skeletons of ancient agricultural populations, and even those of peasant agriculturalists more recently, we see shorter stature than is common among foraging peoples. Short stature can be a sign of increased stress on the body, often related to childhood malnutrition (see Chapter 3).

We also see something called **porotic hyperostosis**: bones that are more porous than they should be. This is a sign of iron deficiency, which triggers the expansion of bone tissues that form red blood cells. Scientists are still in discussion as to what causes the deficiency. It could be nutritional stress, but there is a competing hypothesis regarding immune defenses. It turns out that certain pathogens thrive on iron in the blood; the hypothesis holds that the body withdraws its iron to create an environment in which those pathogens are unlikely to survive.

Whatever its origins, porotic hyperostosis in a group of skeletons signals that bodies were under stress, and more porotic hyperostosis is seen in agricultural populations worldwide. So while there are benefits to the settled farming life, skeletal evidence shows that there also are costs. What we start to see cropping up in agricultural populations is an assemblage of problems representing what we might call the diseases of settled life.

Some of these diseases are written into the teeth. We see declines in dental health in settled agricultural populations. We see it in terms of simple decay due to changes in the diet. Gruels or porridges made of grains and rice and so forth stick to the teeth in a way that forager staples do not, causing more dental caries (cavities). "Virtually everywhere that human populations made the shift to agriculture," writes Larsen, "saw a rise in the frequency of carious teeth" (2006, 93). We may wonder at Larsen's use of the term "virtually"; he himself notes, based on his literature review, that "rice does not appear to be as cariogenic as other domesticated plants, especially maize." In other words, folks who eat more rice should not have as many cavities as folks who eat American corn. Culture does make a difference. Having said that, Larsen goes on to note that there are some indications that even populations subsisting mostly on rice may have had higher dental caries rates than foragers, and in any case their teeth suffered in other, agriculture-related ways.

For instance, we can see general nutritional stress, common to agricultural populations, written on the teeth they have left behind. When people are under nutritional stress the enamel does not get laid down the same even way it does in a healthy, consistently well fed individual. Without ample nourishment, the body does not have enough raw material to contribute to building teeth; pits and lines on the teeth reflect the deficit (see Figure 7.5). This is termed **enamel hypoplasia**, because there is less enamel in some places on the teeth than others (*hypo-* means under or below). We see enamel hypoplasia indexing physically stressful periods. It also can index a struggle with infectious disease which, as the next chapter shows, can plague settled populations. We certainly see more evidence of tuberculosis, syphilis, leprosy—infectious diseases that affect the skeleton—in agricultural populations than among foragers. But for now, let's keep the focus on nutrition.

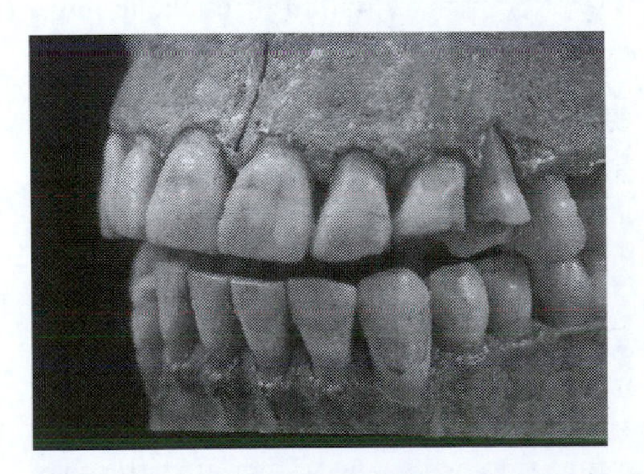

FIGURE 7.5 Linear Enamel Hypoplasia (LEH) caused by periods of biological stress during growth and development, Lower Illinois Valley, Titterington Collection, National Museum of Natural History, Smithsonian Institution. Photo by Arion T. Mayes.

Causes of Nutritional Stress

People who have a settled agricultural lifestyle have poorer nutrition than foragers (Cohen 1989; Larsen 2006; see also Pollan 2006). This links back to a lack of variety, nutrient loss in the foods eaten, and lower nutrient levels in cultivated foods to start with.

In terms of lack of variety, when compared to foragers, settled agricultural populations generally have a much narrower diet. They have come to depend on the products of their agricultural labor for subsistence. If the weather does not support a good crop, or if stored crops rot, agriculturalists are subject to famine. Moreover, the lack of dietary variety even in good times narrows the range of nutrients that their bodies are taking in. Remember, foragers eat lots and lots of different foods. We have a saying that variety is the spice of life; it is not just that: variety is vitally necessary for a well-balanced diet that provides all the nutrients human beings need.

A lack of variety leads to malnutrition among agriculturalists. Nutrient loss does, too: it turns out that many of the major cereals contain within them certain phytochemicals that hinder the absorption of other nutrients. Diets heavy on corn (maize) can foster a niacin deficiency because certain phytochemicals in the corn inhibit niacin's absorption. Wheat eaters often are zinc deficient. A diet highly dependent on rice can lead to vitamin A and thiamine deficiencies.

On top of problems that people might have because of phytochemicals inhibiting absorption, they also can suffer due to food storage-related nutrient loss. The longer we store food, the more time that food's nutrients have to break down—meaning that the same food becomes less nutritious as it ages.

A third factor affecting the nutritional status of agricultural people is the simple lack of wild foods in the diet. Wild foods generally are more nutritious than foods that we have raised domestically. Wild plants tend to have more of their own internal phytochemical protections against predators, for example. Domesticated agricultural plants do not need these as much because humans work to protect them, using pesticides and other tools. It turns out, however, that many phytochemicals that protect plants against predators also have benefits for the human body. In addition to giving a fruit or vegetable more intense flavor, and deeper color, some phytochemicals have antimicrobial effects in the human body just as they do in the plant. They can therefore offer us some protection against illness or infection.

Finally, agricultural crops are lower in nutrients because agriculturalists use the same lands again and again and again. When land is not allowed to lie fallow or helped to regenerate, the soil can become very depleted. Food grown in depleted soil simply is not as nutritious as food grown in rich fertile soil. Plants feed on nutrients in the soil, and if they're taking in fewer nutrients from the soil because it has fewer nutrients to offer, humans will in turn take in fewer nutrients when they eat those plants.

There is a 'wild food chain' as well, of course. A wild plant takes in nutrients from the soil in which it took root. Rich, wild soil fosters more nutritious plants. Animals living in the wild eat these wild plants (and sometimes animals), thereby taking in a wider variety of nutrients than do domesticated animals. This is one reason that wild meat is more nutritious than, say, beef from cattle raised in a feedlot eating today's hybridized domesticated corn.

In addition, meat that comes from wild animals is very lean in comparison to the meat of domesticates. Animals out in the wild on their own get more exercise than domesticates, running as they must to keep away from predators, or running to catch their own prey. Again, although they would have gotten more exercise even 100 years ago than they do now, most domestic animals don't need or get to do so much running or moving, and so they are fatter than wild ones. In addition to being fatter, domestic animals also have a higher proportion of what nutritionists would call bad fat—the kind that our doctors warn us against eating too much of. This is partly why today we have much higher rates of cancers, heart disease, and so forth than we used to.

Even our dental health is in some ways worse now than in the past. It is the case that we have modern fluoridated toothpaste and so forth, but we also have a much, much softer diet than even our agriculturalist ancestors did with their porridges. On top of that, we eat vastly more sugar. The sticky, sugary foods that we eat exacerbate the human tendency to get cavities.

We also get less fiber than early agriculturalists, and than traditional foragers. I talked about fiber in Chapter 6, and its relation to the evolving microbiome. Here I will add that the modern Westernized microbiome has the lowest diversity of species of any microbiome yet examined. It is true that microbiomes are fairly plastic: switching one's diet—purposively or due to migration—can alter its composition radically and quickly. This has been seen, for instance, in East Asian immigrants to the United States: their microbiome diversity decreases notably nearly immediately after arrival (Vangay et al. 2018). Genetics may also play some role in determining which microbes thrive in whose intestines (and elsewhere, such as on our skin); antibiotic use definitely does. Antibiotics kill good as well as bad bacteria. Regardless, little of the scant fiber today's Westerners eat is fermentable or useful to microbes as fuel; that kind is abundant in typical foraged diets and often present in agriculturalists' diets too. It also shows up in the parts of food today's Westerners like to throw away, such as stalks and peelings (Schueller 2014).

Like foragers, then, pre-industrial agriculturalists would have been at lower risk than modern Westerners for the diseases associated with microbial imbalances (e.g., some cancers, diabetes, heart disease, and obesity, as well as depression, anxiety, and even autism; Kohn 2015). That said, they still would have more health challenges than most of their forager forebearers.

A Negative Correlation

With few exceptions, all in all, there is a negative correlation between intensification and good health. That is, as a given population moves from a foraging lifestyle to a settled agricultural lifestyle, its health declines: as intensification goes up health goes down.

Happiness also might decline; or anyhow certain kinds of unhappiness are more evident, perhaps due to increased competition for resources. In some regions, injuries linked to interpersonal violence, as seen in marks left on skeletons, increase over time as agriculturally productive lands grow scarce. Organized warfare also seems to increase.

Forests are cut back to clear new land, erosion occurs, and some species die off due to changes in the ecosystem. Other species, now in increased or closer contact with human populations, share out their parasites and other pathogens. Larsen in fact goes so far as to refer to the Agricultural Revolution, itself driven by environmental changes that served as evolutionary pressures, as an "environmental catastrophe."

While settled agriculture is in many ways a wonderful human invention, it can and often does have a downside in terms of human health (Cohen 1989; Larsen 2006). Many scoffed on first reading Larsen's dramatic description of the Agricultural Revolution as catastrophic, but the long-term, knock-on effects of this change, which we now are recognizing, actually have been just that.

The changes have been so drastic that in 2016 (ten years after Larson published the treatise referenced here) scientists officially labeled the present epoch as the **Anthropocene**. While its official start date is yet to be agreed upon, many view agriculture's emergence as marking its inception. In other words, since we began settling, human activity has become the dominant force shaping the earth's geology and ecosystems.

Industrialized Agriculture

Settled agriculture involves a more intensified relationship to the environment than foraging. With agriculture, humans are doing a lot more to manipulate the environment. When settled farmers began to engage in 'cash cropping' or 'monocropping'—that is, growing just one crop, and lots of it, to sell as a commodity for cash rather than to eat—and when landowners began to use highly mechanized means to do that—**industrialized agriculture** was born.

Streamlining Natural Systems

Domesticating plants or animals inherently involves altering some of the links between them and the ecosystem as it was. The industrial approach to farming requires us to break more of the webbed ecological links that nature had set up environmentally. There are all kinds of links or intertwined activities in natural systems. On the old-fashioned farm, chickens eat worms busy fertilizing the fields. Then, the chickens defecate in the fields, further fertilizing. Cows go out and eat the grass now growing, and they in turn defecate. Flies fly onto these cow patties and lay eggs; chickens eat them when they hatch to maggots, and so on. This is just one of the many, many circles of flowing energy and matter that we find in the complex adaptive system the old-fashioned farm represents. A fancy term for this kind of a system is **polyculture**.

The industrialized food production system strives to be linear. It takes animals, like pigs and cows, and plants, like corn and wheat, well out of the complex systems in which they have evolved in order to streamline their 'production' (see Figure 7.6). In erasing the polyculture of old-fashioned farming, industrial agriculture creates new challenges, and we cannot always predict what these emerging challenges will be. Take, for instance, the invention of synthetic fertilizer, necessary when environmentally provided fertilizer such as chicken dung is removed

FIGURE 7.6 Chickens in an industrial coop on a factory farm. Photo by ITamarK.

from the picture; or take our chemical pesticides. Now, in addition to increased chances for exposure to environmental toxins (poisons found in nature), we gain the chance for increased exposure to **toxicants** (humanly introduced poisons; see Chapter 9).

Or take the fact that feedlot cattle cannot roam and therefore end up standing, sometimes knee deep, in their own waste. Such cattle can succumb to diseases carried in the manure or resulting from stresses that feedlot life engenders. One answer to this is to flood their feed with antibiotics. This has follow-on effects, such as contributing to antibiotic-resistant germs that make humans who eat or handle the cattle very sick (for more on resistance, see Chapters 8 and 9).

The antibiotics might also facilitate the extinction of certain species once prevalent in the human microbiome (regarding which, see Chapters 1 and 6). This and the fact that an industrialized diet is higher in animal fats and simple sugars, lower in fiber, and packed with chemicals not found in a preindustrial diet have been correlated with the lowered diversity in the Western microbiome and with the overproduction of compounds that lead to excess inflammation, unwarranted fat storage, lowered immune function, and other potentially harmful biological activities. The science here is young but with time and more data we will gain a better grasp of how all this happens, and of its health and therapeutic implications, such as for fecal transplantation. Also known as bacteriotherapy, this entails giving fecal matter from a healthy individual with a highly diverse microbiome to a sick individual with a diminished microbiome, with the aim of repopulating the latter and thus rebalancing the host's health. (We aren't the first society to see poop's potential: a fourth-century Chinese handbook describes its use in a soup meant to treat severe diarrhea; Eakin 2014).

In any case, the core point here is that how a population subsists or gets its food (culture) can have a very real bodily (biological) impact, leading to human variation. Some of that impact comes via diet, as shown above; some comes from labor patterns, evidenced in certain kinds of wear and tear on the body, for example. Another portion of the impact comes indirectly: it comes through the way a subsistence type affects the environment when longstanding systems are altered.

Energy and Soil Use

I already have mentioned industrial agriculture's increased introduction of toxicants. Consider, too, that the energy costs of certain kinds of subsistence go beyond the human calories required to find or produce the food. It takes a vast amount of fuel to bring food to the table in a modern industrial food production system. We add fertilizer, which today is made largely out of fossil fuel. We also use gasoline and diesel fuel to transport food, and we use energy to process and package it.

Another way to think about energy costs is to look at loss instead of input. Everything that we eat initially draws its fuel from the sun. The sun shines down on the earth, and the corn grows. The corn grows, it is harvested, and it is made into feed for cattle. Cattle eat the corn, and eventually the cattle are slaughtered to become hamburgers. In that process, energy is actually being lost. By the time we go from sunshine to corn to cow to meat, we have lost 90 percent of the energy that the sun originally had contributed (Pollan 2006).

Beyond fuel costs lies the costs of soil degradation and erosion. When we plant a crop on a piece of land, year after year that land gets leached of its nutrients. Land also loses soil. Our pastures and our range lands are losing soil year by year due to overgrazing and overgrowing.

In sum, intensification helps populations to survive; but it also has costs. These include not only fuel and labor costs, but also costs borne directly by the environment and variously by human populations.

New Food Movements

Before moving on, it is worth noting that food and social activists are gaining traction in efforts to turn things around, in part by relinking polycultural ecological chains that industrial approaches have denied, and taking the lessons of large-scale settled agricultural life as well as of industrialized agriculture into account while doing so. Examples include the biodynamic, slow food, and farm-to-table movements. Each approach has its own emphasis, but all promote the local production of a wide variety of regionally appropriate foods, in season, and with sustainability in mind. All foster multiple and complex relationships between soil, plants, and animals—humans included.

Biodynamic farming in particular emphasizes our mindful stewardship of the earth and preservation of biodiversity. Industrially derived fertilizers and laboratory-produced seeds are shunned in favor of polycultural methods in which farms and gardens exist as their own potentially self-sustaining complex adaptive systems (Biodynamic Farming and Gardening Association 2012).

In general, new food movement efforts favor small-scale enterprise that allows for face-to-face relations. Most keep issues of equity in mind, too, working to ensure that healthful foods are equally available to all people. These new food movements seek to change not only the way we subsist, and thereby our health and so our fitness, but also the way we experience and care for our world.

Intensification and Social Change

The Four Ss

As we have seen, the Agricultural Revolution involved changes to more than simply diet. With agriculture we settled down or adopted **sedentism**. As we became sedentary we also built granaries or silos (see Figure 7.7). We had **surplus** to store—extra food to save for use well after

FIGURE 7.7 "Perfect Granary," an exceptionally well-preserved Ancestral Puebloan (Pueblo III) structure from AD 1100 to AD 1300. Natural Bridges Archeological District, Natural Bridges National Monument, Utah. National Park Service record #42SA6788.

the harvest. The existence of surplus supported **specialization** and a related division of labor among certain segments of the population in areas other than food production, such as the healing arts, or in the crafts of pottery or architecture. A growing but segmenting population also set the stage for **social stratification** or hierarchy. In later chapters we will dive deeply into what these things meant for biocultural diversity. However, first we need a full understanding of them definitionally.

To bring us to that point, we have to talk about something that happens to people when they settle down. They have more babies. Why? For one thing, people do not need to worry anymore about having too many children to travel comfortably because they are no longer walking over long distances. With more than a couple of children, travel on foot can be very, very hard: one can carry only so many at a time. Although nomads tend to have relatively few children then, people who are settled are not so constrained. They also have the need for larger families: more children means more help on the farm.

When families get bigger, social groups or societies get bigger, too. When we look at the ethnographic and historical records, we can see certain patterns in the way that people organize themselves, both politically and socio-culturally, as their populations grow. One of the ways that they do this is through new political arrangements.

Among foragers, people live in small family-based **bands**. Bands may come together also, into **tribes**, in which a number of bands have ties that knit them together. For instance, bands can draw spouses from one another. This often happens when pastoralism or horticulture begins to enter the foraging mix and a few fixed villages, often seasonal, emerge. Sometimes tribes are knit together when representatives from each otherwise generally nomadic band form a cross-linking group, perhaps based on gender or age. This may occur, for instance, in regard to religious or spiritual needs relating to the yearly cycle. In any case, and aside from gender or age distinctions, bands and tribes (whether foragers, pastoralists, horticulturalists, or a mix) are essentially egalitarian groups with very informal governments and decision-making processes; conflict resolution is done between people, who must simply work things out. Peer or cultural pressure is used to enforce decisions and encourage pro-social behavior. Leadership is achievement based; but, however respected, a leader's authority does not entail the overt power to force others to do anything against their will (see Table 7.1).

When people intensify to the point of practicing settled agriculture, political arrangements also change. Government, decision-making, and conflict resolution get more and more formalized. In the political form called a **chiefdom**, hereditary bureaucracy, in which people inherit their positions, emerges. Bureaucrats and people who have the power to enforce decisions now

TABLE 7.1 Four forms of political organization and their implications for selected aspects of life

	Band	*Tribe*	*Chiefdom*	*State*
Government, decision making, conflict resolution	Egalitarian, informal	Egalitarian or achieved leadership, informal	Hereditary, bureaucracy, power to enforce	Centralized, multipart bureaucracy, legal system
Settlement pattern	Nomadic	Semi-nomads, fixed village	Fixed villages	Fixed villages and cities
Division of labor, exchange system	No, reciprocal (gifts)	Not really, reciprocal (gifts)	Some, redistributive (tribute)	Lots, redistributive (taxes)
Add-ins			Public architecture, luxury goods, slavery, literacy	

exist. Indeed, chiefdoms have chiefs. In **states**, which are even larger, generally, than chiefdoms, bureaucracy is centralized, with multiple parts, and a formal legal system exists. While bands and tribes are highly egalitarian, chiefdoms and states are not—just as they have multilayered political systems, they have multilayered social structures. There are people at the bottom of the social structure and people at the top. In other words, there is social stratification; there is a hierarchy of classes. Often, chiefdoms or states have a slave class. There also will be public architecture and luxury goods. These all depend to a large degree on the existence of a surplus, which can be used to support specialists—people who need not labor in the fields but who can (or who are forced to) devote time to non-subsistence-related pursuits. Among the elite, who patronize the artists, we also often find high rates of literacy in chiefdom and state societies (again, see Table 7.1).

Social Evolution and the Organic Analogy

What this brief discussion of political organization encapsulates for us is a pattern of social evolution. That pattern was described for us more than 150 years ago by Herbert Spencer. If that name sounds familiar, it is because Herbert Spencer is the person who really coined the phrase "survival of the fittest," and he coined that phrase in relation to economics. Spencer wrote for the *Economist*, which today is still a major weekly news magazine. As a journalist and scholar, Spencer spent a lot of time thinking about the changes that were going on all around him in England.

What he saw was the impact of the Industrial Revolution. He saw multitudes moving from the countryside into the urban areas to work in the factories, where many people fell by the wayside. The factories were filthy and dangerous; they had a horrible effect on people's health. People suffered from poverty and poor living conditions. There was a sense among the upper classes of uneasiness or anxiety about what was going to happen as a result. The question of how social order would be maintained was very near and dear to the heart of social theorists of the day. Many scholars looked, ethnologically or comparatively, to other societies for answers.

Using a form of the comparative method, Spencer concluded that societies, like organisms, go from a state of being homogeneous to a state of being heterogeneous; that is, they go from a state in which everybody is fairly equal and similarly educated to a state in which people have distinct and different roles in the economy. We know now, of course, that without evolutionary pressure, a band can stay a band forever. However, given a need to intensify, a society will evolve from simple to complex in terms of its social structure; it will go from being homogeneous to heterogeneous. Now, rather than being a **simple society**, with interchangeable parts, it is a **complex society**. This is the core of Spencer's theory of social evolution, which is all summed up in what he calls the "Organic Analogy."

According to the organismic or **organic analogy**, societies are like organisms. Although we now know that evolution is multilinear, in the late 1800s, when Spencer wrote, ladder imagery held sway. Given the climate in which Spencer was doing his thinking, his theory holds that a society will become, over time, more complex (structurally speaking). It will evolve parts with specialized functions; it will evolve, for example, a formal legal system and a formal educational system. All these parts emerge over time as that society evolves.

Compare the evolved society to the human organism. As in a complex society, there are many different parts in the human body. Our bodies are very complex; all of the parts have specialized functions. Some body parts lead (for example, the brain); others follow (for example, lungs do as the brain says). Complex bodies—and societies—must stay organized or they will not survive. In contrast, in single-celled creatures, the inside parts are much less complexly related. In this, the simple organism is like a band society.

Social Cohesion

As Spencer wrote, another social theorist, Emile Durkheim, also was thinking about **social cohesion**—about how societies stick together once they grow big (as they do with an agricultural subsistence base). Durkheim focused on the challenge of what has been termed **social condensation,** which is when larger groups break into smaller factions. In social condensation, a group of people condense or pull together to make their own little group, much as water condenses into droplets on a cold drink bottle (see Figure 7.8). This can happen in bands and tribes; some families simply pull up stakes and move on if the group gets too large or conflicts arise. However, it generally does not happen in functioning chiefdoms and states. People wondered why.

Durkheim's answer focuses on the **division of labor** or, more specifically, extra-familial economic specialization. I make this distinction because every family—even in a band—has some kind of internal division of labor, if only one based on age and gender. However, in a band, or even a tribe, a family or household unit is self-sufficient. Each household unit is, in effect, interchangeable or equal. The same is not so in chiefdoms or states, where extra-familial or household-to-household labor divisions exist. Durkheim's division of labor ensures cohesion as opposed to chaos or condensation.

How does it do that? The division of labor ensures household-to-household interdependence. In other words, single households or groups cannot cleave off to form or start a new version of the society from which they came because they are not independent units. In a foraging group, most skills are equally distributed between households; foraging groups are homogeneous in that regard. Recall, however, that the existence of a surplus among agriculturalists helps to ensure that some people will specialize. Some households will not have to focus on subsistence per se. One could specialize in pottery, another in healing, another in making cloth or making buildings. The model for supporting these specialist households could be communal, or they may trade goods and services for food. Alternately, they may be slaves in the service of an elite (who, again, lives off the surplus).

The division of labor, as Durkheim calls this, means everybody is responsible for a different bit of the labor spectrum. Responsibility for goods and services is distributed throughout

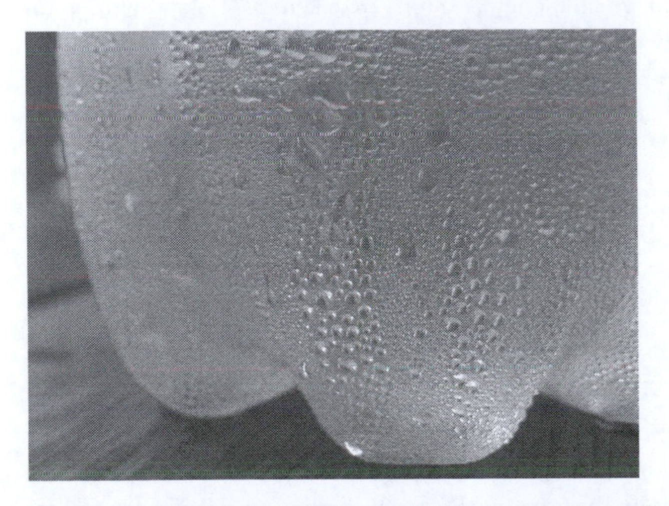

FIGURE 7.8 Water vapor condensing or coming together to form droplets on a bottle, much as people in one society come together to form subgroups. Acdx CC BY-SA 3.0.

the group differentially. If José knows how to make toothpaste, and Jenny wants to brush her teeth, she is very dependent on José. Likewise, if Jenny knows how fix cars, and José needs his car fixed, he is dependent on Jenny for that service. With a division of labor we have interdependence, and that trumps or diffuses any threat of social condensation. Because of the division of labor, people depend on one another in particular ways; from this, social cohesion emerges. Thus, larger societies with a division of labor are in effect self-unified.

The relationships between people in large, complex societies are strong because without them we would struggle, if not perish. They also often are single-use or single-strand, one-dimensional relationships; indeed, they are not really 'relationships' but rather instrumental exchanges between nameless citizens organized around practical needs. They are not person-based but follow the office; that is, the toothpaste maker, auto mechanic, and teacher are important not because of who they are as people but because of what they do (their 'office'); in other words, one toothpaste maker is, in theory, as good in regard to supplying your toothpaste as the next. There is no long-term commitment entailed in purchasing toothpaste from one toothpaste maker—a fact that the use of money for such an exchange makes all the more concrete. Each purchase has a one-time, single-use value. However, the need for exchange creates a strong and cohesive social fabric.

In contrast, in a simple society (a band or a tribe), it is generally the case that the people engaged in an exchange also are kin. They must see each other often and in different roles as well; in other words, their ties are multiplex. Their exchanges have a temporal dimension. That is, they last or endure over time, often through a cycle of exchange actions in which tallies or records may not even be kept mentally because what is being shared is thought of as if a gift. This type of exchange relationship, infused as it is with emotional or expressive value, and meant to be of long-standing duration, entails what theorists call a **reciprocal relationship**. Each positive act is met in turn with another, in continual back and forth movement. Importantly, return gifts given need not be material. They can include help with housework or childcare or hunting or a roof repair. Each gift given contains a little bit of the giver and cements the ties between those doing the exchange, and often between their families or households as well.

With **market exchange**, however, which is the main type of person-to-person exchange that characterizes chiefdoms and states, links between givers and what is given are broken. The use of money signifies this directly, although it also can be true with barter (trading). Chiefdoms and states also use **redistribution**, in which money or goods are taken from one subgroup for use with another, or for use society-wide. Redistribution would be, for example, when the tax collector collects dollars from individuals or households and spends them on things meant for the public good, like education or to build roads. It also can entail hoarding by a self-serving ruler. In any case, the link tax dollars once had with the taxed individual or household is severed with redistribution.

Redistribution of funds or surplus grains or goods in chiefdoms and states often is used to support hierarchy, or social stratification. This means that some groups have more rights and entitlements than others. Some people are members of the elite, and some people are on the bottom of the social heap. Often, food production falls on the backs of those people lowest on the social scale.

Sociopolitical Diversity and Biocultural Diversity

What are the implications of this for us? Diversity in sociopolitical arrangements supports biocultural diversity of various types, from diversity in the kinds of physical stressors that write themselves onto the body, to diversity in the types of diseases to which people might be exposed or vulnerable. We will be exploring such considerations further in the next few chapters. I shall therefore close here with a quick review of the social processes just discussed.

When people settle down to do agriculture, their populations increase. They grow surplus food, and more people plus a stored surplus allows for specialization. When people start doing differentiated kinds of labor, with some of it considered of higher value and some of it of lesser value, the door is open for social stratification, and stratified social systems emerge. Knowledge of these processes helps us better understand the biocultural diversity that came into being with the rise of agriculture and persists today.

8

EPIDEMICS AND IMMUNITIES

This chapter prepares you to:

- Apply basic epidemiology terms and concepts to the question of biocultural diversity
- Characterize the human immune system's role in the body–environment relationship
- Explicate the disease ecology perspective and use it to explain how the agricultural lifestyle contributed to the spread of infectious diseases
- Characterize ways in which local ecologies contribute to population diversity in vulnerability to certain diseases

We human beings have an evolved ability to protect ourselves from the intrusion of unwanted visitors (viruses, bacteria) through our immune systems. Barring individual dysfunction, the basic mechanics of the immune system are the same for everyone around the globe. However, because specific hazards to our bodies vary from place to place, groups of people have over time developed different functional variations in immunity through the process of natural selection. As a result, there is also variation, from population to population, in our vulnerability to certain diseases.

Much of this variation has to do with where a given group settled geographically and how that group therefore fed and organized itself. This (culture) affected human health (biology) and led to diversity between and even within populations. This chapter explores the biocultural interrelationship implicated. To do that properly, we must first discuss the immune system and some of the basic principles of **epidemiology**, the study of disease distribution and its determinants. With firm foundations there, we will go on to examine the relationship between settlement, related patterns of sociality, and the spread—or not—of epidemic disease in various populations.

Remember, our immune systems have evolved, and diseases have evolved right along with them. Some diseases did not even exist in particular places prior to the human occupation of these regions: they exist *because* of us settling there. Adding insult to injury, humans don't just cause new diseases; we also cause more diseases. Epidemics as we know them never could catch hold until our populations grew large enough to support them. Why do large populations support epidemics? The answer has a lot to do with the biocultural changes that came with agriculture. The ecology of epidemic disease is, in many ways, a human creation.

The Immune System

Keeping Out and Dismantling Invaders

The immune system is made up of organs, tissues, and cells that protect our bodies against invaders. Beginning with the skin, organisms and materials that are not meant to enter the body are mostly kept out. At a deeper level, our white blood cells, or **leukocytes**, work to keep intruders at bay. These cells circulate in a system or network of lymphatic vessels as well as in blood vessels, and they are stored in lymphoid tissues and organs, such as in lymph nodes and the spleen. Lymph nodes are found throughout the body, but many are concentrated in the armpits, neck, and groin; the spleen sits above the stomach under the ribs on the left side of the body. The spleen filters the lymph and generates antibodies and lymphocytes, which are one kind of leukocyte.

There are two kinds of leukocytes: **phagocytes**, cells that eat or chew up invaders, and **lymphocytes**, cells that help our bodies remember and recognize previous invaders and support our bodies in destroying them. This is how they work: invading cells are covered in unique substances called **antigens**. A key subset of antigens works somewhat like enemy flags. Sensors in the immune system detect these flags and raise the alert (invader, invader!). Another part of the immune system responds by creating **antibodies**. These are special proteins that can lock, selectively, onto the invaders' antigens and effectively disarm them. They do this by either neutralizing the invader or calling other cells, including sometimes phagocytes, to eat or dissolve the invaders. This, of course, works only if the invaders can be recognized.

Immunization

Immunizations (shots) work by giving our bodies the knowledge that enables them to recognize certain diseases quickly and fight them off immediately if they do invade. Most immunizations do this by introducing a dead or attenuated version of an otherwise dangerous invader into the body. The immunized body, now experienced with that disease, is primed to recognize and react to the real thing and is thereby protected.

Immunization of this kind was first tried in the West in the later 1700s. By that time, some had observed that milkmaids exposed to cowpox seemed immune to smallpox. Smallpox, named after the small boils or skin pustules it often causes, can be deadly; cowpox generally cannot, at least to humans. However, because the germ that causes smallpox is quite similar, from the human immune system's perspective, to the germ that causes cowpox, anyone previously exposed to cowpox could fight smallpox off more effectively than anyone not previously exposed. More milkmaids survived smallpox epidemics for this reason. Edward Jenner, an English surgeon who had taken note of this, carried out the first vaccination on his gardener's son, using pus from a milkmaid's cowpox eruptions. This history is embedded in the word vaccination: *vacca* is Latin for cow.

Technically speaking, there are two forms of immunization: vaccinations use look-alike germs; inoculations use real ones. A smallpox vaccination would have used cowpox while a smallpox inoculation would have used actual smallpox germs. In fact, long before Jenner invented the vaccination, people in China, India, and Turkey used inoculations to stem the tide of smallpox. They would either grind up smallpox scabs and blow the powder up a person's nose or cut a person's skin (sometimes into a vein) and jab in fluid from the pustules of someone whose smallpox case had been mild.

Europeans who had heard of this form of immunization and were willing to immunize at all in fact preferred it over vaccinations. This is because people did not like the idea of mingling cow and human fluids. They would happily eat cattle and even suck beef bones of marrow, but the direct mingling of cattle and human fluids through the vaccination process struck much of the European population as bestial, so for quite some time they would have none of it (see Figure 8.1). Similar beliefs cause some people to reject immunization today. Another source of rejection is misguided faith in an infamous—and fraudulent—study, now discredited, that linked autism to childhood immunization practices (the author was stripped of his medical license once the fraud was exposed).

Four Immunities

Individual immunity brought on through vaccination or inoculation is called **adaptive immunity**. There are three other kinds of immunity, two of which also are seen in individuals. There is **passive immunity**, which infants acquire from breastfeeding because antibodies (and, we now know, certain indirectly protective microbes) are passed along via breast milk (see Chapters 6 and 1). Then there is **innate immunity**, which is inherited. Innate immunity is passed to offspring in the process of natural selection, so it is tempting—but technically incorrect—to call it adaptive. Adaptive immunity instead involves the same kind of adaptation we talked about when we discussed developmental adjustment in Chapter 3. If we remember this little inconsistency, we won't confuse adaptive and innate immunity.

Innate immunity is a bit more complicated than adaptive immunity, too. Whereas adaptive immunity happens to specific individuals through the act of inoculation or vaccination, innate immunity reflects, at the individual level, selective processes that occur over time in a population. For example, if a population is ravaged by a smallpox epidemic, a proportion of people survive. Although some survive because they simply aren't exposed to the disease (the shepherd,

FIGURE 8.1 Cartoon inspired by the controversy over smallpox inoculation; drawn by James Gillray; published in 1802 by H. Humphrey. Library of Congress, Prints & Photographs Division, LC-USZC4-3147.

for instance, who is off herding sheep on a mountaintop when the epidemic sweeps through the village), others survive because they are resistant to the disease for some genetic reason. After the epidemic, then, the percentage of the population that is resistant to the disease is greater than it was prior. Likewise, resistant individuals are contributing a greater proportion of genes or alleles to the gene pool of the next generation because nonresistant people died and cannot do so. We can therefore expect the next generation as a whole to be a bit less susceptible to smallpox, and sure enough, history shows that smaller proportions of the population die from smallpox in generations following epidemic exposure to it. Individuals who were protected genetically and didn't die are classified as having an innate immunity to the disease. Sadly, humans have not yet come close to evolving complete, whole-community innate immunity to smallpox germs: the disease (along with measles, polio, etc.) remained a terrible threat.

This brings me to the fourth immunity type: **herd** or **community immunity**, in which individuals who cannot be immunized—infants, the elderly, the infirm—are protected from a pathogen by others who can. What happens is that once a threshold proportion of individuals (say, 95%; it differs by disease) have gotten immunized, transmission will be suppressed: there simply won't be enough hosts available for a pathogen to get a firm foothold. The more people who vaccinate, the less likely it will be that those who cannot vaccinate (e.g., someone with cancer, a newborn) will be exposed to infection through contact with someone who is ill or infectious.

Herd immunity works well in theory, but there are contingencies. For instance, immunity following most vaccines wanes, so sustaining the necessary level of protection through all relevant transmitting age groups can require repeated doses over the life course and not everyone keeps up with these. Also, the individual costs of immunizing (e.g., time, money, risk for injury) may be misperceived as too high once the proportion of immunized individuals has gotten high enough to make a disease rare. Further, even if threshold level vaccine uptake rates can be maintained, outbreaks still can happen if nonrandom pockets of people do not vaccinate.

Immune System Deficiencies

In addition to diseases that overcome otherwise healthy immune systems, auto-immune disorders can compromise our immunity. For instance, an immune system that lacks certain kinds of lymphocytes, or lacks enough of them, cannot effectively fight disease. Certain medications can weaken the immune response in this way, as can germs such as HIV, the human immunodeficiency virus. By contrast, sometimes one's immune response can be too strong, even turning its weapons on the body itself. This can happen, for example, when the body over-reacts to common environmental materials and develops rashes, swellings, or other symptoms well known to anyone who has suffered from allergies.

Sometimes, allergies are the result of the training—or lack thereof—that the immune system receives when young. A study conducted by Erika Von Mutius in 1999 helped fuel a major shift in our understanding of the immune system. Von Mutius compared allergy and asthma rates in what used to be East and West Germany, hypothesizing that those in the East, which was historically poorer and dirtier, would have higher rates of these ailments. To her surprise, she found exactly the opposite to be true.

Children living in polluted areas of East Germany actually had lower rates of asthma and allergic reactions than did those in 'clean,' modern West Germany. Why? East Germans used more daycare; their children were around numerous other children. They also were around more animals. Based on these differences, Von Mutius built upon an idea developed by David Strachan in 1989 called the **hygiene hypothesis**, which holds that hyper-sanitation leaves the immune system with so little to do that it is primed to jump on the first bit of flotsam it encounters, no matter how harmless it might really be. Conversely, the immune systems of children exposed to more infectious or otherwise 'bad' microbes have greater tolerance for the irritants that trigger asthma and allergies (Public Broadcasting System 2001).

In keeping with discoveries regarding the microbiome, Graham Rook proposed, in 2003, a new version of this hypothesis that flips the lens to focus more on 'good' than on 'bad' microbes, or on what Rook termed 'old friends.' In the **old friend hypotheses**, because our immune system did most of its evolving "in an environment of mud and rotting vegetation" (as Yolanda Smith puts it), our immune function is dependent on continued contact with the microbes we were thereby exposed to. In Chapter 6, I mentioned some of the ways some microbes help us to fight disease, such as by attracting pathogens or exuding elements toxic to them. Without these 'old friends' our immune system cannot be optimized (Smith 2018).

SUPPLEMENT BOX 8: WORMS, ALTERED STATES, AND THE DAWN OF MEDICINE

Having helminths (parasitic worms) can compromise a person's nutritional status as well as cause cognitive problems, intestinal issues, and even death. Helminths have therefore played a fundamental role in shaping the evolution of our immune system as well as our nonimmunological defenses, including avoiding contaminated objects. They also shape our development directly, such as by guiding the body to invest in self-maintenance via immune function versus investing in growth.

Although flush toilets, soap and running water, and the hygiene practices developed in tandem with those do wonders to minimize the number of worms we each carry, they are not our only defense. To deal with helminth infections, we evolved to self-medicate using plants. Plants contain compounds meant to support their own growth, maintenance, and reproduction—and to defend themselves from plant-eating predators and pathogens or germs. Those self-defensive compounds are, effectively, toxins; and many species of animal, including chimpanzees, exploit them on instinct to treat or guard against worms.

One special class of plants' defensive toxins work by disrupting the central nervous systems of herbivores and pathogens. These neuro-toxins are often psychoactive when ingested; that is, they alter one's perceptions, moods, or consciousness. These compounds include not only the caffeine in coffee and tea but also the nicotine in tobacco, cocaine in coca, tetrahydrocannabinol (THC) in cannabis, arecoline in betel nut, and so on. In other words, they include many compounds that we might today label as 'recreational drugs.' They also happen to be largely anthelmintic: they help get rid of worms.

Recreational drug use is documented in nearly every culture. Could it be possible that it evolved as part of a broader, naturally selected system of health maintenance—one that later gained cultural meaning and grew into what we know as medicine today?

Casey Roulette and colleagues (2014, 2016) tested the first part of this hypothesis among Congo Basin foragers who have, historically, been heavily infected with intestinal worms and who also use lots of tobacco and cannabis. This isn't the kind we buy at the corner convenience store or dispensary. These foragers also neither vaped (which introduces all kinds of other hazards) nor injected drugs intravenously (bypassing the nausea and vomiting that provide a measure of protection against overdosing). Their substance use was markedly nonindustrial.

Roulette's team collected saliva and urine to quantify tobacco and cannabis use, and stool (fecal) samples to count the number of eggs present from three different worm species. What did they find? Cannabis and tobacco were both negatively correlated with helminth infections. That is, infections were lower when use rates were higher. This supports the hypothesis that, in locations where intestinal worms remain endemic, bioculturally-regulated recreational drug use helps treat helminth infection.

In a separate study Roulette's team also found that a treatment group (foragers who received a commercial anthelmintic to remove their worms) smoked less tobacco and cannabis two weeks following treatment than a placebo control group (one given a fake anthelmintic). This suggests that helminth infections might moderate drug use behavior, although how that happens remains an open question. Further findings indicated that tobacco and cannabis use actually could prevent infection: one year after treatment with the anthelmintic, reinfection levels were lower among those who smoked more.

Taken together, these data suggest that recreational drug use might be part of a larger strategy of nonimmunological defense against helminths. The data may even help explain why our bodies are 'rewarded' when they consume these neurotoxins: those more responsive to their psychoactive effects may have used them more and the increased fitness this offered may have ensured that the genotype whose human carrier was more sensitive to being pleasured this way was found at greater frequency in the next generation's gene pool.

That said, evolution is full of trade-offs, as is development. Using too many psychoactive drugs can be terribly damaging. Moreover, some worms can be good: wiping them out completely seems to prime the gut for inflammation, and so to increased autoimmune disease rates. How does this work? It seems that the more worms we have the more mucus our intestines generate. The defensive barrier that mucus provides protects the intestinal wall. When the amount of mucus is low, it seems, the immune system goes into overdrive, attacking oneself versus only invaders (Jabr 2010). In other words, our relationship with worms does have a benefit. Nonetheless, with worms, as with cannabis and tobacco (and antibiotics of all kinds), 'everything in moderation' is probably the best approach.

Infection

Even the healthiest immune system cannot always fight off intrusions. One kind of invader that causes trouble is the communicable or infectious disease germ. Simply put, in the context of human society, an **infectious disease** is a disease resulting from the presence and activity of a pathogenic microbial agent that can be communicated from one person to another.

Agents, Hosts, Environments

Germs—bacterial, parasitic, viral, or otherwise—act on the human body to make us sick; they generate pathology in the body. Because germs actively do this, epidemiologists call them **agents**. Agents need **hosts**: they need bodies on which to live. Out there alone in the world, agents cannot survive. On or in a host, they can—but agents make host organisms sick.

How is a disease or germ transmitted or communicated to a host? Often this happens directly from host to host. Humans are great germ communicators because we are so social. We constantly reach out to shake hands, give or receive hugs and kisses, and touch each other socially. In some sense, our very sociality is our downfall when it comes to infectious disease.

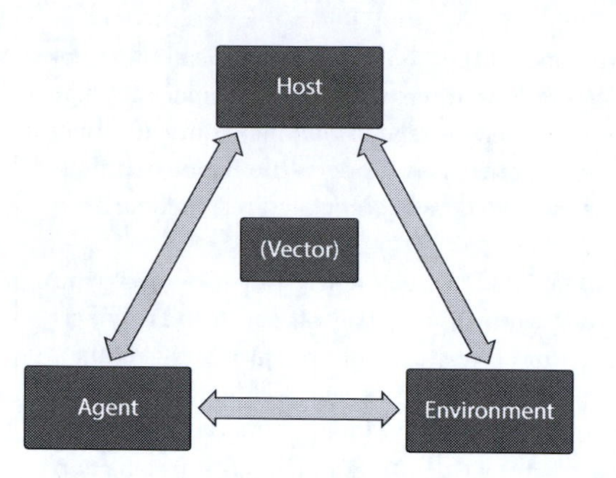

FIGURE 8.2 The epidemiological triangle demonstrates the multidirectional relationships between host, agent, and environment.

For their part, the germs help ensure their own transmission through the symptoms they provoke in us, such as sneezing. In the case of tuberculosis, for instance, a cough and the droplets that come out of the nose and mouth facilitate the spread of germs. In diseases that lead humans to form a rash, the fluid in pustules, which should eventually burst, is the vehicle of transmission. The germs in that fluid are gambling on the chance that someone else will come in contact with them. If the student sitting next to you has a rash on his arm and he unintentionally brushes against you, or if someone who just sneezed into a hand turns (with that hand) a doorknob that you are about to touch, the agent will have hitched a free ride to the body of its next potential host.

The agent cannot always get to the host on its own. Sometimes, a **vector** connects them. Examples of common vectors are ticks, fleas, and mosquitoes. Agents also may get to hosts via an **environmental reservoir**—something in the environment that holds the germ, for instance soil or water (or that doorknob). While a vector could bring certain germs from a reservoir to a host, an environmental reservoir generally holds certain germs until the host happens to come in contact with the reservoir. Epidemiologists also sometimes speak of **reservoir hosts**: living organisms that are both hosts—infected themselves by the agent—and reservoirs that can communicate their disease to other living beings, directly or via a vector. A diseased rat, for example, may bite and infect that pesky child next door; or it may harbor fleas that do so.

Whether it entails a vector or not, the infection process can be seen as based in the relationship between a host, an agent, and the environment (which is, of course, quite humanly created). This trio often is termed the **epidemiological triangle** or triad (see Figure 8.2).

Epidemiological Vocabulary

Infectious disease is not the only kind of disease out there. Noninfectious diseases exist, too; for example, cancer and diabetes. A noninfectious disease is not caused by an agent or germ. We should also distinguish between chronic and acute diseases. A disease that comes on very fast and then (if it doesn't kill its host) goes away fast is **acute**. The common cold and 24-hour stomach flu are good examples. By contrast, a **chronic** disease, such as HIV/AIDS, lasts for a long, long time. One way to remember the difference between these two types of disease is to note that chronic stems from the name of the Greek god of time *Chronos* (think of the word 'chronological').

These are terms we need to know to foster our exploration of the relationship between biology and culture as expressed through health. Another term that we need to know is **endemic**. An endemic disease is one that is native to a given population or area. It has been there for a long time and exists in balanced coexistence with the host population. 'Demic' comes from *demos*, also used in 'demography,' and even 'democracy'; *demos* means 'the people.' The prefix 'en-' means inside, so an endemic disease is one that's inside the population—one that the people in that habitat have simply learned to live with. Malaria, for example, is endemic in certain areas.

On the other hand, an **epidemic** disease is initially not inside a community or people but rather comes down upon them—and spreads very quickly and very efficiently among them ('epi-' means upon or over; e.g., the epidermis or skin lies over our bodies). In epidemics the agent usually moves from host to host—jumping among people, spreading itself very quickly, and infusing the population with the disease that it brings. Polio, for example, swept over the United States just after World War II, sickening huge numbers of people, leaving many paralyzed and many dead.

A **pandemic** is basically an epidemic that has a global toehold; it is an epidemic that's not just sweeping over one village, local region, or nation, but reaching all around the world. As we learned from the spread of the H1N1 (so-called swine flu) virus late in the first decade of this century, something that is pandemic or even epidemic does not necessarily kill the majority of the population. Technically speaking, these terms refer not to severity but rather to how far and fast a disease spreads. However, traditionally when people use the terms epidemic and pandemic, they do refer to acute diseases with high mortality rates.

The science of epidemiology emerged in response to some of the epidemics spreading through Europe, specifically England and France, during the 1800s. Just about any introductory epidemiology text will cover the London cholera epidemic, recalling the way Dr. John Snow cleverly traced it back to a water pump on Broad Street in London. How did he do that? He identified and mapped out the households in which people were getting sick (see Figure 8.3), figured out where their water came from, and found a positive correlation between getting water from the Broad Street pump and contracting cholera. It took some persistence, but finally he did convince public health officials to close the pump. The number of cases fell almost immediately.

Settlement and Sickness

This story brings us back to our discussion of how settlement patterns and subsequent changes in social organization (culture) lead to diverse human health outcomes all around the world between and even within populations (biology). Recall that bacterial and parasitic infections were a big problem for foragers. They continued to be a problem for settled agriculturalists. Other health problems persisted too: traumatic injuries such as those that can be sustained during abnormal childbirth, from long falls, or from fighting during warfare or in the context of domestic violence. On top of all that, however, people began to suffer, and suffer greatly, from epidemic diseases—bubonic plague, tuberculosis, smallpox, cholera, measles. Why were epidemic diseases such a problem once people settled down? Three key reasons were malnourishment, more reservoirs and vectors, and mercantile trade.

As we learned in Chapter 7, many settled agriculturalists were not that well-nourished. Sometimes whole groups were malnourished, but often groups within groups, particularly the lower classes, were more malnourished than the rest. When a person is malnourished, the immune system cannot respond as quickly as it should; malnourished people are not as well protected as they would be otherwise.

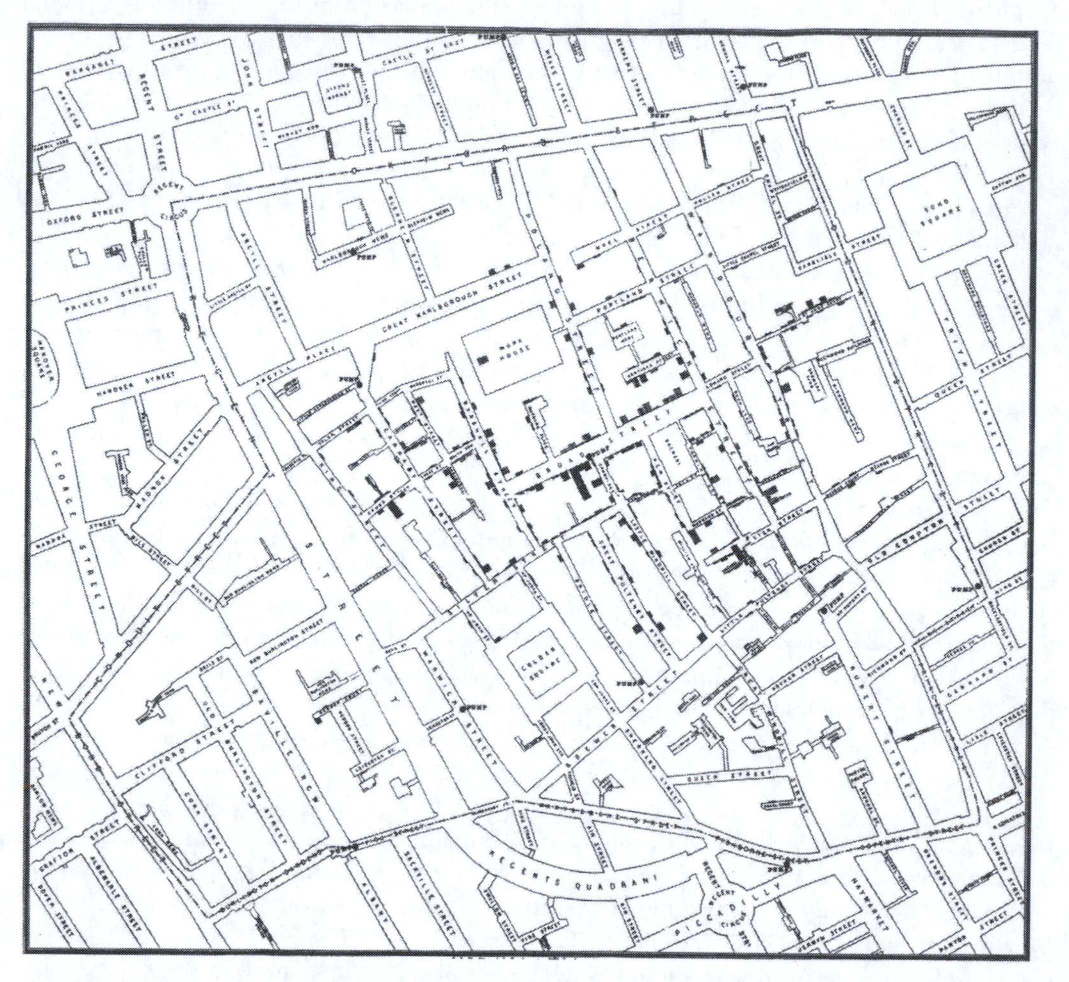

FIGURE 8.3 Original map made by John Snow in 1854 for "On the Mode of Communication of Cholera." Cholera cases are highlighted in black.

Also, when people are sedentary, there simply are more reservoirs. First, there are more people to act as reservoirs, which fosters host-to-host communication, such as in the case of tuberculosis (TB), which is borne in the droplets emitted when an infected person coughs. Second, settlement is followed—at least in successful agricultural groups—by stored surplus food. Food stored in silos or granaries attracts vermin, such as rats, implicated in bubonic plague.

Rats themselves are not actually bubonic plague vectors for human infections; that is the job of the fleas that the rats carry. The rats are only a reservoir host. In fact, while controlling the rat population does serve a purpose (see Figure 8.4), some theorists argue that it was not until the host rat population was itself decimated by bubonic plague that people became susceptible. At that point, fleas jumped off of dying rats in search of better (that is, living) hosts and found humans. So if dead rats begin filling an area, the humans who live there should be worried: whatever killed the rats might be brought by the dead rats' ex-fleas to new human 'homes.'

Bubonic plague, or black plague, is known as such because infected people would often develop buboes—big lumps like giant boils. These big, often blackening lumps generally form

FIGURE 8.4 Professional rat catchers in 1900 in Sydney, Australia, display their work. Photo by John Degotardi for the Department of Public Works, Australia. CC-BY-2.0.

first in the groin, in armpits, or on the neck. These locations are home to lots of lymph nodes where white blood cells—infection fighters—cluster. The buboes turn black when they have gone septic or necrotic—when they are full of decay. If a bubo bursts, the pus that escapes will communicate the infection to others; whole towns sometimes succumbed (see Figure 8.5). In the past doctors commonly opened buboes, believing this would let out the bad juices (here is culture at work), thereby spreading plague germs (and here is biology).

However, the real culprit in spreading the plague was trade. The spread of the black plague in the Middle Ages tracks exactly to known shipping and caravan routes. Rats carrying fleas with plague got on ships; ships went from port to port; and rats went from town to town, bringing with them their fleas. Since then we have made trade much, much faster, beginning with train and then truck routes over continents and then with airline routes that link continents together by mere hours. Germs are jet-setting, too.

Disease Ecology

A Germ-Centered Approach

When we examine infection from the perspective of the agent—from a germ's point of view—we can understand how infection, as horrifying and grievous as it may be to people, is essential for the survival of each agent species. Disease ecology offers a germ-centered approach that we can use to situate biocultural diversity in a systems context.

Disease ecology focuses on the immediate or proximal environment, including a germ's immediate habitat, and on the context in which such germs are spread. Disease ecology is a narrow framework in the sense that it takes a very close-up, on the ground or in the weeds look at disease transmission. Were we to move from talk about trade routes to explore how different

FIGURE 8.5 Plague scene ("Epidemics Die Pest"). Image from the History of Medicine collection, National Library of Medicine, 101405382.

societies are linked together—and how geopolitical relations, power structures, political economy, and so forth fostered a disease's spread—we would be taking a helicopter perspective—one too broad to be termed 'disease ecology.' Think of a camera operator on a movie set. The camera would be observing events at a wide angle, drawing back to take everything in. Conversely, a disease ecology perspective calls for the close up; the camera lens narrows the focus right down to the germ, the disease, and the mechanics of how infectious disease spreads. This is the angle we take when we talk about how the immune system may be weak in a malnourished individual, allowing for infection. We also take the disease ecology perspective when we talk about the number of reservoirs in a farming settlement or focus on livestock, which we shall now do.

Zoonosis

When people first settled down to farm, forests covered much more of the earth than they do today. However, farming requires cleared, often flattened land and so, when in the way, swaths of forest had to go. Geographic features were also removed or altered; this aspect of niche construction for agriculture ranged from tossing rocks away to diverting rivers. Given the human-made or anthropogenic alterations to their respective niches, human–animal contact in some instances increased.

For instance, when farming came to equatorial Africa people broke open the forest and cleared land (Robbins 2012). This may have supported the emergence of varieties of malaria that make people sick: it both provided more standing water for mosquitos that carry malaria and brought humans into closer contact with chimpanzees, who had long carried their own strain of the disease (see Cormier 2011). In Bangladesh today, habitat encroachment has resulted in increased human–bat contact. Hungry fruit bats raid humans' date palm sap containers, contaminating their contents with the potentially deadly Nipah virus, which then infects humans who drink the raw sap (Robbins 2012).

Beyond illustrating 'One Health' in action (this model was introduced in Chapter 1), these two cases exemplify zoonosis. A **zoonotic** disease is one that originates in an animal species. Zoonosis always was possible, but through the environmental alterations it required, intensification broadened this possibility. When we started to raise livestock, we further increased this risk: cows, pigs, and chickens are among the many domesticated animals that have been associated with the spread of disease among human populations, sometimes in epidemic proportions.

Normally, a germ that can infect a particular species cannot necessarily infect another because each species is a little bit different. In this, speciation is a bit like sexual reproduction, in being a source of variation (see Chapter 2), although the latter serves only the single species by ensuring diversity in the gene pool while the former serves life in general by ensuring diversity at the level of the genus and beyond. The passkey a germ needs to enter and infect one species can be very different than the passkey to enter and infect others. For example, certain germs are well-fitted to thrive within (infect) pigs, but because of physiological differences between pigs and human beings, often those germs cannot infect us.

This status quo of germs and hosts holds only until the day that a germ mutates in just the right way to gain a toehold in the human body. This is what seems to have happened with malaria long ago and, more recently, with the H1N1 ('swine flu') virus. Through this simple evolutionary process, a new humanly infectious disease is born. The longer a human group has handled or otherwise dealt with a particular animal, and the closer that physical contact has been—the more bodily contact those animals have with the humans—the higher the likelihood that one day a zoonotic mutation will happen.

Such a mutation is necessary, but of course not sufficient, for a new zoonotic disease to become epidemic. Remember, epidemics require a large population to support the transmission of a disease from generation to generation. There must be survivors. In a small population, if everybody is wiped out, so, too, is the particular pathogen that killed them, as there are no more hosts to keep it alive (unless, of course, it can jump the species barrier again). In a large population, however—particularly in one that reproduces sexually and thereby exhibits variation in the gene pool—chances are that some people will be immune to the new germ. They will go on to have children. Few traits are 100 percent inheritable, and not all immunity is innate anyhow, so children not born immune will provide the germ with a whole new host population.

Smallpox

This kind of scenario played out again and again with smallpox so that, over time, European populations in close contact with cattle were, in aggregate, less vulnerable to smallpox than populations that were never exposed. A traveler from a cattle-raising community might bring the smallpox to a town without cattle and inadvertently wipe it out entirely. Likewise, a traveler—a merchant, conquistador, colonialist, missionary, or any other—might carry smallpox to another nation or continent. The result could be absolute decimation.

Native Americans, of course, had never been exposed to either smallpox or cowpox prior to the invasion of the continent by European parties. Following European colonization, Native Americans, lacking any form of immunity, innate or adaptive, to the smallpox germ, died in droves, despite various precautions (see Figure 8.6). In the Northeast this period has been called The Great Dying.

Some scholars have wondered why the opposite did not take place—why the invaders did not sicken and die from indigenous American diseases. Although the Europeans had cattle, people

FIGURE 8.6 Burning of a Navajo hogan occupied previously by a smallpox victim, near Indian Wells, Leupp Indian Reservation, Arizona, c. 1890–1990. National Archives and Records Administration.

living in the Americas had herd animals as well; in South America, for instance, they had llamas. There were huge differences, however, in the ways that these populations related to their livestock. For example, domestic llama herds were smaller than European cattle herds. Further, people in South America did not milk their llamas, nor did they live very near to them, so they never established that kind of physically close contact that the milkmaids and other Europeans had with cows. In Europe, cattle often provided a source of body heat for the household; they might live in a house's ground floor. Europeans were just closer to their animals—to their skin, their feces, their milk. The South Americans and their animals were not so close. They did not have the opportunity to foster zoonotic diseases equal to smallpox; they did not coevolve with a disease to which they would be a bit immune but the Europeans would not (Diamond 1997). Given the geopolitical context of the time, this example of biocultural diversity had huge implications for global history

Recap and the Red Queen Redux

With settled life, and concurrently larger populations, disease is more likely to spread. No longer dispersed, agricultural people live in close quarters and very nearby to large numbers of other people. They tend to live near their own waste as well. Agricultural systems result in and are sustained by food surplus, which attracts vermin and therefore supports more disease. Living with livestock has the same result. Polluted water supplies compound health challenges. While foragers generally roam about the countryside, leaving waste behind, in permanent settlements health problems pile up.

To better comprehend the chain of effects that come with settlement, let's extend the Red Queen metaphor. As we saw in Chapter 2, the Red Queen keeps moving because the world around her is moving all the time. Human culture and human biology work much in the same way, adapting to new environmental pressures or challenges. Importantly, many of the challenges humans face are in fact humanly created. Changes in social organization and subsistence patterns foster the spread of disease. In changing the way in which they made a living, our ancestors made much more intensive and extensive changes to the environment than they ever had as foragers. In turn, they had to change—change their cultures (for instance, inventing such things as housing that would last more than a season and synchronizing their belief systems to their new diets) and their bodies (for example, to digest lactose, or to resist new diseases). Our ancestors had to adapt to the effects of the changes they brought about; so do we, and so do other species. The world around us is in constant motion.

Cholera

Cholera is a horrible intestinal illness brought about by a bacterial pathogen that became very well adapted to thrive within large, settled human populations. Cholera epidemics have swept through the so-called Old World numerous times in the past. By the 1800s, cholera had penetrated the New World, too, as the reach of Old World powers grew. Typically, 50 to 70 percent of those stricken in an initial epidemic of classic cholera will die (Joralemon 2006, 33).

Cholera's main initial symptom is the sudden onset of intense unabated diarrhea. This brings about immense dehydration: the body is excreting so much it cannot replenish its fluids, which is why it dies. Diarrheal excrement, of course, is helpful for the cholera germ because, if uncontained (and containing it can be quite difficult due to the nature of the illness), it serves as an excellent way of moving that germ between one host and the next. Water or raw foods can provide a reservoir; if the germ gets into another host before it expires, the cycle will continue. Ceviche, a raw shrimp and fish dish, is a very good conduit for cholera transmission if the ingredients come from water infected with the cholera pathogen.

The bacteria that cause cholera do so by excreting a substance that, to human beings, is very toxic. The substance disrupts the regulation of water and salt across the wall of the intestine. Under normal circumstances, nutrients (including salt) and fluids move across the lining of the intestine. One of the biological mechanisms that regulate this process is a particular kind of **transmembrane conductance regulator**, or TR. TRs move (conduct, transfer) things across the intestinal membrane, regulating the flow (see Figure 8.7). When this regulation process is thrown off kilter by toxins secreted by the cholera-causing bacteria, the affected individual will be sick, not only because fluids and nutrients are insufficiently moved out of the intestine and into the body, but also because fluids are pulled out of the body from outlying tissue and into the intestine. All this then comes out in the form of diarrhea, leaving the body malnourished and dehydrated.

Interestingly, there is variation across the global population in the number of TRs that people have. People with a greater number of TRs have a much harder time with cholera than do those with fewer. Those with fewer are less vulnerable to choleric diarrhea: they do not have as many regulators available to become infected by the cholera toxin and break down in the first place. They therefore will not have as many TRs pulling fluids out of the body into the intestine and creating diarrhea and will not suffer as much as a person who has more regulators. However, although protected in this way against cholera, people with lower TR counts need cool climates. If the climate is hot, they will suffer from other health challenges because of a related inability to sweat well to keep cool (the TRs implicated in cholera occur in places other than just the intestine).

It turns out that the TRs we are talking about here—which actually and more specifically are called cystic fibrosis transmembrane conductance regulators (CFTRs)—are themselves regulated by a particular gene that has various alleles, one of which works much like the sickle cell

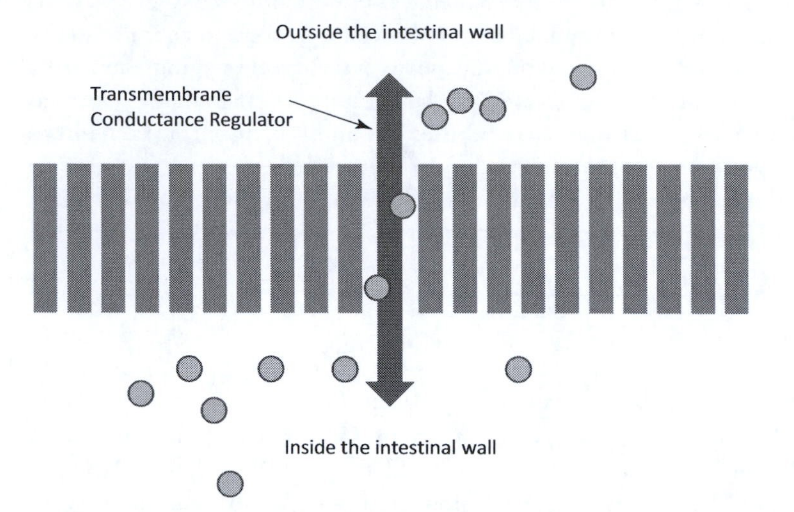

FIGURE 8.7 Transmembrane conductance regulator (TR); TRs regulate the flow of targeted materials across cell membranes.

allele that we learned about in Chapter 2. That is, it is partially expressed when inherited from one parent—and it seems to confer an adaptive advantage when cholera is a threat (for the sickle cell allele, the equivalent disease is malaria). The allele's partial expression includes the generation of fewer CFTRs.

This allele is more likely to be found in white people from northern Europe where not being able to sweat very well generally poses little problem. The allele is thought by many to have evolved in these populations as part of a balanced polymorphism that protected the majority against death by cholera, albeit at the cost of sacrificing a few members of the population to a disease that inheriting two copies of the recessive gene would cause. That disease is cystic fibrosis (CF).

With some CFTRs missing, a heterozygous person encountering a cholera outbreak would be advantaged. This person would have fewer functioning CFTRs to be negatively affected by the cholera toxin. Indeed, heterozygous individuals may have been more likely to survive the cholera epidemics of the 1800s; that may be how the allele became so prominent in some groups. There is some indication, too, that heterozygous individuals are more likely to survive typhoid fever and TB. The latter caused 20% of all deaths in Europe between 1600 and 1900 and so protection from it could certainly also explain the high prevalence of the CF allele in white Europeans (MacKenzie 2006). Indeed, there is probably no 'one' reason for the persistence of CF: like so many other aspects of human experience, inheritance of low-functioning CFTR alleles is without doubt a very complex process.

In any case, anyone homozygous with the recessive allele thereby lacks healthy, functioning CFTRs. For this individual, real trouble ensues. Fully expressed, the recessive genetic disease CF disrupts the body's ability to absorb nutrients, breathe, and sweat. Cystic fibrosis was, prior to the development of modern treatment regimens, generally fatal in childhood.

The Germ's Point of View

I have been telling this story through the disease ecology lens. The existence of this protective recessive allele reflects evolution in action. We made a change in our environment so that we were more likely to drink water containing excrement fouled by the cholera bacteria (we gained more exposure to typhoid and TB also through our changed lifestyle, but to keep this example simple let's stick just with cholera for now). The changes we made did benefit us initially, for instance in terms of giving us somewhere easy to put our waste; but eventually our dirtier water benefitted the cholera pathogen more than us. For our population to survive, we had to change something in response to that. One kind of adaptation could have been cultural: we could have invented sanitation, as we eventually would do. Before we did so, however, genetic adaptation went to work. The gene pool happened to have a pre-existing variation in the polymorphic CFTR gene. One CF allele became more advantageous to carry when cholera came onto the scene. In the face of cholera, it was favored by natural selection.

These kinds of small, microevolutionary changes are happening all the time, both in us and in the world around us. For instance, the bacterium that causes TB has recently evolved ways to resist being wiped out by particular pharmaceuticals; that is, multi-drug-resistant strains have emerged. In a nutshell, particularly when TB treatment is sporadic or incomplete, the few TB germs that are resistant to a particular drug will survive it, reproduce, and repopulate, making the next generation better fitted to survive in an environment flush with drugs that easily killed ancestral strains.

Not all evolution leads to increased or enhanced resistance. Another way pathogens evolve is to become able to infect a new host, for example through zoonosis. Yet another direction is toward **attenuation**: if a pathogen's impact on its host becomes sufficiently minimized, hosts carry it around longer. If germs had intelligence, attenuation could be classed as a very smart move. The super-virulent disease that kills off all its hosts thereby sabotages its own existence; but a good host is a living host, so a well-adapted pathogen has a stable life with lots of living hosts to inhabit.

Many scientists think that the common cold was once a very virulent disease that attenuated; it became weaker in its effects and that was beneficial to the germ itself because it allowed the germ to spread much more easily. We see attenuation happening right now in a particular strain of cholera. Once people realized that water sanitation was key to stopping the spread of cholera, the classic strain was placed at a disadvantage. It also was disadvantaged by public health efforts to get people to stop eating ceviche and undercooked foods and encouraging the consumption of only boiled or disinfected water. Our cultural adaptations placed real environmental pressures on the cholera-causing pathogen. As a counter-response, classic cholera, which is very, very deadly, seems to have evolved one strain, the El Tor strain, which is better adapted to thrive within the confines of water sanitation.

El Tor causes less mortality and morbidity in humans, and as a result is excreted by human hosts for a much longer period of time (classic cholera victims die fast). Moreover, El Tor lasts longer outside of human beings when in water, and it can survive even in moderately cooked food. To top it off, people do not have any post-infection immunity after recovering from this new strain. In sum, it does not make us as sick as classic cholera does, but it can survive better and longer and infect more people. From a disease ecology point of view, El Tor is well-adapted to its present environment (Joralemon 2006, Chapter 3).

The Pivotal Role of Niche Construction

In the future, in addition to efforts to eradicate epidemic and pandemic health threats, we might further explore the possibility of culturally manipulating features of the environment that support attenuation in problematic pathogens. We also may be able to track, in real time, evolution not only in the pathogens that trouble us but also in the human genome. In either case, we should be wary of unanticipated secondary effects. A small change in one area—in a gene, in a pharmaceutical concoction, in how we treat water or process waste, and so on—can have a huge impact elsewhere in our global system.

Through the cholera and the other examples in this chapter, we have seen that viewing human biology as static or unaffected by the environments that we have created to live in blinds us to the very real and very important connections between our physical bodies and the large, complex system that we are part of as inhabitants of the earth. We have seen how the various susceptibilities different populations have to particular diseases is a result of diverse interactions between the varied environments we have created (cultures) and the bodies (biologies) with which we inhabit them.

9

POLITICAL ECONOMY OF HEALTH DISPARITIES

This chapter prepares you to:

- Explicate the political economy perspective and use it to identify and explain the varied, differentially distributed impact of social stratification on human health (epidemiological polarization)
- Describe the key components and main aim of primary health care (PHC) and demonstrate the value of PHC in combating epidemiological polarization
- Explain the need for a social justice approach to combat structural violence

In this chapter, using the political economy perspective, we examine the differentially distributed impact of social stratification on biocultural diversity, as reflected in population health profiles. In disease ecology, explored in Chapter 8, the focus is on evolution—in fact on co-evolution, as seen for example in the evolution of resistance in germs assaulted by culturally invented drugs. In disease ecology, we focus on **proximate** causes—factors in the immediate vicinity of infectious processes or things we see when we take a close-up look. In the framework offered by the **political economy** perspective, which we now shall adopt, we look from a distance, with a wide-angle lens. We consider **ultimate** or distal causes—causes far removed but that precipitate an event.

The political economist asks questions such as "Who benefits from this situation?" "Who is profiting?" or even "Who is maintaining their gains, maintaining their position, or keeping the status quo as it is (that is, in their favor)?" The political economy perspective asks about power relationships; it asks about authority. Much of this power and authority is tied up with the economy—with resources including money, and whoever has it.

The lens of political economy reveals the many ways in which those with less power and authority end up carrying a disproportionate share of a society's ill health. In other words, it explains certain kinds of biocultural diversity as rooted in social-structural arrangements. Before getting into social structure, however, I should introduce a few more epidemiological concepts.

Epidemiological Variation

An **epidemiological profile** is the disease profile or picture of a given group. It identifies what diseases and other health challenges group members are experiencing or have experienced, and to what degree. It also attempts to say why.

Epidemiological profiles can shift over time. Recall that if a foraging group becomes agricultural, its epidemiological profile shifts to highlight epidemic diseases. When agriculturalists intensify, epidemiological changes also follow.

In the late 1800s, the Industrial Revolution was well underway in many until-then wholly agricultural societies. With it came not only the rise of factory-based industries but also the demise of ways of life that had, until then, been longstanding. People migrated to the cities in large numbers. Living conditions for workers were often unhygienic, uncomfortable, and crowded. This was a boon to the infectious diseases that thrive in such circumstances. However, as the Industrial Revolution became the industrial way of life, a new epidemiological profile emerged. In that profile, the diseases that bothered people most were not so much infectious diseases with high mortality rates; rather, they were noninfectious or chronic diseases with low mortality but high morbidity rates: people lived longer, but with longer lives came the diseases of old age.

This shift (from more acute, communicable diseases and high infant mortality rates to more chronic, noninfectious diseases that show up later in life) has been termed the **epidemiological transition**. To focus on this particular transition as 'the' transition of course ignores the importance of the forager-agriculturalist transition—and of more recent transitions such as the globalized transition to Western diets (see Chapters 7 and 12)—but 'the' transition is, anyhow, what epidemiologists have called it. As we shall see in this chapter, the simplicity implied in the label and its dualistic (either–or) and narrow focus conceal great complexity and broader patterns related to human biocultural diversity—complexities and patterns that here we aim to explore.

Soap and Hot Water

What underwrote the epidemiological transition? In a word: sanitation. People began to wash their hands and take other steps such as building better sewerage systems. Why did they not wash their hands previously? For one thing, running water was not always easy to find; even when it was available it often was cold—sometimes (e.g., in winter) very. Moreover, people did not know that hands could carry pathogens from person to person.

One day, however, an obstetrician named Ignaz Semmelweis realized that the midwives in a hospital where he worked were having much better outcomes than the medical students. Semmelweis's data showed that the medical students were actually killing off birthing mothers because they were going straight from doing autopsies and working with cadavers to delivering babies, bringing germs with them. Even so, it took the medical community a long time to come around to Semmelweis's suggestion for hand-washing. Culturally, it did not make sense. The germ theory of disease had not yet gained credence, and people at that time did not know that microscopic pathogens could exist. People for the same reason did not think twice about drinking water if it looked clean enough, or about drinking from a shared cup.

Although we'd been living in our own germ-friendly filth for quite a long time, people in the later 1800s were forced to confront this because the factories that they built were so obviously polluting and tenement housing was so flagrantly awful. That's not to say that cities previously were not redolent because many were—but by the late stages of the Industrial Revolution there had been developments in science that helped people see what they previously could not. Moreover, democratizing political changes and a belief in citizens' rights that had been underway already in Europe for a few hundred years now came to the fore, as seen for instance in the French Revolution. Friedrich Engels's classic 1844 study of the horrible conditions under which the lower working classes in Manchester, England labored and the slums in which they lived spurred further activism, as did Jacob Riis's late 1800s photographs of squalid US tenement life (Engels 1970; Riis 2004; and see Figure 9.1).

FIGURE 9.1 "Five cents a spot": unauthorized immigrant lodgings in a Bayard Street tenement, New York. Photo by Jacob A. Riis, c. 1890.

The sanitation movement was really a social movement. Social justice campaigns that emerged out of the realization of how awful workers' lives were did a lot to clean up the cities and otherwise promote sanitation. Yes, the government was interested in stemming any potential for worker unrest that foul living conditions might inspire. Yes, capitalists were interested in keeping the cities attractive to rich shoppers. However, it was really the social justice campaigners who underwrote the epidemiological transition, and they did so in the name of the working class.

Polarized Profiles

Sanitation led to a shift in the overarching epidemiological profile away from high levels of mortality from infectious disease and lots of death at young ages toward a profile in which people lived longer and later died from noninfectious diseases. Only, not every population made the transition.

Many groups have maintained a pre-transition epidemiological profile. These groups comprise whole countries—poor countries—as well as the poorer classes in rich countries. This antithetical division between the haves and the have-nots, or the transitioned and the non-transitioned, forms what Paul Farmer termed "The Great Epi Divide" in his campaign to defeat it (Kidder 2003, 125). The more common label is **epidemiological polarization**.

Social stratification, which arose as a byproduct of the Agricultural Revolution, is a very important factor underwriting epidemiological polarization. Indeed, when we refer to biocultural diversity today as measured in terms of differential health outcomes or different epidemiological profiles, we are really talking about the consequences of holding different positions within local and global social structures. Health inequalities highlight parts or features of the social structure that divide societies like fault lines, or which exist as gaps—as places where people can fall through the cracks.

When looked at as wholes, better off societies (such as US society) have much better health outcomes generally; they have made the epidemiological transition. Societies at the bottom of the scale (such as Haitian society) are not so well-off health-wise.

We also see epidemiological polarization within societies. I am reminded of it every day when I see people waiting on a well-known street corner in my town for someone to drive by and offer them work. Then there are waged laborers who are exposed through their jobs to contaminants and hazards. Agricultural workers in the United States have a very different health profile than, say, chief executive officers (CEOs).

Indeed, in all hierarchical societies, health outcomes are worse overall for those on the lower rungs; and they grade or incline upward as one's socioeconomic status climbs. Scholars variously term this phenomenon the social gradient of health, or the **health–wealth gradient**.

Helpful in summary, the health–wealth label does mislead slightly if taken literally, because money alone cannot explain the health disparities seen. For one thing, the more income inequality a society has, the worse off its members will be overall for outcomes that measure health and well-being. For instance, the United States may be the wealthiest of wealthy nations but its average life expectancy is among the lowest and its obesity, teen pregnancy, infant mortality, social distrust, and drug addiction rates are among the highest in this grouping (Wilkinson and Pickett 2009). Why? Large masses of have-nots bring down averages; but also, the stress that culturally created inequity can provoke can be very damaging biologically (see Chapter 10).

Moreover, one's socioeconomic class intersects with other demographic variables such as gender, age, sexuality, and race or ethnicity. The convergence of various such dimensions of identity is termed **intersectionality**. Intersectionality often is used also to refer to overlapping, interlocking systems of discrimination or oppression that support the political-economic status quo and undermine nondominant groups. The barriers and enablers entailed can affect one's well-being drastically.

Multi-Drug-Resistant Tuberculosis

If all this still seems a bit abstract, never fear: Paul Farmer (2005) has demonstrated concretely the advantage of viewing and responding to biocultural diversity through the critical lens of political economy. A favorite example for him entails multi-drug-resistant tuberculosis (MDR TB), in Russia. There, as a result of political and economic changes, many people have been thrown into poverty. Many have ended up in prison for crimes committed in circumstances of severely limited access to legal, paid employment by which to feed one's family.

What happens in jail? People live in very close quarters where they are exposed to TB, a bacterial infection of the lungs that generally leads to death if untreated, but which, being slow-acting, has lots of opportunity to spread. Russian prisons simply have not been able to afford to provide sick prisoners with the right kinds of medical treatment on a consistent basis. Over time, a highly drug-resistant strain of TB has evolved. Prisoners infected with this strain can spread it to prison staff members and, on release, to outsiders. This is one reason that Farmer's colleague Alex Goldfarb calls prison an "epidemiological pump" for the MDR TB crisis (Kidder 2003, 232).

When we consider TB from a political economy point of view, we do not ask so much about what a TB germ does once it has entered a person's body or how it is spread per se; that is the purview of disease ecology. When we look at things from a political economy perspective, we ask which individuals are exposed, why they are exposed, and why they are vulnerable to a problem anyhow. What in their political-economic or social context set them up for infection?

With TB, for example, we ask who is at risk for incarceration. We ask who is malnourished, too, because the better nourished people are, the less vulnerable they are to infection. We ask about who has access to the right medicine, and why. We ask why one class of people get TB and another does not.

I mentioned that the prisons do not have the money to buy the right kind of medicine for incarcerated patients. "We just don't have the money" is, as Farmer says, a very common phrase in explanations of health inequities. What Farmer shows and what many theorists also argue is that "we" do have the money. We just prefer to spend it on other things—on luxuries for our pets for example, or fancy cars, or status handbags, or any number of nonessential consumer goods. There is in fact enough money circulating in the world. The problem, then, is in its unequal distribution; it is that priorities for spending do not extend to the provision of appropriate pharmaceuticals to all who need them (Farmer 2005; Kidder 2003).

Structural Violence

Sometimes certain social groups are cut off from access to health care or put into positions that make them vulnerable to exposure, such as to TB. Sometimes they are put at risk or in certain conditions, such as via the occupational hazard of inhaling pesticides, that can lead to cancer and so forth. The harm brought on by this kind of vulnerability is a form of what Johan Galtung (1969) named "structural violence." In **structural violence**, the shape of a given social structure harms or is harmful to the people who occupy certain positions within that social structure.

Some social structures cannibalize themselves, using up their own human resources in structurally violent acts. This is quite obvious when we consider a cotton or sugar plantation fueled by slaves—individuals treated as human resources: purchased, and easy to replace when they burn out or break. In reality, of course, slaves are people—people who, because of their position in the social structure, come to harm. They are consumed by the system as it strives to keep itself going.

A contemporary US example of cannibalization entails so-called poor whites. Ever since their forebears first came to the United States (many as indentured servants), poor whites have been willfully disenfranchised and manipulated by those in power (that is, rich whites). Concurrently, poor whites have been charmed by the social structure's ideological foundation, which promises that anyone can move up to the top with enough hard work. This kind of (mythical) meritocracy is inherently competitive, allowing those on top to, effectively, divide and conquer those groups whose labor they would exploit.

To this end, in the Old South, the land- and slave-owning elite encouraged racism in poor whites, who effectively fought the Civil War on the slave owners' behalf. When slavery was abolished, poor whites became a key source of cheap farm labor in the South and on the prairies. Those living in the mountains of Appalachia (a mostly white, mostly Christian region centered in West Virginia, Kentucky, Virginia, Tennessee, North Carolina, and Georgia) labored in coal mines. In the North, steel barons (who also used poor whites to fight the war) worked them in industrial processing and manufacturing facilities.

While the coal industry remained a big consumer of these human resources until recently, sharecropping (essentially, feudalism) had lost its appeal for land owners by the later 1930s, due largely to the perceived advantages of tractor farming. The appeal of mechanization was intensified by the fact that the unsustainable farming methods meant to increase yields had, coupled with drought, culminated in a dust bowl. Displaced poor white sharecropping families became another cheap source of migrant farm labor (see Figure 9.2). John Steinbeck memorialized this turn of events in *The Grapes of Wrath*.

(a)

(b)

FIGURE 9.2 (a) Cotton sharecropper Floyd Burroughs, photographed in Hale County, Alabama in 1936 by Walker Evans. (b) Destitute pea picker Florence Owens Thompson, age thirty-two, with three of her seven children, photographed in Nipomo, California by Dorthea Lange. Library of Congress, Prints & Photographs Division, FSA/OWI Collection, nos. LC-DIG-ppmsc-00244 and LC-DIG-ppmsca-50236, respectively.

Forty years later, poor whites who had entered the working and even middle classes through factory jobs in the North also began falling on hard times: deindustrialization related to factory owners' profit-boosting preference for outsourcing labor needs decimated local economies and turned what was known until then as the Steel Belt into the Rust Belt (see Supplement Box 9).

SUPPLEMENT BOX 9: DEATHS OF DESPAIR

In the late 1990s, mid-life deaths by drugs, alcohol, and suicide began to rise in the United States. The first big cluster was seen in the early 2000s, in the Southwest. Within a few years, increases had spread into Florida, Appalachia, and the West Coast, and then the problem lodged in the Northeast. High mid-life death rates now are seen nation-wide, in rural areas and urban centers, among men and women similarly. The only way that these deaths do seem to discriminate is by race and class: they center on non-Hispanic white Americans with a high school degree or less. This group now has a higher overall mid-life death rate than any other—a distinction that, once discovered, shocked many: the overall mortality rate for said whites was 30 percent lower than that of blacks as recently as in 1999; but by 2015, it was 30 percent higher. What happened?

In attempting to figure that out, Anne Case and Angus Deaton have looked at context data. Here, correlations are telling. The increased rate of deaths due to alcohol, drugs, or suicide has been accompanied by measurable increases in mental illness and social isolation. As might be expected under such conditions (see Chapter 10), physical pain also increased—much of which was doctor-treated with medicinal opioids, such as OxyContin®. Pointing to the latter's rapid initial uptake as a result of physicians prescribing in primarily white states, Helena Hansen and Julie Netherland (2016) note that regulations and marketing strategies "gave US White patients the 'privilege' of unparalleled [opioid] access" (p. 2127). High levels of dependence followed, accordingly, and the number of deaths by opioid overdose skyrocketed for the white population.

Yet, rather than being forced into interventions in line with the 'war on drugs' or similar to those aimed at crystal methamphetamine or crack cocaine use (read: drug use by people of color), pharmaceutical opioid control measures took a lenient stance—one more in line with how powder cocaine users (read: whites) had been treated by the law. That is, rather than arresting and incarcerating those addicted to pharmaceutical opioids, authorities crafted drug give-back programs. Good Samaritan laws popped up to protect those who might call for emergency help. Moreover, when whites began turning to heroin for an opiate hit, policy-makers called for reduced sentencing and expanded treatment access in their communities—something not done where people of color are in the majority.

This response and the problems with it notwithstanding, rather than to stop at opioids in explaining the substantial upward trend in deaths among working class whites, Case and Deaton point to broader issues that opiates, like alcohol and other drugs, as well as suicide, can be used by individuals to address. They point to decreases in economic and social well-being measures, and in marriage and labor force participation rates. They argue, with plenty of federally collected data, that pain, distress, and social dysfunction have been accumulating in the lives of the people affected. The heyday of white, blue-collar workers ended in the early 1970s and since then, fueled also by the 2008 financial crisis and collapse of the housing market, the previously pervasive sense of status and belonging was lost to members of this group.

Their deaths, then, are what Case and Deaton have termed "deaths of despair." The two propose an explanation "in which cumulative disadvantage from one birth cohort to the next—in the labor market, in marriage and child outcomes, and in health—is triggered by progressively worsening labor market opportunities… for whites with low levels of education" (2017, 397). The labor market has deteriorated and the opportunities white workers expect to find there as well as those that a robust labor market might provide (e.g., the opportunity to support a family and a middle-class lifestyle) simply are not there. Deaths of despair express the distress this disappointing situation causes by turning it inward, onto oneself (although, inevitably, with alcohol, drug, and suicide as the cause of death one's family and other relations also suffer; and outward-facing despair can be problematic too, if it turns jingoistic and racist instead of toward socially just civic activism).

In the end, Case and Deaton highlight "the collapse of the white working class after its heyday in the early 1970s, and the pathologies that accompany this decline" (439–440). Whether and how mid-life death rates among these whites will restabilize, and at what level, remain open questions. Answers depend on our ability to approach diverse forms of drug use objectively and democratically, and our capacity to fix the problems that push varied groups of people toward varied forms of self-medication to begin with.

Before moving on, let's be sure that we are all clear on exactly what social structure means. Societies are made of groups of people, and all of the people in a society have some kind of a relationship with one another. The **social structure** emerges from those relationships, just as a building can emerge out of the relationship of wood, nails, and concrete. Put together one way, a ranch house gets built; another way, an A-frame cabin; still another way, a two-story townhouse (see Perry 2003, Chapter 4). Each structure emerges out of the relationships between the bits of building material. Likewise, if we fit people together in one way, one kind of social structure will emerge; if we fit them together in another, we can have a different social structure entirely. Think of the difference between an egalitarian social structure and, say, a feudal society governed by a monarchy. Part of the difference lies in the impact of each social structure on the well-being of particular subgroups, which in turn depends on how those subgroups—the structure's pieces—are related.

In the slave-fueled plantation, the horrific impact on slaves seems obvious now, which is why I began with that example. However, structural violence (like any effect of the social structure) generally appears, at least to those who benefit, unremarkable in the moment. It is understood as part of the natural order. In this way, slave deaths and maltreatment were passed over by members of slave-owning societies as nothing problematic. Likewise, today, many play rhetorical tricks to keep ultimate causes for slaves' descendants' suffering hidden. For instance, some write people's problems off to just plain bad luck or individual, proximate factors; some make sweeping racist generalizations. Some blame victims rather than pulling back to get a wider-angled view of the structurally promoted aspects of the status of African Americans and other people of color still disadvantaged today. We further minimize disadvantage when we refer to people of color as 'minorities' even where their numbers are large: however unintentionally, the implication can be that a group is inconsequential. For that reason, some speak instead of 'minoritized' communities.

For a second concrete example of how we recast structural violence let's return to poor whites. Remember, most white immigrants came to this country as indentured servants or refugees of one kind or another. Most were landless, indebted, and otherwise disadvantaged from the start. Indeed, they have long been seen and treated (and even labeled) by advantaged whites as society's 'trash.' This is cast as their own fault: poor whites remain poor, it is said, because they are inherently lazy and degenerate.

To help prevent this degeneracy from polluting the upper class, in the early twentieth century poor whites were targeted for both birth control and sterilization by wealthier whites, who also instituted Better Babies and Fitter Families competitions emphasizing the hereditary value of 'good breeding' and the dangers of procreating with people of so-called lower standing (see Figure 9.3). These programs were administered largely at state fairs, where livestock already were judged this way. Additional tactics for diverting attention from the true nature of the structural violence oppressing poor whites included—then as now—dominant norms placing them in parallel and in competition with other disenfranchised groups, hobbling their ability to see let alone resist their own exploitation and consumption by the social structure.

To this point I've used words like 'cannibalize' and 'consume' to describe how structural violence depletes the lower classes. Another way to think about structural violence, and the biocultural diversity it creates, is to consider that each individual bears some of the burden of supporting society, whatever its shape. Wherever one is positioned social-structurally, one must bear some weight. How much weight one bears and where it is borne on the person will differ depending on where a given individual is located in the social structure. Those lower down or at the bottom foundations bear the biggest structural burden; thus, they suffer from more wear and tear. Put another way, society allocates more risk to them (see Schell et al. 2005). Moreover, people who are further down in a social hierarchy have much less **agency**—they have much less ability to impose their will or make choices than the people who are at the top. To add insult to injury, they are not even given the same kinds of choices to make as people at the top of the heap.

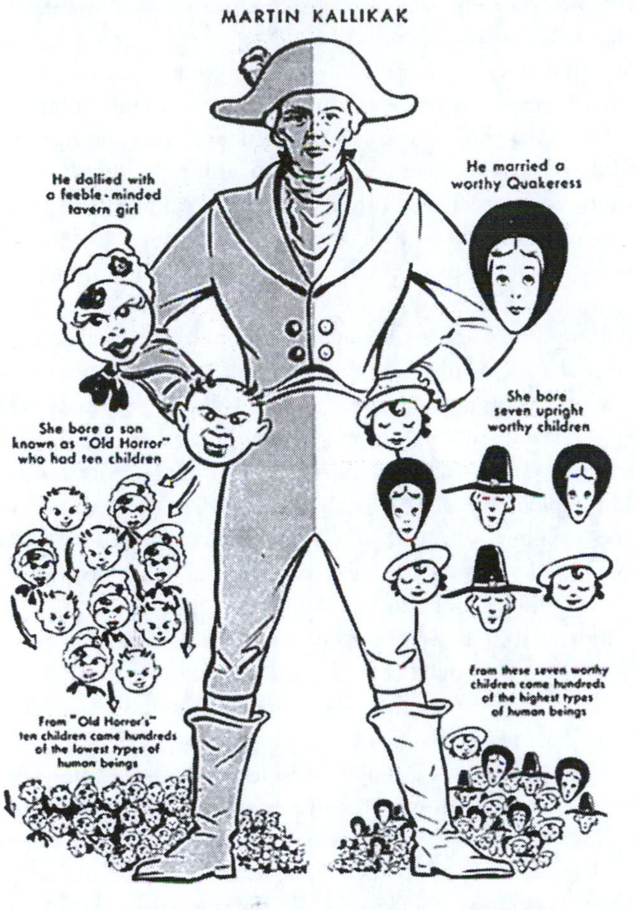
MARTIN KALLIKAK
He dallied with a feeble-minded tavern girl
He married a worthy Quakeress
She bore a son known as "Old Horror" who had ten children
She bore seven upright worthy children
From these seven worthy children come hundreds of the highest types of human beings
From "Old Horror's" ten children came hundreds of the lowest types of human beings

FIGURE 9.3 A caricature of the Kallikak family from a 1950s psychology textbook by Henry Edward Garrett. Modern research indicates that there is nothing accurate about the descriptions offered here, which take a racist approach to (racialized) class differences.

In structural violence, the forces of history have come together to structure a society in a particular way, and to link it up with other societies in a similarly historically particular way. If we think back to the changes brought on by the rise of agriculture and work our way forward through history until the present day, we can see that most structural arrangements track back to economic issues emerging from the organization of subsistence activities. All the historical economic forces that are bundled up in a social structure get written into the bodies of the people who occupy different positions within the social structure in diverse ways depending on the positions occupied (note that here 'bodies' includes minds, despite the dualism that the English language tries to force upon us; for more on this, see Chapter 10).

While there certainly are some costs to the people on the top of any social structure, it is at the lower levels that inequity is most pathogenic (disease producing). Moreover, in keeping with the intersectionality framework, mentioned above, more than class is implicated. For instance, gender, too, has long been a part of this process; and as societies became more heterogeneous, race and ethnicity entered the equation. The more complex a social structure is, the more complicated and even convoluted structural violence can be. Just pointing to structural violence generically as a cause of harm, then, is not enough; the successful application of the concept demands specifics regarding hows, whats, whens, whys, and so on.

We also must acknowledge the roles of related classes of socially structured violence. Building on Galtung's concept, and on the work of Pierre Bourdieu and Nancy Scheper-Hughes, Philippe Bourgois (2001) highlights political and everyday violence. Political violence is overt and includes military repression, police torture, and harm inflicted by public mobs or militias. Everyday violence includes normalized, socially structured interpersonal brutality, such as child beatings, date rape, hazing rites, and even microaggressions: insults to individual members of marginalized groups contained in overtly civil discourse (as, for instance, in the phrase "Where are you *really* from?"). Like structural violence, which some argue microaggressions are just another expression of, everyday and political forms of violence both reinforce and are made possible by the social structure.

Returning to structural violence as such, we should note that it is not just a local phenomenon; frequently it reaches around the world. This is because each social structure exists within a larger one, and poverty in one location in the world often supports wealth in other places. This is the case, for instance, in the Haitian countryside area where a dam (the Peligre Dam) was built to power foreign-owned industry in the capital city. The peasants living where the dam was to be built were not consulted, and in the end they were forced to flee when water suddenly (to them) began to rise and cover their homes and gardens—all for the short-term benefit of Haiti's foreign (including US) investors and urban elite (Kidder 2003).

Losing one's home, getting tuberculosis, getting a dental infection and losing one's teeth due to a lack of affordable dental care, having a hand chopped off in a meat grinder at work that had no proper safety mechanism—these kinds of facts represent more than just individual bad luck. Such events do not happen randomly; they happen to particular people because of the positions those people occupy in the social structure. This is not to say that individuals should not take responsibility for themselves. Of course they should. However, many occupational hazards exist not because of worker actions but because employers are not protecting workers. Even when a person working two jobs has an accident at work due to tiredness, something else also is to blame—and that is the system that makes working two jobs necessary for this person to get by.

Structural Violence in Historical and Contemporary Perspective

Structural violence is not a new phenomenon; archaeological evidence demonstrates that it has been around since the rise of agriculture at least. We know that, compared to groups lower on the intensification scale, agricultural groups are larger and more stratified. They have more inequality. With more inequality we see different subgroups being exposed at different rates to different hazards. We see different subgroups being allowed at different rates to access health care and healthy lifestyles.

We can see 10,000 years of health disparities written into the archaeological record if, for instance, we look at what skeletal and material remains reveal. In Chapter 7 we considered some of the signs of nonspecific nutritional distress. While pervasive in whole societies after agriculture was adopted, such signs were more pervasive in some subgroups than others. Some archaeological sites include separate burial grounds, and we can see from the skeletons of the elite how well-nourished they were in contrast to the working class or the slaves. We can also see different types of occupational stress and injury on the bones that we find. These data show that health inequity related to social structure has been around for a long, long time.

Today, we have masses of data on class-based inequities in many nations. In the United States, much inequity shows up along so-called racial or ethnic lines too. For instance, in terms of deaths due to diabetes, we have much higher rates among Hispanics and blacks than whites.

The difference is actually getting bigger—and it is not the only health problem that plagues blacks and Hispanics more than whites. There are many, and often they feed each other in a synergistic fashion.

Syndemics

This is my cue to add one last word to our epidemiological vocabulary, and that word is syndemic. We've had endemic, epidemic, pandemic; a **syndemic**, in contrast, entails not just one health problem but a cluster of problems that work together, reinforcing and often exacerbating each other synergistically. One example of a syndemic would be HIV disease, or AIDS, occurring hand-in-hand with TB, and both of those reinforcing each other in a given population. Moreover, where we have a very malnourished population, we will have people at a higher risk for being infected with HIV and TB.

What can we do when we see a syndemic? We can help in the immediate situation. We can provide drugs, bandages, stitches, and nutritional supplements, for instance. Yet, according to an insight credited to Albert Einstein, "No problem can be solved from the same level of consciousness that created it." We have to step back from proximate concerns to try to understand the situation in its broadest context because until we understand and address the ultimate causes of the situation, the facts that have created it will not change and it will remain an ever-present problem (see Figure 9.4). The only real way to prevent or control a syndemic is to prevent or

FIGURE 9.4 Poster from the Aboriginal AIDS Prevention Society illustrating persistent and syndemic health problems. Image from the History of Medicine collection, National Library of Medicine, National Institutes of Health, 101451231.

control the forces that lead to that cluster to begin with. Again, those forces are political and economic; they have to do with power relations—and these have to do with money and its circulation. Power relations must be altered to prevent or control syndemics.

Sustainable Health

Health is an indicator of how well populations are adapted to the environments in which they live. While the human species may seem to be thriving, evidence shows that it is not fully optimizing its potential—at least not across the board. This becomes obvious when we compare the epidemiological profiles of various groups.

In many cases, people are struggling; the systems in which they live are generating problems for them—problems expressed in the form of poor health outcomes. Variation in the distribution of these problems is tied to variation in living conditions. This variation is, in turn, tied to the political economy, both locally and globally.

The existing political economy was humanly created. Humans created the local and global systems that exist. If humans created them, humans can change them. This may sound daunting, but as Franz Boas's eminent student Margaret Mead is famously claimed to have said, "Never doubt that a small group of thoughtful, committed people can change the world. Indeed, it's the only thing that ever has."

Two approaches to health improvement that have been very successful are primary health care and the social justice approach, both of which we are about to discuss. Used in combination, they are valuable in combating epidemiological polarization and the structural violence that sometimes brings it about. They are valuable in combating the negative kind of biocultural diversity (that is, deficient health statuses and outcomes) so that the positive kind can flourish.

Both approaches are a legacy of health programs that were instituted after World War II when the United States got heavily involved in overseas aid provision. Through organizations like the US Agency for International Development (USAID), US experts and staff members conducted many health programs all around the world.

Initially, most such programs failed. As we came into late 1960s and 1970s, people started to wonder why, with so much expertise, and money, the programs were not working. Anthropologists and other social scientists were called in by the government to study the situation. They concluded that the programs were failing because they were imposed from the top down. There was no community involvement and no cultural connectivity.

Social Soundness

In 1975 USAID instituted an initiative called **social soundness**. This initiative called on all aid groups to take up the specific goal of community involvement—of connecting with community leaders and members and working to create a good fit between programs and cultures. At the same time, leaders and aid workers all around the world were noticing that many of the health issues that they were trying to address stemmed from poverty. They realized that these problems would not actually abate until poverty itself was alleviated. They saw that in many cases solutions were technologically simple, and cheap. They entailed, for instance, providing clean drinking water, immunizations, and mosquito nets.

Such solutions sat at the core of a very important meeting convened in 1978 by the World Health Organization (WHO) in Alma Ata, Kazakhstan—a meeting attended by people, including leaders, from all over the world. At this meeting, people declared that we would achieve health for all by the year 2000. This led to an initiative that we still have in place today, now called "Healthy People 2030." The overarching aim remains to achieve health through simple solutions. One of those simple solutions is primary health care.

Primary Health Care

The prime mover of the epidemiological transition, where it has happened, was sanitation. The prime mover in the primary health care approach is *prevention*, or preventing problems before they happen. Key routes to prevention involve water, food, and adaptive immunity.

Primary health care first entails providing people with safe drinking water. Even in the United States, this was not always assured: the Safe Drinking Water Act of 1974 was brand new when people met at Alma Ata to begin to define this new direction.

Nutrition is the second component of the primary health care approach. This does not mean dropping famine relief supplies from a plane. It means helping people to achieve a sustainable nutritious food supply.

The third component in the primary health care approach is immunization. This helps ensure that people do not get sick in the first place.

Some proponents of the primary health care approach go even further than to champion prevention via clean water, good nutrition, and immunization. For instance, Partners in Health, a nongovernmental organization dedicated to serving the poor, also recommends free health care and education for the poor, community partnerships that address basic social and economic needs, and that we serve the poor through the public sector. This is exactly the kind of approach that many people talked about in Alma Ata (see Figure 9.5).

Social Justice

Partners in Health adopts a **social justice** approach, which promotes equitable distribution of basic human rights such as the right to healthful living conditions. Social justice proponents promote equal opportunities for equal outcomes among all social groups so that one population

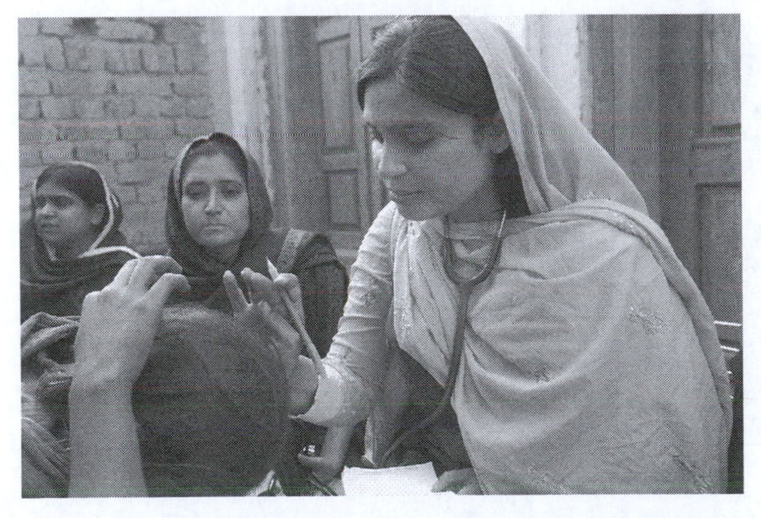

FIGURE 9.5 International Medical Corps doctor examines a patient at a mobile health clinic in Pakistan. UK Department for International Development, CC-BY-2.0.

is no more privileged than others. There are many groups in dire need of social justice—not just in poor nations but also in poor or otherwise disenfranchised segments of our own country.

Distribution of Environmental Contaminants

Environmental inequities explain one kind of the biocultural diversity produced when social justice is not a priority. Take lead exposure. As noted in Chapter 7, lead poisoning in children interferes with nervous system development, and so can cause developmental problems—if not death.

Figure 9.6 shows demographic and epidemiological data from New York City (the racial and ethnic categories the city used conform to US Census norms). The first bar in each set represents the percentage of all children under eighteen years of age who belong to the particular ethnic or racial category listed. The second bar represents the percentage of lead poisoning cases among children when sorted into the same ethnic and racial groupings. These are cases that required environmental intervention, which meant clearing lead paint and lead pipes and so forth from their houses.

Notice that the bars of each type are of different heights for most groups. For example, while 9 percent of New York's child population is Asian, Asian children account for 18 percent of its lead poisoning cases. Likewise, while 29 percent of the children are black, black children account for 42 percent of those poisoned by lead. Conversely, while 24 percent of children in New York City are white, white children account for only 5 percent of those suffering from lead poisoning. The distribution of lead poisoning by racial and ethnic group does not match the distribution of race and ethnicity in the population at large.

To try and answer the question of why not, maps can come in handy. Had we a map of New York City showing where poverty is concentrated, and a map showing where the identified groups are concentrated, and a map showing where lead poisoning cases are concentrated, we would easily come to the conclusion, based on overlap in the maps, that poverty and lead poisoning are positively correlated, as are poverty and race or ethnicity. Poor people of color are much more likely to be exposed to lead. The housing stock is old. It is not well taken care of:

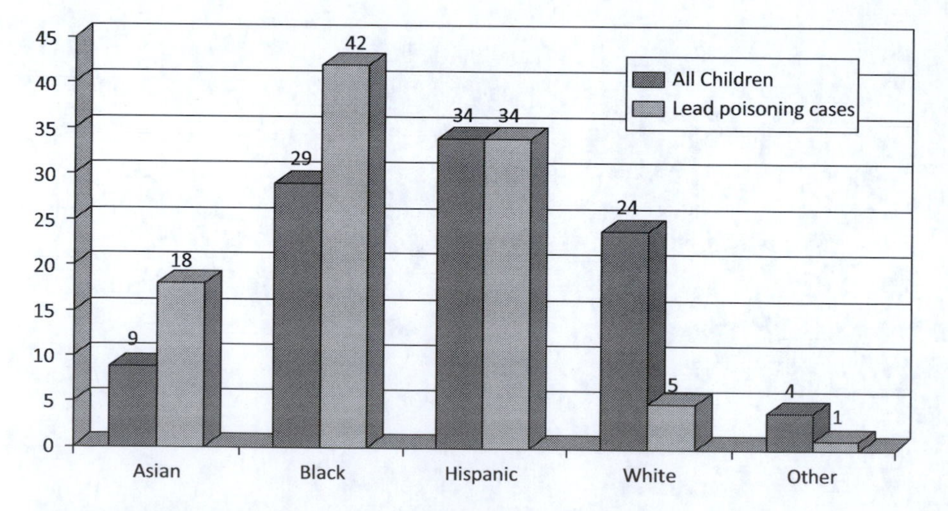

FIGURE 9.6 Percentage of children younger than eighteen years of age, and percent of lead poisoning cases among children that required environmental intervention; New York City, 2004. Created by author from data shown in Mullin and Duwell 2004.

landlords do not regularly repaint, or replace pipes. This means that children in these areas, who already are disadvantaged in terms of their impoverishment, malnourishment, poorly performing schools, and so forth, now have the added disadvantage of lead exposure on top of everything else.

Exposures are not limited to urban settings. Rural Americans suffer, too, when industries pollute the natural resources on which they rely, or when they work in these industries. Take, for example, poor whites in Appalachia, where coal mining has, historically, been a major industry (see Figure 9.7). Work at the coal face is dangerous. Along with other hazards, coal miners face a high risk for pneumoconiosis (CWP, also known as black lung disease). Tighter restrictions on the amount of respirable or breathable mine dust have been shown to lower this risk substantially where they have been enacted. However, industry objections and the lack of other economic options for workers, particularly as mines are shutting down, make mandating such restrictions universally a difficult prospect.

Rural Native Peoples also have suffered due to corporations' profit-optimizing choices. For example, until they were designated by the government as so-called superfund sites and targeted for clean-up, General Motors, Reynolds Metals, and Alcoa factories discharged contaminants into the St. Lawrence River, upstream of Akwesasne Mohawk territory (which spans New York in the United States and Ontario and Quebec in Canada). The contaminants included polychlorinated biphenyls (PCBs), known carcinogens that also disrupt the endocrine system, potentially causing reproductive, growth, cognitive, behavioral, and other irregularities. Due to such contamination, fishing, hunting, and other outdoor pursuits that are central to Mohawk identity have become problematically dangerous.

To help mitigate the situation, the Akwesasne Mohawk community devised a project in partnership with university and government experts. After laying ground rules informed by the social soundness approach, they measured the contaminant load on youths through blood tests and so on (Schell et al. 2005, Schell 2012). These data were plotted against near-term

FIGURE 9.7 Kentucky coal miners (Jenkins, Kentucky; 1935). Photo by Ben Shahn. Record number LC-DIG-fsa-8a16947, Library of Congress Prints and Photographs Division.

developmentally significant measures (e.g., height, weight, thyroid function, onset of menstruation, cognitive function) rather than specific diseases, because diseases can take a long time to manifest and because some diseases can be predicted based on the developmental indicators anyhow.

Findings confirmed that developmental challenges and abnormalities co-occurred with high pollutant levels, as did reproductive abnormalities (earlier sexual maturation in girls, reduced testosterone levels in boys), depressed thyroid activity, and obesity (Schell et al. 2005, Schell 2012). Continued exploration of the data has helped broaden our understanding of how PCBs and other measured **toxicants** (humanly introduced poisons) work on the human body.

The community also has been collecting context data and working on interventions. Effective interventions depend on both identifying the riskiest activities for the Akwesasne Mohawk and engaging in responsive planning that prioritizes Native rather than Western concerns. Seen in this light, telling people to avoid activities central to Mohawk identity ("just stop fishing"; "don't use wild herbs") clearly cannot work: loss of key subsistence activities will entail a loss of culture that would further destabilize the Akwesasne Mohawk Nation (Schell et al. 2005, 2012).

Parasite Stress, Iodine, and IQ

In keeping with the concept of structural violence, a group's socioeconomic position or its location within the social structure determines, to some degree, that group's allocation of risk—such as for toxicant exposure. Yet, a group's present socioeconomic status often also is the outcome of risks experienced earlier. Likewise, this generation's exposure-linked problems will reduce the opportunities of subsequent generations, compounding a risk's toll. In this way, social distinctions reproduce themselves (Schell et al. 2005).

For instance, when cognition is hampered life chances spiral downward: without average cognitive skills, school is hard. Without an education, securing gainful employment is tough; without gainful employment, staying out of poverty is difficult. Being poor increases one's chances—and one's children's chances—for exposure to environmental contaminants and various diseases too; and risks are multiplied for people of color and those otherwise already marginalized. Exposures are thus self-reinforcing (see Figure 9.8)—so much so that their multigenerationally persistent outcomes may be taken by outsiders as proof that a given group's cognitive capacities are predetermined genetically when these limitations are in fact humanly imposed.

Many things besides PCBs and lead can dampen development—even just carrying a higher burden of disease can do it. Indeed, whatever problems there may be with defining and measuring human intelligence quotients or IQs, the synergistic link between higher disease burdens (including malnutrition) and the lower IQ scores of certain populations is by now well established. We know less regarding exact causal mechanisms and intra-group variation and resilience. Still, the recent discovery of a correlation between IQ and life years lost to 28 infectious diseases for 192 countries is indicative. The scientists who made it were testing the "parasite-stress hypothesis," which holds that fighting a high burden of childhood infections caused by viruses, bacteria, and things like tapeworms that live, parasitically, on the human body can be such a stress—can divert so much of a child's energy—that normal cognitive development cannot take place (Eppig et al. 2010).

If the hypothesis were wrong, you might expect no correlation between IQ and the disease burden index the researchers created. Yet a strong inverse correlation did exist. The highest average IQ measured was in Singapore (108), followed by South Korea (106). Third place was a tie, between Japan and China (105). Of English-speaking nations, both Australia and the United States averaged 98, Britain averaged 100, and Canada 99. All of these nations have low levels of

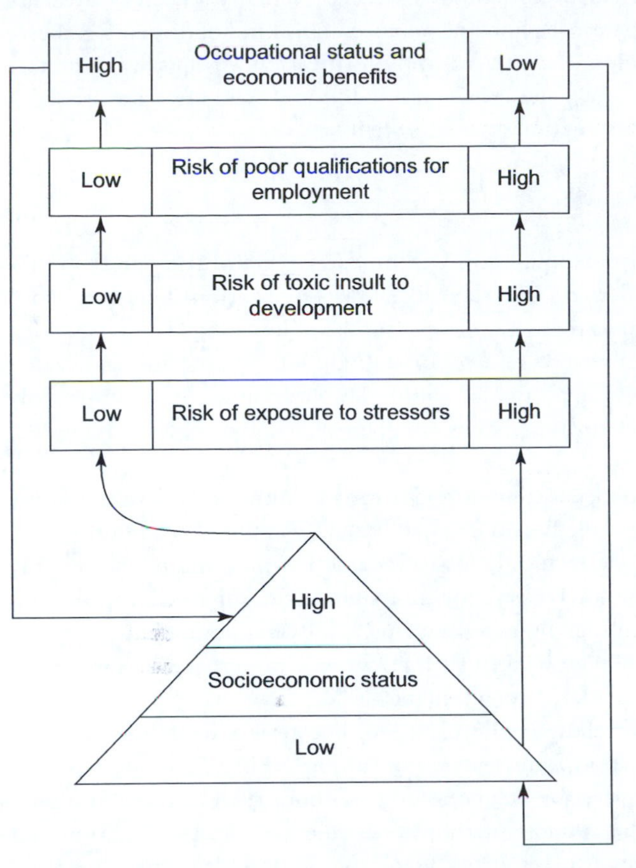

FIGURE 9.8 Schell's model of risk focusing on the relationship between socioeconomic position, toxic exposure, and socioeconomic mobility. Published in Schell and Denham 2003 (adapted from Schell 1992). Used with permission.

infectious disease, relative to countries whose IQ averages were lower—and the lowest was 59. Sure enough, the disease burden for that lowest group was 4.45 as per the study index; the highest group's, Singapore's, was 2.67 (*Economist* 2010a; Eppig et al. 2010).

Even the simple lack of access to iodine is associated with impaired cognitive development. Most of us get plenty of this micronutrient via iodized salt, but about one-third of the global population does not get enough of it. The implications are enormous: "When a pregnant woman doesn't have enough iodine in her body, her child may suffer irreversible brain damage and could have an I.Q. that is 10 to 15 points lower than it would otherwise be" (Kristof 2008).

Knowing what we know now about the (negative) synergy between IQ and malnutrition, disease burden, contaminant exposures, and so forth, the word 'syndemic' should be coming to mind. Importantly, then—and even leaving out the impact of culture on IQ score (which can be great; for instance, those who do not come from the culture that devised a given test are unlikely to score high on it)—we can see clearly that diversity between populations in terms of measured IQ averages is not genetically determined.

Put another way, reports that people living in certain areas or people who are members of certain groups have "lower IQs" on average than people from other areas are, in fact, true—but they do not mean that populations are *inherently* more or less clever. When put into context, they only show that population-to-population IQ differences, like so many aspects of diversity within the human species, are to a large degree humanly created. This kind of diversity is therefore not at all a good candidate for being used as a 'natural' justification for oppressing certain populations or limiting their access to, say, education—although those who are ignorant about the link between IQ and environment sometimes would have it be so.

Zoning

We already have seen how a people's overexposure to environmental contaminants (and so forth) and the ramifications thereof, has much to do with their social marginalization. These are products of a political economy that demands that some people subsist as low-wage laborers or that their region's natural resources to be exploited, tainted, or wasted if others are to amass wealth. It also stems from a political economy supportive of zoning laws that allow environmentally problematic businesses—which would not be allowed in 'better' neighborhoods or wealthy counties—to operate in poor ones.

For instance, zoning laws permitted metal stripping and plating facilities to exist in untenably close proximity to residential housing in the Barrio Logan section of San Diego (see Figure 9.9). This neighborhood has long been home to mostly lower income Latinx families. Yet plating operations (many more in the past than now) release contaminants, including heavy metals, into the air. Those who live near them are thereby exposed. Plus, a lot of contaminants go home on the clothes and in the hair of people who work at such places, meaning that workers' family members, too, are exposed to the occupational contaminants.

Barrio Logan's residents were used to the plating facilities, but after a while they noticed that they were experiencing higher rates of miscarriage, breathing problems, headaches, and other physical ailments than people in nearby neighborhoods without such facilities in them. Residents got together and, under the auspices of the Environmental Health Coalition and eventually with help from the California Air Resources Board, they collected and analyzed data, secured an intervention from the city, changed some laws, and got some of the plating facilities to close (Environmental Health Coalition 2004).

Of course, similar stories do not always have such happy endings. Plus, environmental pollutants from the plating facilities were not and are not the only problems that Barrio Logan's residents have faced. They and residents further south also have been exposed to chemicals from nearby shipbuilding yards. Further, in Barrio Logan the housing stock is old, and so residents

FIGURE 9.9 A contaminant-generating plating operation that was shut down through community efforts due to close proximity to dwelling houses in Barrio Logan; photograph courtesy of Environmental Health Coalition.

have been exposed to lead as well as to asbestos (a known carcinogen). There also are many 'brownfield' sites there—sites where contaminating activities have taken place in the past and which now host schools or other facilities, sometimes without having undergone the environmental remediation necessary to make them healthy places to be.

Proponents of the social justice approach find this way of treating the poor unconscionable. Be that as it may, the preceding examples make the influence of the cultural side of our biocultural existence—of economic forces and social and political decisions—manifestly evident.

Community-Driven, Culturally Responsive Change

To close this chapter, let us look to a case study offered by the Tohono O'odham Nation of Arizona, who presently have one of the highest rates of diabetes in the world. This case study highlights the human creation of diversity, particularly in obesity and diabetes rates. More than that, it demonstrates the link between social justice and health while charting the ramifications of structural violence for the Tohono O'odham people; showcases the enactment of social soundness, as per Alma Ata; and exemplifies the importance of prevention, as promoted in the primary health care approach. Importantly, too, the programs that ultimately may heal the Tohono O'odham Nation emerged from within that group—not from without. They are truly community based.

Several generations ago the Tohono O'odham lost their right to the region's water when the river fueling their farming lifestyle (see Figure 9.10) was diverted to service nonnative farmers upstream. At that time, and in fact until the 1960s, no known case of diabetes existed among the

FIGURE 9.10 Tohono O'odham woman winnowing (cleaning) wheat c. 1907. Library of Congress, Prints & Photographs Division, Edward S. Curtis Collection, LC-USZ62-123312.

Tohono O'odham. However, when they lost their farm-based livelihood and the cultural ways that went with it, the group also lost the means to good health: physical activity, a balanced diet, and a hopeful outlook (California Newsreel 2008a; and see Schell 2012).

Today, more than half of all Tohono O'odham adults have been diagnosed with type 2 (adult onset) diabetes—as have increasing numbers of Tohono O'odham children. Because of their unnaturally elevated diabetes rates, the Tohono O'odham have become favorites for biomedical research. However, rather than to simply submit to genetic testing in support of the biomedical community's individualized, gene-based understanding of why they are more likely to have diabetes than other populations, or to otherwise offer themselves as human test subjects for outsiders' research projects, some members of the Tohono O'odham Nation formed Tohono O'odham Community Action (TOCA), a community-based organization dedicated to re-creating a sustainable, healthy culture (see Tohono O'odham Community Action 2010).

Co-founded by Terrol Johnson and Tristan Reader, TOCA promotes cultural renewal, and the empowerment this entails, as key to wellness. TOCA has had four primary program areas: basket weaving, traditional foods, youth and elder outreach, and traditional arts and culture. All are intertwined. The traditional foods program area, however, is perhaps most directly related to diabetes abatement. It emphasizes healthy eating and seeks to re-teach traditional farming skills, which were lost to many when the river was diverted. This it does through school programs and other outreach efforts. Concurrently, TOCA strives to revitalize the cultural aspects of Tohono O'odham foodways, using native traditions in the context in which they were meant to be used rather than as, for instance, performing a ground-blessing ceremony on a stage. TOCA also helps to support traditional farming efforts by creating markets for the foods so grown, and with a café.

Rather than simply endorsing and working to ensure access to the current clinical model of diabetes care, which of course is important, TOCA emphasizes a community-based model of holistic diabetes intervention. It celebrates the community's farming-based roots, all the while casting the community as part of a larger subsistence system. Ultimately, TOCA aims to address the structural causes of unemployment and poverty and so of ill health among the Tohono O'odham people. Advocacy, like that leading to the rehabilitative Arizona Water Rights Settlement Act (which went into effect in 2008 but was over thirty years in the making; Archibold 2008), has been a crucial part of this process.

Populations Embody Their Social Positions

We've seen in this chapter how power relations within and between social groups underwrite biocultural differences. In complex societies, and in the global political economy, a group's structural position has consequences—consequences that sometimes last into succeeding generations, affecting life chances (and, sometimes, genomes and epigenomes). As Nancy Kreiger says in the film *Unnatural Causes: In Sickness and in Wealth* (California Newsreel 2008b):

> There's one view of us as biological creatures; we are determined by our genes [and] what we see in our biology is innately us, [it is] who we were born to be. What that misses is that we grow up and develop .... We interact constantly with the world in which we are engaged. That's the way in which the biology actually happens. We carry our history in our bodies.

Biology isn't a given: it *happens*, and culture plays a key role. As this chapter has shown, structural position is part of the history written into our bodies. It is part of what underlies human biocultural diversity.

PART III
Meaningful Practices

In the first part of this book, we learned about and applied systems thinking to understand bodily adaptation to the physical world, both at the population level, via genetics, and at the individual level, via developmental plasticity or epigenetic mechanisms. We saw, too, how individual developmental adjustment and epigenetic change, on the one hand, and population level change, on the other, can be connected.

Once we came to understand how humankind emerged as such, we examined the impact of geographic variation on specific traits. Then, still using the complex adaptive systems framework, we focused in Part II on the social structures humans created as they sought to exploit new areas of the world, and to do so in new ways, such as via settled agriculture.

While society (or social structure) and culture are two sides of the same coin—with 'society' describing the relationships between available roles, and 'culture' justifying those relationships and roles—our focus in Part II was on the social side of things. We did talk about culture, but only in terms of what it contributes toward human survival directly, such as in providing means for provisioning food, clothing, and shelter. We did not explore the cultural knowledge or belief systems that, for instance, support paying tribute to a chief of divine lineage, or submitting to feudalism, agreeing to rule a nation state, or serving as a shaman. We did not explore the ways in which people, however organized, attempt to manipulate supernatural forces to bring on better harvests, ensure procreative fertility, cure a tumor, or encourage love. We did not examine various cultural knowledge and practices relating to kinship, gender, age, or other attributes that can be used to differentiate groups of people.

Yet all these things entail biological and cultural interaction. The more cultural side of things is limited to some degree by the biologies we are endowed with, but at the same time our biologies reflect cultural actions—actions taken (consciously or not) in response to cultural meaning systems. In some instances culture creates the biology it desires; in others, effects on biology are secondary or unanticipated. Sometimes meaning systems are so deeply internalized that our bodies respond in ways that are beyond our awareness.

Chapter by Chapter Overview

Chapter 10 deals with meaning directly as it characterizes the link between social structure, mind, and body through a discussion of population-level variations in stress and its long-term effects. Stress is meaning-laden; different events may be perceived as stressful or not in different cultures. Indeed, different cultures may entail different levels of stress altogether—and different sources for resilience. Further, diverse cultures have certain circumscribed health conditions that occur in them but not elsewhere, or that they perceive distinctly; certain treatments also might work in one culture but not in another. The role that culture plays in interpreting, experiencing, treating, and even generating particular symptom sets or sicknesses, such as *susto*, a form of soul loss known in certain Latin American populations is examined.

Chapter 11 encourages readers to explore the role of day-to-day, belief-fueled cultural practice in creating forms of diversity thought to be 'natural,' such as gender-related differences in how we carry our bodies. To help grasp this, after we distinguish sex from gender, briefly exploring sex as a biological process, we consider cultural roles that cross-match sex and gender or refuse binary distinctions altogether, and we explore the cultural features that support diverse role expectations. We then characterize and apply the 'rites of passage' framework framework to demonstrate how expectations for manhood and womanhood are culturally constructed.

Chapter 12 extends the analysis to include body ideals, with an eye to their biological impact. We explore cross-cultural and historic differences in weight ideals in relation to food supply and examine weight-related health challenges in light of our evolved metabolic biology. Then we identify and describe the functions of body modification and decoration such as tattooing cross-culturally, and we discuss the main dimensions along which such practices vary.

Beyond how people are shaped, who one's relatives are can differ cross-culturally. Chapter 13 describes the three most basic kinds of kinship, paying particular attention to cross-cultural variation in the ways humans create consubstantiality—that is, kinship due to shared substances, of which blood is only one. We also explore diverse understandings of when, how, and if life begins, and examine how these various understandings affect practices such as abortion. Finally, the concrete, physical consequences that ideas about kinship have for family-building and thus population demographics are examined.

Central Lessons of Part III

In sum, Part III of the book pushes us to give equal weight to the cultural end of the biocultural equation by taking into account how some of the deepest-held ideas different populations have—ideas about, for instance, social roles, gender, the ideal body, and kinship—affect human biocultural variation. Part III guards us from settling for a focus that is solely material by demonstrating the value of a perspective equally sensitive to the corporeal impact of ideas and meanings.

The book concludes by reaffirming that biology and culture co-create each other and the human experience; holism and systems thinking are useful; variation is good; and we are all connected. Areas ripe for future inquiry are also noted, and causes for optimism highlighted.

10

STRESS, MEANING, AND HEALTH

This chapter prepares you to:

- Characterize the interconnection between social structure, mind, and body by identifying of class- and race-linked variations in allostatic load (chronic stress) and by explaining its causes and effects
- Characterize the interconnection between culture, mind, and body through explication of diverse cultural models for interpreting, experiencing, treating, and even generating certain disease states
- Characterize the interconnection between culture, mind, and body through discussion of the Meaning Response in terms of self-healing and resilience

The last section's chapters concentrated on the relationship between biology and culture by highlighting how the various ways that human groups sustain and organize themselves lead to biocultural diversity. Beginning with this chapter, our inquiry broadens to include, more directly, cultural meaning systems. Without losing sight of the co-determinate nature of biology and culture, we will first explore chronic stress and its diversely distributed causes and effects. Then, we'll take on culture-specific sicknesses and self-healing.

The Stress Response

Technically speaking, **stress** is one's immediate physical response to environmental pressures. The stress response entails reactions of the nervous system, such as when the brain releases certain neurotransmitters; the hormonal system, which releases stress hormones; and the immune system, in which some parts go on hold so that energy can be diverted for self-protection from the stressor.

It may sound surprising that hormones are involved in the stress response, because so often we think of them only in association with masculinization or feminization. However, they serve a broader function than just that. Hormones regulate the relationship between the various organs of the body. When a person is under stress, those relationships must change, because the stressor demands a different kind of response from those organs than everyday living demands. Exams, for example, stress students and teachers alike. Other stressors might include the threat of a layoff at work or that a snowstorm will close the road home. Being attacked by a lion is stressful, too, as are famines, fires, hurricanes, tornados, earthquakes, and so on. In each case, the body is mobilized to act quite differently than it does under normal conditions, and hormones play a role in fostering that mobilization.

Selection for a Strong Stress Response (Fight or Flight Reaction)

Why does the stress response exist? As humankind evolved over time, it was selected for because it was helpful. You have likely already heard of the **fight or flight reaction** that the stress response entails. This prepares the stressed individual immediately, via internal changes, for meeting a challenge with physical force or running away from it as an aid to survival. Immobilization can also ensue, for instance just prior to fight or flight; so another name for the reaction is 'fight, flight, freeze.'

Imagine a forager, quietly digging up tubers, inadvertently disturbing a rattlesnake. If the forager can act faster than normal in an effort to escape danger—which is what happens if the stress response triggers flight—the forager gains an adaptive advantage. That is, a forager who survives when others would not is more likely to have more offspring, thereby contributing more genes to the next generation's gene pool. The frequency of genes supporting a stronger fight or flight reaction increases.

The Biology behind the Supercharge

Biologically speaking, what happens to cause an individual's body to so quickly provide the energy needed to climb up a tree or flee or otherwise act out of the ordinary in a way that will foster survival? What causes the supercharge? The mind registers the threat and the body responds, churning out the necessary biochemicals, including hormones and neurotransmitters. One biochemical central to the stress response is cortisol. There also is a key category of biochemicals called catecholamines, which includes epinephrine. Epinephrine used to be called adrenaline: it is generated in the adrenal glands, which sit on top of the kidneys (see Figure 10.1).

Exactly what happens when someone's cortisol level and so on shoot up? We all know something of what happens because we all have been there. Each and every reader has likely experienced the fight or flight syndrome. From my own life, a very particular scenario stands out. I was writing a book (not this one) and had just finished Chapter 3. I began tidying my desk and then, in a wicked instant, deleted the entire chapter. I later recovered it, but within that first moment of disaster I instantly felt as if a quart of stress hormones was coursing through my veins. My heart rate went up, my blood pressure surged, and sugars in my bloodstream mobilized, potentially providing my muscles a source of instant energy. Indeed, blood rushed to my muscles. This would have been terribly handy had I needed to run away from a bear or fight off an intruder. Sadly, no such action could be taken. I could not run up a hill or climb a tree—well I could, but the stressor was not something I could counteract by doing that. My body simply had to reabsorb everything (I did go to the gym later, which provided some relief).

How Stress Varies

As in my experience, most of us these days do not find ourselves stressed by challenges such as meeting up with grizzly bears, lions, or mad dogs. Still, we encounter similar stresses, such as when a car pulls out in front of us quickly, without warning. This kind of stress is, like an illness

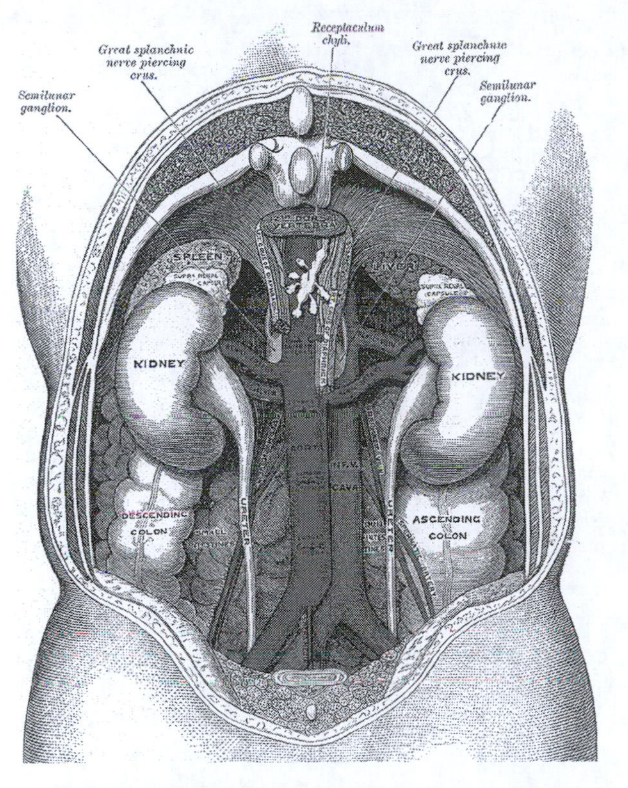

FIGURE 10.1 Adrenal glands, located on top of the kidneys (as seen from behind). Epinephrine is made here. Image based on lithograph plate from *Gray's Anatomy of the Human Body*. Philadelphia: Lea & Febiger, 1918.

that hits swiftly, acute. However, many contemporary stressors are chronic. They persist over time. Even my assumed loss of that chapter was part of a chronic stress onslaught having to do with professional expectations. The contrast between chronic and acute is thereby a question of the level of analysis; and more simply, it is a question of duration.

Another variable is the source of the stress. It may be a permanent fixture on the landscape or it may be removable through certain actions. It may be work stress that one is powerless to do anything about, for example, or it may be a mad dog that actually can be shot (see Figure 10.2). In other words, stressors can be more or less constant, meaning that in some cases the stress response is always set to 'on'.

Chronic Stress: Allostatic Load

Individuals can and do respond to stressors with varying amounts of resilience, and habituation is a possibility. Yet, generally speaking, when a stressor is constant and irremovable, a moment will come when the body gets so overloaded by its own stress response that it can in fact begin to break down, or undergo system failure. The fight or flight reaction that is so helpful when we really do have to either fight or run away fast can backfire or become dangerous when unabated.

FIGURE 10.2 A mad dog on the run in a London street: citizens attack or flee it as it approaches a fallen apple seller. Colored etching by T. L. Busby, 1826. Wellcome Collection. Library no. 663745i. CC-BY-SA 4.0.

One way to think about this is to picture a dump truck or trash chute that just keeps spilling garbage on a group of people. One dump is stressful enough, but dumping that does not stop—chronic stress—is downright debilitating. The process by which stress does its damage is not as simple as being crushed by garbage's weight. Rather, it has to do with recalibrating bodily set points to try to stave off system failures that the onslaught of stress otherwise might bring on. This onslaught also is known as the allostatic load.

Technically speaking, **allostatic load** is the "cumulative multi-system physiological dysregulation that results from exposure to challenges over the life course and that places individuals at greater risk for poor health" (Gersten 2008, 532). In other words, our normal regulatory processes (for example, metabolism of sugar) are disturbed or discombobulated, in unhealthful ways. When a particular habitat includes certain long-term environmental pressures or chronic stressors, the population living there will bear a high allostatic load.

A heightened allostatic load can lead to long-term immune suppression, infertility, atherosclerosis (hardening of the arteries), and heart disease among other things. It can promote abdominal obesity by affecting the pattern in which the body stores fat. When fat accumulates in the gut, around the organs, rather than in the buttocks and thighs, it is more likely to harm the vital organs it thereby surrounds. In addition, coupled with the fact that people under chronic stress have extra sugar available in their blood, it can underwrite the development of diabetes. Hypertension is also a possible outcome. In these ways, chronic stress writes or inscribes itself into the metabolic and cardio-pulmonary biology of a given population.

Chronic stress also seems to tax the brain. It may even do so in a sinister way that perpetuates cycles of poverty (recall Figure 9.8). Working memory—the ability to hold informational bits in your brain for immediate use, such as the date when a term paper is due or who won the battle of

Waterloo—correlates negatively or inversely with allostatic load. That is, the more chronic stress one is under, the fewer items one can remember at any one point in time. Moreover, because poor (stressed) children's memories do not work as well as their less-stressed middle- and upper-class counterparts, they get lower marks in school, end up similarly poor as adults, and raise their own children in likewise poor circumstances. The synergy between nature and nurture here is clear: brain biology helps determine social status and social status helps determine brain biology (*Economist* 2009).

Of course, memory problems and disease responses such as in diabetes are exacerbated by other contributing factors, such as lack of sleep due to working two jobs or crowded living conditions, or poor diet. Nevertheless, from a medical viewpoint, they stem directly from the recalibration of once-healthy set points to allow stress-induced levels of certain biochemicals or responses to persist.

In fact, scientists named ongoing stress the allostatic load in reference to **allostasis**: the process of re-creating what we've learned to call 'homeostasis' by changing the body's initial set points or 'factory settings' in order to accommodate chronic stress (as in allocare, *allo* means other). Regaining balance in this way is good in the short term because the failure to maintain homeostasis itself can cause death. Ultimately, however, if the new balance is unhealthy, long term damage comes about.

Accordingly, common measures of allostatic load also are indicators of long-term damage. They include high blood pressure, a large waist-to-hip ratio, high urinary epinephrine and cortisol levels, and so on. Such indicator scores generally are combined in studies to create what's called an 'allostatic index.'

New science regarding the allostatic mechanisms by which stress does its damage is both fascinating and groundbreaking. However, the link between stress and ill health has long been known. In the American context (we will get to some others later), for years studies have demonstrated a positive correlation between poor health outcomes and household stress levels (indexed by measures of unemployment; communication problems; whether a member has a chronic illness; a recent experience of divorce, death, or desertion; and so forth). For example, Schmidt's review of the literature back in the 1970s found that high household stress correlated with a steady increase in streptococcal illness, and a rise in antibodies. High stress also was related to increased risk of stroke, angina or chest pain, and concern with one's own and one's children's somatic (physical) symptoms. Pregnant women with high stress and low levels of social support had a pregnancy complication rate of over 90 percent (Schmidt 1978).

Likewise, divorce has long been known to be unhealthy because of the stress it entails in certain contexts. Where divorce is not condoned, or when only one party seeks it, stress is heightened. Of course, divorce does not have such an impact in cultures where it is nonproblematic or where other arrangements, such as serial monogamy, are the norm.

Another source of stress is occupational. I mentioned the health–wealth gradient in Chapter 9. One reason for it is that many jobs on the low end of the ladder are what theorists call **high-demand, low-control** jobs: jobs in which employees are ordered around and have little control over their work (restaurant staff, assembly line workers, fruit pickers). In short, holding a subservient job in a Western industry is bad for one's health; and this translates into differing job-linked epidemiological profiles. In keeping with the concept of intersectionality (see Chapter 5), in the United States, many people in jobs linked to worse health outcomes are of color, less educated, and poor. In wealthy countries with more inequity, outcomes are even more compromised (Wilkinson and Pickett 2009). This is partly to do with the stress engendered by the perceived need to keep up while concurrently facing social-structural barriers to self-advancement—a topic I soon return to.

Stress in Cultural Context

Responses to Events Are Culturally Relative

Many people believe that contemporary humans bear a much higher stress load than our foraging ancestors did. However, our ancestors are long gone; we have no direct way to test this hypothesis with, for example, comparisons of salivary cortisol levels or aggregate allostatic load indexes. Even if we did, such comparisons would not tell the whole story.

This is because peoples' responses to events are culturally relative. The events and issues that create high and chronic stress for me as a member of the professorial occupational subculture of academics frequently are considered idiotic by members of other occupational subcultures (including some members of my own extended family, which adds to the stress). Conversely, not having an up-to-date wardrobe is no problem for me while it may be for some of my students, just as not having a particular tattoo might be problematic for someone whose tribe expects it but not for someone whose social network does not. Many expectations are stressful only when unmet in particular cultural contexts.

One time in Jamaica, where I conducted my initial dissertation fieldwork, I was doing some laundry. While hanging out my clothes to dry, I walked across the little river that my old wash-water had made in the dirt. You might have thought that I had cut myself with a straight razor considering the consternation this caused for Grandma and Auntie May, who were nearby at the time. They admonished me fretfully regarding the implications this might have for my health due to what might rise from that water, where it might go, and what it might do. Walking over wash water would be very stressful for a person brought up in their tradition, due to the understandings referenced. Conversely, for me, at least until I learned to fear these things, crossing a wash water stream was no big deal.

Another example illustrating the relativity or, put another way, the **cultural construction** of stress concerns physical expectations for old age. To say that something is culturally constructed is to say that it exists as it does only because one's culture teaches that it is so, and our views on aging bear this out. In cultures that valorize youth, people want to look young. At the least, they dye their hair; those with more resources and further commitment to the ideal get facial and other surgeries. Those who are most committed to the ideal do whatever is within their means to achieve and maintain a young-looking exterior. They feel ashamed and anxious when gray hair roots or wrinkles show.

In other cultures, youthful looks are equated with immaturity and a lack of achievement in life. In such cultures, the aged instead are held in high accord (see Figure 10.3). People with gray hair are respected. If you don't look old and worn out when the expected time for that comes, people may talk. They may assume, as they do among the Hua of Papua New Guinea when older people are too vital, that you have not been contributing to the community at the expected level—which, if you had been, would have led to proper aging (see Meigs 1997). In such a context, an older person naturally endowed with a youthful face and body would feel worried by and about their appearance and others' reactions to it. In such a culture, youthful-looking elders might eat less overall in an attempt to look frail; certainly they would eat fewer foods known to vitalize. They might expose their skin to extra sun or abuse their teeth to simulate the wear and tear of the average aging process.

Another example of how stress is culturally constructed concerns teen sexual activity. In communities where virginity is sacred, the teenager who has had sexual intercourse, even if only one time or by force, may live in fear of being found out. Or, in contrast, take a community for which the youthful initiation of sexual interaction is the norm. Here, a young person who either has not yet managed any sexual relations or a young person who is disinterested

FIGURE 10.3 Toposa elder in South Sudan. Photo by Steve Evans, 2011. CC-BY-2.0.

might feel worried and anxious. Concurrently, a homosexually oriented teenager living in a culture to which adolescent homosexual activity is unremarkable (such as in traditional Polynesia) would have no sexuality-related worries. A narrowly heterosexual male youth living where male homosexual relations are prioritized and heterosexual ones thought highly polluting (such as in some Melanesian groups) or where other non-heterosexual relations are prioritized may, on the contrary, feel very stressed indeed.

Here I should note that social media can intensify stress in that it adds a multiplicative dimension to one's experience of social scrutiny. Related to this, it enables—and Western culture encourages—users to present themselves as if 'living the dream' by fulfilling cultural expectations exceptionally successfully. Although such self-presentations may be pure puffery they leave many social media users feeling that their own (or real) lives are sadly lacking. Poor sleep due to night-time screen use also affects mood and cognition. All this adds up to notably high suicide and depression rates among users, particularly adolescents (Twenge 2017). While some today just assume that youth are constitutionally prone to mood disorders, historical and cross-cultural comparisons confirm that our current mental health profile—like the diverse profiles of other populations—is an artifact of the complex interaction of culture and biology.

Resilience

Just as sources of stress may differ, so do sources of **resilience**—of strength in the face of stress. Resiliency behaviors enable those who otherwise might pack it in under pressure to stay the course. They include spiritual practices, such as magical or religious acts undertaken in an effort to control the uncontrollable; humor; use of physical outlets such as sports activities; and reliance on networks of friends and relations for social support.

The size or extent of a social support network is not in itself a source of resilience. Having many friends or relatives interfering in your affairs can in fact be stressful in its own right, as anyone who has lived in a small village or with a large extended family knows (and which undermines the assumption that our ancestors were by definition less stressed than we are today). Resilience comes instead in the buffer resources provide by a large social network. These may include information, childcare, financial support, or assistance in requesting supernatural help, for instance through sacrificial rites or prayer meetings.

The hormone **oxytocin** also supports resilience by dampening the stress response and promoting trust and empathy, although not unconditionally. Part of the childbirth (uterine contraction) and lactation processes, oxytocin also affects the amygdala, a part of the brain involved in emotions. The hormone ebbs and flows in tandem with reproduction but it also can be induced through skin stimulation, as happens with hugs and backrubs—when they are welcome. Peer massage programs, instituted in some northern European schools to increase classroom calmness and enhance children's ability to focus, seem to capitalize on the way that positive touch stimulates oxytocin production (Blair 2012). Such programs will not work in touch-phobic cultures or where massage would itself be a stress-inducing event. This fits well with experimental scientists' observation that when they administer oxytocin in controlled studies its impact is affected by immediate context and prior life experiences (Olff et al. 2013)—exactly as the biocultural paradigm would predict.

Spending time away from the humanly built environment, such as by camping or hiking—also seems to heighten resilience. The Japanese call the practice *shinrin-yoku* or forest bathing; it induces significantly lower blood pressures and heart rates as well as increased perceived calmness and well-being (Hansen et al. 2017). Other countries also promote time in nature, or 'nature therapy,' with reference to health-related outcomes. England's National Health System (NHS) Forest Project, for instance, specifically lists lower cortisol levels as among the many therapeutic benefits of green space in its promotional material (http://nhsforest.org/benefits).

Stress Inequities: The Case of Race

To deepen our understanding of stress we must now apply the comparative method directly. We can compare, for example, the stress experiences and allostatic loads of diverse societies, or of members of different religions. There are many directions we could go in setting up such comparisons, but the direction in which I would like to turn now is back toward social structure—and then forward to culture, which provides ideological support for the structure that a given society may have. I want to highlight social structure because, as the last chapter indicated, some social positions entail more long-term stress than other social positions. To bring culture directly into the picture, the positions I want us first to think about here are racialized ones.

To begin, let us take, as an example, the case of Kim Anderson, described in the *Unnatural Causes* documentary "When the Bough Breaks" (California Newsreel 2008c). Ms. Anderson is a successful lawyer, having graduated from Columbia Law School in 1984. She was well-paid and in good health in 1990—the same year she had her first child. Also and by the way, Ms. Anderson self-identifies as African American.

Given Ms. Anderson's occupational status, one might expect excellent outcomes for her and the baby. True, infant mortality rates are twice as high for black people in America as for white ones (MacDorman and Mathews 2008). At the same time, many African Americans also are poor (see Barr 2008, 134–168). We know from Chapter 9 that occupying a lower class position in society—having a disfavored location in the social structure—is tightly linked with poor health outcomes due to inequities in access to health care and to unhealthful living and working conditions.

Yet, it turns out that even babies born to highly educated African American mothers like Ms. Anderson are at a higher risk for premature birth—and death—than is true for the population as a whole. Infant mortality among these moms is two and one-half times that of equivalently educated white American mothers.

If the difference were racially generated (a hypothesis that presumes human races exist, but see Chapter 5) we would expect to see high infant mortality rates in all populations with African heritage. Yet research shows that this is not the case. One study, for instance, compared the newborns of African American women, white American women, and African immigrants to the United States. While the African immigrants' babies and the whites' babies birth weights were generally the same, the babies of the African American mothers were different. They were about one-half pound lighter. This, for a baby, can be a big deal. Furthermore, after one generation of living in America, outcomes became significantly less favorable for the African immigrants as well. So low birth weight and high infant mortality rates among blacks are not racial or 'natural' outcomes; rather, something environmental—something cultural—seems to be at work (see also David and Collins 1997).

Kim Anderson is, as noted, African American. In regard to lifestyle, as she tells it, "People would think I'm living the American Dream: a lawyer with two cars, two-and-a-half kids, the dog, the porch, a good husband, great family. I've always been lucky to have good health. Always ate well. Exercised. Never smoked." During her pregnancy, "I did all the right things. They told me to take vitamins; I took vitamins. They told me to walk. They told me to eat vegetables. They told me not to drink." Yet more than two-and-one-half months before her due date, Ms. Anderson went into labor. When her baby was born, she said, "I heard her cry, I said, 'Thank, God.' But she was so small. [You could] hold her in the palm of your hand" (California Newsreel 2008c).

Why did things turn out that way? Stress. Remember, Ms. Anderson is black, just like the other African American mothers whose pregnancies did not go full-term, and just like the daughters of African immigrants who went on to try to build families in America. Thereby, every day of her life, she has a prospect of experiencing racist treatment.

African Americans are more likely than others to experience prejudicial treatment, for instance in being charged more for equivalent automobiles, being denied mortgages, and receiving fewer job interviews. In a study done in 2004 in Milwaukee, Wisconsin, black and white men (actors) were sent to apply for 350 entry level jobs, all with exactly the same resume. The black men fared worse. Moreover, black men with no criminal records were less likely to be offered jobs than white men with a felony conviction (David Williams, in California Newsreel 2008c).

Work from 2010 in a similar vein demonstrated that Craigslist ads for iPods shown being held by black hands received 13 percent fewer responses and 17 percent fewer offers than those in which the iPods were in white hands. The dollar value of offers received for the iPods displayed in black hands was 2 to 4 percent lower. Buyers corresponding with black-handed sellers also displayed lower levels of trust: they were less likely to share their names or to accept a delivery at their mailing addresses (Doleac and Stein 2010). A lifetime of such discrimination—of living in a racist culture—surely can add up.

At least in the United States (and in many other places, too), Ms. Anderson and others who belong to minority groups live chronically stressful lives because the world they live in is racialized. So, while about half of all white women report that they never think about their race, about half of all black women think about their race at least once a day; about 20 percent think about it all the time (Camara Phyllis Jones, in California Newsreel 2008c).

When Ms. Anderson walks into a shop, or a bank, or just down the street, the first thing many people notice about her is that she is black—and because in our world racism is layered onto racialization they adjust their actions toward her according to stereotypes held regarding the 'black race.' In the documentary cited, Ms. Anderson recounts a time when she walked into a shop, thinking to buy some jeans. The clerk started following her around, as if she was going to steal something.

Now maybe the clerk wanted to advise Ms. Anderson of a sale or keep her away from wet paint. Nevertheless, based on her past experience of racist treatment (which, as we saw above, research confirms is quite real), Ms. Anderson felt otherwise—in the flesh, quite literally. Describing her response in the face of racism, she made physical reference to the anxiety it provoked: "Your stomach just gets like so tight. You can feel it almost moving through your body; almost you can feel it going into your bloodstream" (California Newsreel 2008c).

Racism can, in fact, result in heightened stress in those who experience it. We already have established that chronic stress writes itself into the body, for example altering heart function and insulin-related processes. The bodies of people under chronic stress are therefore more prone to diseases such as hypertension and diabetes. They also are more prone to premature labor, which contributes greatly to infant mortality rates. Prematurity is likely because some of the hormones generated in stress also are involved in triggering the start of labor. If the body begins pregnancy with an already high load of stress hormones, it may take less time to reach the tipping point at which contractions begin (California Newsreel 2008c).

Importantly, low birth weight (if survived) has durable developmental and epigenetic impact. For instance, it is associated with high blood pressure and early signs of diabetes in African American children and teens; and it places adults at an elevated risk for related conditions such as full-blown diabetes and, ultimately, cardiovascular disease (CVD). Indeed, nearly half of all African Americans have some form of CVD. As if this were not enough, low birth weight in one generation has implications for the next: one's compromised health status in turn compromises one's offspring's well-being, much in the same way that maternal stress physiology affects infant birth weight to start (Kuzawa and Sweet 2009).

As the foregoing suggests, premature births and low birth weights represent just some of the many health problems more prevalent in African Americans and linked to racism's stress. Other minority group members also experience problems, helping to demonstrate that the stress of racism (culture), and not race itself (biology), is at work. For example, one study showed that after the September 11 attacks on the United States by Al-Qaeda, pregnant women with Arabic names had a sudden surge in low birth weight deliveries.

Mounting evidence shows that health problems such as hypertension, diabetes, and infant mortality—long known to occur at higher rates among black people than among white people in the United States—are in many ways the result of our racist culture. In an interesting twist, the national tendency to see racism as an individual, aberrant practice rather than an institutional one or a feature of the present social structure means that victims must often prove that (or are forced to question whether) racist treatment occurred, intensifying the stress already experienced (see Wyatt et al. 2003; see also Mays et al. 2007; California Newsreel 2008b; California Newsreel 2008c).

Individual versus Population Risk: Stereotypes versus Generalizations

Importantly, all the problems we are talking about are population-level problems. That does not mean that they do not affect individuals; they surely do. In fact, I just used a person-focused case study to give life to the issues under discussion. However, the figures provided and the extrapolations regarding risk are not about individuals. They are about *groups*.

It is logically problematic to go from a population-level statistic to an individual. Just because an individual is a member of a given population does not mean that he or she will have the same outcome that is seen at the population level. This is not only because that individual belongs to lots of other groups, too, which muddies the relevance of the statistics. It is because logically speaking—mathematically speaking—individual risk and population-level risk are two different things.

To make this a bit clearer, we might take, for instance, population-level statistics regarding a college course on biocultural diversity. The instructor might tell students that the class average was 75 percent (a C) on a given test. The average score holds for the group as a whole, but it tells single students very little about their individual scores. A student's score may be much worse—or, for you, having gotten this far in our book, much better. A student can compare his or her individual score to the average—but that student cannot guess his or her own score based on that average alone. Likewise, you may hear that members of a group you belong to have a 25 percent chance of developing hypertension, but there is no way to jump from that figure to predicting your own odds for the disease.

When someone takes a population level statistic and applies it across the board to any individual belonging to that population, that someone is stereotyping. As Geri-Ann Galanti (2008) has noted, a 'stereotype' is an end point. A 'generalization,' however, can be seen as a starting point. For instance, based on generalizations we make from the population-level statistics we have, we can move forward and implement public health programs; in turn, these can raise the health status of entire populations.

Cultural Consonance and Role Incongruity

Cultural consonance, a construct developed by William Dressler (see Dressler and Bindon 2000), is the degree to which one's lifestyle fits with the lifestyle that one's culture recommends and that one thereby aspires to. In this, the term 'consonance' is somewhat like concordance: it means things align. When the fit is bad—when we have not achieved the things that we had in mind to achieve by a certain point—things that our culture has taught us to want to achieve (such as having earned a college degree, or being married with children, a dog and a white picket fence)—cultural consonance is deemed low. We therefore feel stressed. More concretely, health outcomes suffer.

Dressler and colleagues have examined the link between health outcomes and cultural consonance in several cultures (including in Brazil, the Caribbean, and Alabama) via clinical biological measures, symptom self-reports, and ethnographic methods. This body of work has demonstrated the link between high aspirations, low levels of the resources necessary to achieve them, and poor health outcomes such as high blood pressure. It also has shown how community resources for resilience can serve as a buffer.

The concept of cultural consonance includes role congruity, which occurs when one lives up to the expectations entailed in the role one has been placed in or has elected to take on. **Role incongruity** is when this does not happen. For instance, students may experience role incongruity if they have some idea of what a good student does and achieves but cannot seem to get these things done or get their grades up. These students feel a gap between the role as they want to play it and as they really are playing it out. Another example might be when parents cannot quite live up to their internalized cultural expectations of what makes a parent 'good.' When there is a gap, there is role incongruity; and with this as with low cultural consonance more generally, there is stress.

Dissonance can be felt, too, when one has the outward trappings of success but not the real basis for it, as when a person who seems to be living well is really deeply in debt, or when someone with a good job actually does not have the education necessary for it.

People who are either under pressure to keep up appearances or accustomed to respect and now get none feel stressed. People likewise feel stressed if they reject cultural requirements but still feel judged by those with whom they must live and work. Feeling stressed over a long period of time can trigger disease processes, including sometimes mental breakdowns.

Culture-Bound Syndromes to Channel Stress

Our examination of the impact of chronic stress makes it clear that the physical body (biology) and the mind, which interprets events (culture), are intimately connected. Beyond stress overtly recognized as such, culture gets into the body to create and sustain biocultural diversity via what have been termed **culture-bound syndromes**. These are culturally named conditions. Each entails a variety of symptoms, most of which would otherwise (in the mainstream West) be written off as 'stress' and many of which may not have any biomedically related explanation for concordant or concurrent (syndromatic) expression.

While culture-bound syndromes initially were studied as unique and taken simply as indicators of cross-cultural diversity among humans, more recent work highlights their universal features. For instance, many entail what is termed **somatization**—the projection of mental attitudes or concerns onto the body so that they are expressed as physical symptoms. As a simple example, think of the school child who gets a stomachache every time that his or her teacher schedules a test. Anxiety, distaste, annoyance, and the like also can be felt in the head, and neck; we do experience pain or become tense in these areas (we somatize) when such feelings cannot be otherwise expressed.

In addition to their high likelihood of featuring somatization, culture-bound syndromes often provide an outlet or channel for stress brought about by role incongruity or a lack of cultural consonance. In doing this, cultures provide stressed individuals with safety valves of a sort, while alerting others to their distress and so enabling others to mobilize, to assist. As Ann McElroy and Patricia Townsend note, "Rather than becoming and remaining labeled as mentally ill, the person is considered to be a victim of witchcraft, soul loss, severe shock, vengeful ghosts or other forces. After recovery, little stigma is attached" (2004, 280).

Traditional Examples

For example, in many Spanish-speaking societies, an individual who experiences a sudden fright may have her soul or some of his vital essence scared or taken away. This person becomes *asustado*: he or she is a victim of *susto*. Susto's symptoms include loss of appetite and weight, apathy, withdrawal, and depression; nightmares, swollen feet, diarrhea, general pain, and headaches also have been reported. Some equate susto's symptoms with those of post-traumatic stress disorder (PTSD). While asustados experience higher rates of organic disease than others, this may most often be an outcome of their vulnerability; no pathogen or lesion causes the primary condition. An exception would be cases triggered by hypoglycemia or low blood sugar; not coincidentally, some susto-susceptible societies link susto with diabetes—a point I return to later. For now, however, it suffices to say that most episodes seem to be brought on by the stress of role incongruity.

Many susto sufferers experience some kind of crisis of inadequacy just prior to the susto attack. They may have been socially sanctioned, perhaps laughed at or shamed; their household partners may have been cheating or using household savings for personal gain; they may lack necessary resources to feed the family, which may be in debt—in short, something will have made them feel unable to carry out ascribed social roles. The discrepancy or gap between their actual or socially viewed performance and their own expectations for a job well done leads to their excess of stress (Rubel et al. 1984).

Another common example of the culture-bound syndrome is *pibloktoq* or "Arctic hysteria." Individuals affected by pibloktoq—mostly women—suddenly enter an altered state in which they shout, make animal noises, sing, brandish knives or harpoons, tear off their clothes, throw things, run away, roll in the snow, jump into icy water, and so on. It often takes many people to rescue, hold onto, or subdue an individual in the throes of pibloktoq. Prior to an attack, they may exhibit anxiety, irritability, confusion, or depression. Afterward, they may sit still and tremble or sleep.

The embodied stress of role incongruity can underwrite pibloktoq just as it can susto. Culture-bound syndromes like these enable the expression and socially supported management of this kind of stress. That said, when such stress is more than individual—when it affects broad swaths of a population or seems as if epidemic—and when it is seen during certain windows of history but not others—a deeper dive into its causes is warranted.

Contact as Catalyst

Pibloktoq is not, it turns out, a longstanding, wholly indigenous part of Arctic life. The written record of pibloktoq cases spans 1892–1928, with the bulk having occurred in the very early 1900s. Most sufferers were working for expeditions led by Admiral Robert Peary, a driven, daring explorer who expected the most of his crew—and of the Inuit families he hired as assistants. In exchange for valuable trade goods Inuit men hunted and drove the dogsleds while women sewed warm clothing and boots (see Figure 10.4). Intimidation or awe also likely spurred them to sign on: not only was Peary bold and self-assured; expedition members carried guns and did, after all, arrive as if from another world. It is in the context of this kind of colonialism that pibloktoq must be examined.

Peary's crew included men trained in science and medicine. They treated the syndrome when they could with morphine or mustard water (Dick 1995). The former is a narcotic; the latter, a caustic agent long used medicinally to induce vomiting and to expel poisons or ill humors of various kinds. The idea that medicine could end an attack points to a biological understanding of its mechanisms.

Biologically oriented researchers have suggested that pibloktoq could have been caused by calcium deficiencies or vitamin A overdoses. Both are most likely in the winter. For one thing, the light–dark cycle in winter is linked with biological 'desynchronization,' so that more calcium is lost through the urine than normal; for another, the shortness of food supplies in winter can lead people to rely more on organ meats high in vitamin A, like liver (the fat of arctic animals and fish also are high in this vitamin) (McElroy and Townsend 2004, 281–283; see also Chapter 7, this text).

Notwithstanding, pibloktoq peaked in October—not wintertime. October's weather is characteristically rough, raising the likelihood of hunting trip mishaps. Also in this month

FIGURE 10.4 Inuit women sewing on ship's deck; National Archives photo no. 401-XPS-7-1, from the Robert E. Peary Family Collection: Photographic Prints of Subjects Relating to Polar Expeditions of Adm. Robert E. Peary, 1886–1939.

the sun disappears, the ocean freezes over, and sea mammals migrate away. Numerous ethnographic reports cast October as a depressing time for the Inuit, met with increased shamanistic drumming, dancing, chanting, and trancing. In hindsight, perhaps one-third of pibloktoq cases reported were just expressions of shamanic activity (Dick 1995).

Although as a religious practice shamanism would have occurred with or without the explorers, it may be that Peary's expeditions intensified the need for it, and of stress in general for the Inuit. Peary implemented hierarchical, militaristic order on his expeditions while the Inuit were used to a more egalitarian life. Further, families were not always allowed to stay together. Members missed one another. They worried more than usual for one another as well, because some of Peary's decisions, such as sending men on hunting and exploratory missions when the ice was not solid, seemed ill-considered.

Women stayed on Peary's ships or in his base camps. The less hospitable the expedition camp was, the more pibloktoq episodes were recorded (LeMoine et al. 2016, 10). At the least hospitable camp, in addition to limited food and so on, there was no blubber for the lamps women brought. They had to rely on kerosene or alcohol, and to burn that they needed to repurpose tin cans. The tin-can lamps were hard to work, smoky, sooty, and odorous. This might not have been so bad, but traditional lamps are critical to Inuit women's identities. Well-tended, they provide smoke-free light and heat for the home, food, and water, and to dry clothing. There is a spiritual aspect too: in tending their lamps, women tend the household's soul. The inability to fulfill this role expectation would have been deeply stressful (LeMoine et al. 2016).

On top of everything, Peary encouraged his crew to engage sexually with the women, regardless of the women's desires or matrimonial situations. He saw this in practical terms, as an aid to crew "retention" (Peary, as quoted in Dick 1995, 16). Accounts indicate that harassment could be accompanied by force at times, and that alcohol was used to aid coercion. Trauma was no doubt often induced.

Like the longstanding autumnal concerns of Inuit people, role loss and sexual predation could have been answered with shamanic rites such as trance activity. It also might have provoked shows of resistance; as may have other situations when expedition masters' demands became too disagreeable: some pibloktoq accounts cast sufferers as if willful children refusing commands. A third option was the dissociative trance, through which people channeled, in a form that was culturally intelligible to them if not to Westerners, the cumulative biocultural stresses of hunger, dangerous working conditions, the inability to fulfill culturally ascribed roles, and sexual predation—all in the context of an imposed power structure that placed them at the very bottom (Dick 1995; LeMoine et al. 2016). All three kinds of episodes—shamanistic trance activity, acts of resistance, and dissociative states—were labeled as pibloktoq by observers who did not care or were unable to make contextualized sense of them (see Supplement Box 10).

SUPPLEMENT BOX 10: OUTSIDER DEPICTIONS OF *PIBLOKTOQ*

The stress of subjugation may be a key driver of many of the culture-bound syndromes described for non-Western peoples. That is, contact may make role expectations harder than usual to fulfill. The domination entailed may force people into untenable situations and make life in general more precarious.

When first-hand observers of a culture-bound syndrome are involved in exploiting indigenous peoples or lands, however, their accounts may not directly address how domination shapes what they are seeing. Other kinds of records or other parts of observers' diaries and memoirs must be inspected to see the biocultural connections, as Lyle Dick (1995) and Genevieve LeMoine and colleagues discovered for *pibloktoq* (1995, 2016). Nonetheless, it is worth rereading observers' accounts to get some sense of what their ethnocentrism did allow into their field of vision.

To that end, bearing in mind how blind reporters could be to their own roles in fostering the deprivation and trauma expressed in pibloktoq, here are a few excerpts from Dick's 1995 compilation. As Dick notes, these accounts do sometimes reference stressors such as food deprivation and physical illness, being away from home, family separations, and dangerous hunting or traveling conditions. Sexual predation is not explicitly obvious in the descriptions but other records make clear that it was not uncommon.

> While the *Windward* [a ship] was in winter quarters off Cape D'Urville, a married woman was taken off with one of these fits in the middle of the night. In a state of perfect nudity she walked the deck of the ship; then, seeking still greater freedom, jumped the rail, on to the frozen snow and ice. It was some time before we missed her, and when she was finally discovered, it was at a distance of half-a-mile, where she was still pawing and shouting to the best of her abilities. She was captured and brought back to the ship.
>
> *(Account #3, Robert Peary, 1898 regarding Inalu)*

Note the contrast between "freedom" and "capture" in the foregoing account. This comes up also in the idea expressed in the next excerpt that a pibloktoq sufferer "wants her way and cannot get it." Note too how the condition was medicalized, and drugs used to subdue the "patient":

> One evening I was reading in my room when I heard a number of voices, making a good deal of noise. I walked out to the alleyway, when I met the Doctor going out on deck,

> I asked him what the trouble was, he said that one of [the women had] Pibloktoo, the form of hysteria which is more prevalent among the fair sex than among their partners. I would diagnose it as pure cussedness. She wants her way and cannot get it. The crying and yelling they make would cause you to think that they were going to do violence to them-selves. The Doctor sized up the situation and calmly said bring Madam to the surgery, the patient was brought to him and in the severest tones possible, but with great dignity, he lectured her, and injected a liberal portion of mustard water into her arm, and made her drink some of it. This had a desired effect, she immediately became sane again and was restored to her right mind. I must say that the Doctor lost a good deal of patronage for he did away with Pibloktoo.
>
> *(Account #20, Robert Bartlett, 1905; sufferer's name not given)*

The next account reveals more figuratively the explorers' condescension:

> We heard a commotion for'ard and word was passed that "Buster's got piblokto." The "Scientific Staff" promptly proceeded to take such observations as the weather and place of the forthcoming performance permitted… . There was the lady in a pool of ice-water, breast-high, right under the bow, looking like an inebriated fish treading water, singing like a siren, and banging her hands together: "Yah! Yah! A yah yah! Yah!" … Many were the opinions as to the best way of getting the lady to ice or dry land. None of us was stuck on pretending she was a bird's egg [treasure] and wading after her, so some fellows got up in the bow and tried to lasso the heifer, but, as we were not cowboys, she was perfectly safe… Finally the old girl was saved, though not until she'd given a good imitation of a cat arguing with fly paper.
>
> *(Account #31, George Borup, 1909; regarding Alnaya)*

The foregoing accounts all concerned women. Men's attacks generally occurred away from the base. Hunting accidents were of course part of Inuit life prior to contact; but being put into danger by one's employers' lack of familiarity with Arctic hazards entails a different order of stress, as the following account suggests:

> Tukshu, on a block of ice, was scarcely half-way across the open lead, when with a roar like the discharge of artillery, the floe he had just left broke into three parts. An upheaval of water followed, the pan upon which Tukshu was broke apart, plunging him into the sea … Tukshu seemed lost, but in some manner he succeeded in reaching the main ice and was hauled upon it. The other Eskimos began at once to beat the water, quickly forming into ice, out of his bearskin trousers, while he pulled off his wet kuletar and donned a kopartar. Then I gave him a small drink of whisky from my flask, and he began running up and down to warm himself. I do not know whether it was the whisky, or the excitement attendant upon his narrow escape, but suddenly Tukshu went problokto, and nearly two hours lapsed before he was sufficiently recovered for us to begin our retreat.
>
> *(Account #29, Harry Whitney, 1908)*

Taking historical context into account opens our eyes to how culture contact shapes indigenous people's suffering. Here let us reconsider the folk assertion that susto causes diabetes. Although diabetes sufferers in susto-susceptible societies seem to have only slightly elevated susto rates compared to their peers, stress does correlate, epidemiologically, to diabetes

(Baer et al. 2012). So does poverty. In this light, the role incongruity implicated in susto is part of a disenfranchisement complex reflective of a colonial legacy of socioeconomic disadvantage and dominative power relations. At the population level susto, like pibloktoq, and likely other culture-bound syndromes too, is the embodied expression of the trauma of structural violence.

Modern Culture-Bound Syndromes

There are culture-bound syndromes in so-called modern cultures, too; here, also, power seems key. Take, for instance, road rage, which entails the sudden onset of extremely aggressive behavior among (mostly male) drivers. Its occurrence has been documented in the United Kingdom, Holland, and the United States (McElroy and Townsend 2004, 286). It may be an expression of the stress engendered by the gap between being treated or received on the road as one believes that one should be treated or received, or it can be an expression of perceived incongruity in other areas of life. It may be similar to syndromes such as *amok,* a temporary kind of insanity occurring in New Guinea, Indonesia, and Malaysia, mostly among young men. A victim of amok goes suddenly frenzied and runs about, trying to injure or kill anyone in his path. The English expression 'running amuck' stems from colonialists' knowledge of this condition.

Two more modern examples are anorexia nervosa and menopause. Anorexia nervosa is an extreme form of self-starvation conjoined to an intense fear of being fat. On the surface, it can be linked with an overabundance of cultural messages regarding the value of a slender body, particularly for women in the white middle class (in other groups, being plump can be taken as a sign of vitality and good health). However, at a deeper level, anorexia nervosa also involves rebellion against the social relationships represented in food-sharing. It has been analyzed as an enabling condition of teenage girls who, particularly in white middle-class families with controlling parents, may be having difficulty individuating or otherwise feeling empowered over their own destinies.

Menopause happens further along in the life cycle. Clinically, and across all cultures, menopause entails cellular changes in women's ovaries, related hormonal changes, and the eventual cessation of menstruation. These occur universally, in all women, but the degree and speed of the changes, particularly hormonally, vary broadly, being affected by diet, number of births, lactation durations, tobacco habits, exposure to endocrine disruptors (e.g., PCBs; see Chapter 9) or hormonal birth control, and so on. Population-level variation in such factors combined with differing gender and other cultural expectations underwrite the diverse ways menopause is experienced in various groups (Melby and Lampl 2011).

In Western culture, 'menopause' refers to a syndrome including mood, body temperature, and sleep disturbances. Menopause in this sense, which does not happen everywhere, can be understood not only in tandem with the Western diet, birth rate, and so on but also as a culturally acceptable way of diffusing or grappling with the stress of the gap between the cultural equation of femininity with childbearing and youthful beauty, on the one hand, and the impact of aging on the other. The menopausal woman in this cultural context may not be the (young) woman she has been enculturated to wish to be.

If this interpretation has any truth to it, in cultures where postmenopausal women are culturally valued contributors, or where social status increases postmenopausally, the syndrome should not be seen. Indeed, although the average age of fifty is remarkably universal for onset of menopause, a survey of thirty societies conducted before the recent and potentially confounding surge of globalization revealed that in only two societies besides the United States was menopause seen as a major physical and emotional loss (Bart 1969).

Culturally Interpreted Disease States

Until this point, my focus has been on syndromes expressed in behavioral and mental states, even when somatization is identified. However, some syndromes have more narrowly physiological signs and symptoms. The convenient distinction between body and behavior is, of course, ethnocentric. It also flies in the face of this chapter's contention that body and mind are interrelated. That said, it still may be helpful here to explore some ailments that are (or seem to us to be) more body-bound—ailments in which symptoms themselves are overtly physical rather than attitudinal. These include conditions brought on by cultural practices, such as late nights spent staring at a computer screen or long spans of time spent sitting in front of one. They also include bodily problems found around the world in one form or another, but in each place with a distinctly cultural twist.

Empacho provides a clear-cut example of this kind of culturally relative physical malady. Empacho, which is most common among people of Latin American cultural origin, entails a kind of stomach distress—the kind that happens when food gets stuck to the inside of one's stomach or intestine. The food can become attached through suction, much as soft sandwich bread can adhere to the roof of the mouth. In empacho, the food stays stuck, however, sometimes having trapped other substances between it and the wall of the belly or wherever it has lodged. Whatever has stuck stays stuck long enough to start to rot. The whole conglomerate generates gas and swelling, leading to a lack of appetite that can be accompanied by headache, diarrhea, and vomiting.

The sensory feelings that I am describing can be felt by anybody, anywhere in the world. We all get stomach upsets sometimes. The feelings reflect universally measurable biochemical processes and concomitant signals and impulses traveling through the nervous system. Yet, depending on culture, people interpret the symptoms entailed quite differently. Those who grew up in a Latin American culture will have been provided, since childhood, with the cultural construct of empacho. Knowledge of the condition provides a person with a mental tool through which to interpret the signals of abdominal fullness and so on—and with ideas on what to do to treat them. With empacho, the goal is to loosen the stuck or impacted food so that the stomach is emptied again and ready for regular use. For instance, in Guatemala, people report resorting to stomach massages, over-the-counter remedies containing sodium bicarbonate (an alkali and so an antacid), lemon juice, purgatives, and herbal teas, some of which also are purgatives. Another option is to take some olive or cod liver oil or to ingest something ashy, such as a burnt tortilla (Weller et al. 1992). Other cultures have their own culturally influenced responses (see Figure 10.5). A mainstream white middle- or upper-middle-class American, for instance, might interpret those same signals in terms of an acid imbalance. Sodium bicarbonate or something similar may figure in his or her remedy scheme (bicarbonate is an ingredient in Alka-Seltzer).

Note here that there is some convergence in terms of medicinal strategies. Sometimes similarities stem from the diffusion of cultural practices: ideas do spread. Other times, people discover, independently, what works, and alkaline substances, including ash, do neutralize stomach acid around the world.

However, diverse material or environmental options (what kinds of treatments actually are available) or varied views regarding an ailment's cause can lead to very different forms of treatment across cultures. Further, one culture may subscribe to 'homeopathic' views, in which symptoms are treated by intensifying them, with the idea of helping the body to do what it intends to do anyhow to get rid of the problem (fighting fire with fire); another might take an 'allopathic' standpoint, in which the symptom is countermanded (fighting fire with water). So, for instance, someone with a fever might be wrapped up in a warm blanket or otherwise helped to further heat the body in one culture, while in another he or she might be given a cooling bath.

FIGURE 10.5 Beecham's Pills, a laxative treatment made from aloes, ginger, soap, and several minor ingredients, came onto the market in England in about 1842. They continued to be manufactured until 1998.

Beyond differences in treatment plans, or in the interpretation of symptoms from population to population, the very experience of stomach upset is bioculturally diverse. I did say that at the biological level the signals would be the same worldwide; that is true. However, because at the interpretive level a stomach upset might carry such a different meaning from one culture to the next, it actually can be experienced very differently. In a sense, when comparing the stomach upsets of, say the Guatemalan peasant and the Wisconsin-born Anglo American, we are comparing apples and oranges, or tortillas and kaiser rolls.

Yet another experience will emerge for someone whose culture provides not a naturalistic (for example, food or stomach-acid based) explanation but rather a personalistic one, based on how one's relations with others are faring. In such a context, a stomachache could signify social rejection, homesickness, an angry neighbor, or any number of problems within one's social network. Treatment plans will focus not only on fixing the stomach, but also on—and sometimes instead of—fixing one's social situation. Treatment will be not individualistic but rather community based.

Mentioning personalistic treatment here helps drive home an overarching theme of this book: folk traditions notwithstanding, it is unhelpful, at best, to try to explain scientifically any human experience as either wholly biologically grounded or completely cultural. Our language sets us up to make this mistake. We have no unified word for the mind–body, and so tend to see ourselves dualistically even when trying not to. Yet, because biology and culture (or nature and nurture, or body and mind) are actually a unity, only a unified, integrated approach can provide real answers, and real understanding.

Placebo Effect

The Meaning Response

At this point, we must consider the potential objection that much of what passes as healing is not actually 'real' medicine. A community ritual held for a person with a gastrointestinal ailment is not what 'really' heals it. Rather, the educated reader may say, something used in the ritual had clinical efficacy; the ritual itself made no difference; or the illness just ran its course although people misattributed its disappearance to the concurrently timed ritual. A common cold, for instance, eventually will get better on its own with or without a healing ceremony.

Another explanation for why people believe in ritual healing is that they are taken in by the **placebo effect**: a measurable, observable, or felt improvement in health that is not directly attributable to biomedical treatment. The placebo effect is generally ascribed to self-healing attributable to knowledge or belief—to the cultural meaning—that an inert substance or an act that is not part of the biomedical tool kit—carries. The placebo itself, traditionally defined, is something seen by biomedical doctors to do nothing. Examples include laying on of hands or other forms of faith healing, and use of charms (see Figure 10.6a and b). However, because an effect cannot logically be named after or caused by something that by definition has no effect (that is, a placebo), Daniel Moerman (2002) suggests that we should label this kind of healing 'the **Meaning Response**' instead.

Whatever called, this response or effect is well-utilized in many cultures. Even in biomedicine, cures or improvements often are created with inert substances given to patients who believe them to be medication. For example, as much as 50 percent of analgesic pain relief is due to a placebo response (Watkins 1996). Even a prescription itself can have a placebo effect; as Norman Cousins (1981) noted years ago, leaving the doctor's office with one in hand is a major part of the expected healing ritual. So can just being told that a pill will do this or that: in a study designed by Fabrizio Benedetti, patients were given pain relief drugs either with or without being told about it; those who knew about the analgesics did better (Moerman 2012, as cited). This study

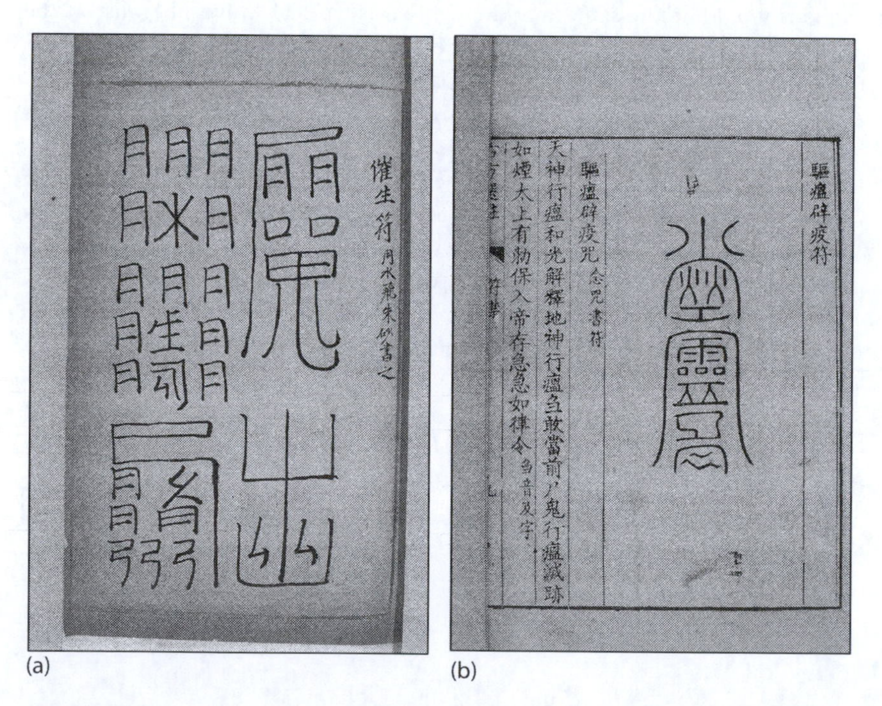

FIGURE 10.6 (a) Charm to be inscribed in cinnabar on yellow paper and attached to doctor's hat to protect against infection when examining victims of pestilence (1737). (b) Charm for accelerating labor and delivery, to be written out using cinnabar and burnt to ashes over a lamp; ashes should then be administered to the laboring mother, mixed with water. Wellcome Collection. Creative Commons Attribution, CC BY 4.0.

has been replicated, with the same results, in three other areas of biomedical treatment since, including for anxiety and Parkinson's disease.

Although we do not know exactly how most placebos make a difference, we can be certain that what works as a placebo is culturally relative. A placebo's success relies on the patient's socialization, beliefs, and trust in the healer. As Jamaicans say, belief can cure, and belief can kill—although placebos with negative effects are more correctly called 'nocebos.'

Fear Kills

Some of the earliest work regarding nocebos was undertaken by Walter Cannon, also one of the first to help define the biochemical pathways for the stress response. Cannon was interested in explaining 'unexplained' deaths among those who believed they had been hexed. After reviewing a number of cases in which people had died after learning that someone had cursed them, Cannon realized that their deaths had to do with the stress response.

While to cultural insiders the hex was the root of the problem, Cannon explained this kind of death from the biomedical perspective, attributing it to the traumatic stress brought on by intense and unabated fear. He argued that when epinephrine (then called adrenaline) runs high, blood vessels constrict, lowering the amount of circulating blood; vital organs are therefore not well enough oxygenated to support the body. Cannon was more or less correct, although newer work on allostasis, such as we already have discussed, has broadened our understanding of what is really going on. In any case, as Cannon showed, death is in fact a certain result when circulating blood levels are low and are exacerbated when victims stop sleeping due to worry, stop eating due to belief in one's sure demise, and experience social rejection, for instance, by relatives already digging a grave and singing funeral dirges (Cannon 1942).

A preternaturally early funeral is a form of social death, quite literally. In other contexts and more broadly, **social death** entails treating a person as a nonperson or with a very cold shoulder. This can be carried out even for whole groups of people, such as when one population wishes to subjugate another. Being targeted for social death is hugely stressful to say the least, particularly when coupled with other forms of harassment or oppression.

Meaning Matters

Proximate, biomedical causality such as Cannon described notwithstanding, it turns out that, ultimately, a person's culturally directed interpretations of his or her situation are what underwrite biomedically inexplicable—but physiologically measurable—healing (or pain reduction, or death). To make this clear, let me recount one more classic example (Blackwell et al. 1972, as described in Moerman 2002, 47–51).

About forty years ago, fifty-seven second-year medical students came to class to hear a lecture. At the start of the lecture they were told that scientists had discovered two new drugs, a stimulant and a sedative, and that they would be part of an experiment to test the new drugs. The students were randomly divided into four roughly equal sized groups and then given pills—which were inert. The groups respectively received either one pink pill, two pink pills, one bluish pill, or two bluish pills. The students were told only that the pills were sedatives or stimulants; they were not told which pill was which.

Before taking the pills, the students took each other's blood pressure and completed a self-report mood survey. After taking the pills, they listened to a one-hour lecture, as medical students commonly do. After the lecture, they again took their blood pressures and reported how they were feeling.

What did the findings indicate? They showed that one quarter (26 percent) of the group that took one pink pill felt less alert after the lecture, while two-thirds (66 percent) of the group that took one blue pill felt so. The pink pills seemed to inoculate or protect the students against lecture-linked boredom much better than the blue pills did. Furthermore, students who got two pink pills were more protected than those who got one pink one, and those who got two blues were less protected than those who got one: doubling the dose seemed to intensify the effect. Remember, however, that the pills were inert; they did not have any kind of pharmaceutically driven impact on the body. Everything that happened to the students, then, stemmed from the meaning response.

As members of mainstream American culture, the students had learned to associate blue with coldness, sluggishness, nighttime, and sleep; they had learned to associate pink, as blue's opposite, with warmth, heat, and activity. The students associated the colors of the pills they had taken with the states of slowness and quickness.

Note that in other contexts or cultures these colors can mean something quite different. For instance, Italian men find blue (in Italian, *azzuri*) invigorating: blue is the color, and name, of the national football team of Italy. So blue pills do not put them to sleep (Moerman 2012).

Notwithstanding, given the US students' cultural assumptions, the pills' color-related meanings worked as the students expected them to. Through the mind and through the brain they affected the body. The connotations of the numbers of pills also played a role. We live in a culture where more is better, so again, apparently naturally and without effort, but in a way that really was very cultural, the students assumed that if they had double the dose, they would have double the effect. Of course, being second-year medical students interested in helping people with necessary pharmaceuticals, they also assumed that pills must have some kind of an effect to begin with, which in itself supported their experiences of the pills' impact.

Even knowledge about price can support healing. In a US-based placebo pill pain reduction study involving electric shocks, subjects given the placebo but told that it was rather expensive ($2.50 per pill) fared better than subjects given the placebo but told it was only 10¢ per pill (Waber et al. 2008). The adage, "You get what you pay for," or the assumption that items that cost more are better, may have been at work.

Similarly, studies regarding the role of the built environment in facilitating or impeding healing in hospital settings have supported the use of features such as garden views for patient rooms. Roger Ulrich (1984) followed gallbladder surgery patients who were put in hospital rooms either looking over a grove of trees or looking out at a brick wall. These were comparable patients, who had had comparable surgeries. The tree-view group had shorter stays and used less pain medication than those who were looking out at the brick wall. This will come as no surprise if you recall our discussion of Japanese forest bathing in relation to resilience. In any case, this study has been repeated over and again, in diverse forms and ways, with confirmatory findings. For instance, patients who stay in sunny rooms use 22 percent less pain medication (Ulrich 2006).

Their improvement, the improvement of others like them, and the improvement of those treated by nonbiomedical healers (such as curanderas, shamans, and priests) seems to be the result of the treatment's meaning, or of some meaning attributed to the context in which the treatment is given. That is, a pill's color, name, shape, or even the dress and demeanor of the person dispensing it, the extra attention entailed in its provision, or what the patient is told to expect can underwrite a degree of healing, or pain reduction, or the like—sometimes enough so that researchers comparatively testing a new drug or procedure against a placebo cannot legitimately or legally claim the drug or procedure to be in itself a significantly better source of healing.

Leveraging Knowledge

As Moerman (2002) notes, rather than dismissing the placebo effect as 'unreal,' we should find out what specifically in or about the treatment supports self-healing. If we knew more about the connotations that help make placebos so effective in various cultural contexts, including those that support ritual nonbiomedical healing—if we knew more about the mind-body, culture-biology connection—we could begin to leverage the placebo effect to optimize the meaning response and improve patient outcomes. Similarly, if we knew more about allostasis and the stress response, we would be better equipped to stem its onslaught.

Meaning—culturally constructed meaning—acts directly on the body, triggering a helpful or a pathogenic response. Yet, belief has its limits. Some diseases are much easier to cure through a meaning response than others. On certain conditions, it works fine, but on others—cancer, HIV infection, malaria—its impact is palliative at best. Humans are wonderful and amazing beings, but nobody lives forever, even with culture's help.

In any case, in the meaning response, and in chronic stress, cultural meanings are embodied. Just what diverse populations embody will differ depending upon the cultural context in which each population has been raised. We saw, for instance, how events may be perceived entirely differently in distinct cultural contexts or for different social groups, leading to high stress—and all that that entails—in one setting or group, but not elsewhere. We even saw how racist ideas underwrite measurable differences between so-called races. Culture's power in shaping biology, or in triggering evolved biological responses in ways that vary from population to population, should not be underestimated.

11

CULTURE IN PRACTICE

Embodying Gender

This chapter prepares you to:

- Explain how culture is diversely embodied in its practice (e.g., in cognition, emotion, carriage)
- Demonstrate the difference between sex and gender with reference to differences in sex development and diversity in gender practices across cultures
- Explain the conceptual basis, power source, and cultural functions of transgressive or anomalous gender roles
- Characterize and apply the 'rites of passage' framework to explain how social transitions—including into culturally constructed gender roles—are accomplished

Introductory anthropology classes often present culture as a set of rules that people follow or as the superorganic 'thing' that emerges when we follow those rules. In advanced classes, however, and beyond, many argue that culture does not actually exist—except for when we put it into practice. It comes into being that way: culture is a processual, event-bound, emergent creation. It emerges through actions we undertake with the use of our bodies.

Thus, a deeper look at the 'practice approach' is warranted. We do this here, taking as initial examples bioculturally diverse emotion, cognition, and carriage practices, then drilling down into gender. After skimming variation related to sex, our focus will land squarely on how cultures (their members) create and enact gender itself—how they envision and practice masculinity and femininity. Finally we will explore how rites of passage are used to help children make the transition from being children into adults who rightfully can lay claim to, and must bear the responsibilities of, full status as members of a particular gender category.

Why Study Culture in Practice?

I would not go so far as to say that culture exists only in the 'doing' of it, because each culture certainly also pre-exists most actions that express it: culture drives people to do as they do. Still, it is helpful to think about culture as a process versus a thing because, among other benefits, this approach helps us understand culture's stickiness or how it becomes such a habit. Practiced actions, repeated frequently enough, can come to seem so natural that we may not be able to conceive of any other way of doing said actions. Also, having one way of doing an action can predispose us to doing other actions in particular ways, as in a chain reaction. What is more, for better or worse, prominent action patterns can be highlighted by people who wish to argue in favor of reductionistic biological determinism as, for example, in regard to how boys versus girls throw a ball.

While we previously considered practices directly linked to notable health challenges, in this chapter we will think about less obviously linked ones, such as wearing high-heeled shoes. In certain communities, wearing high heels assists people in claiming femininity. High heels also affect the anatomy of the foot and leg if worn regularly, in a way that can be drawn upon by those wishing to disparage women's walking and running abilities. What happens is that the Achilles tendons of long-term high-heel wearers become notably stiffer and thicker than those of people who wear flats. This apparently helps offset forced and measurable calf muscle fore-shortening; it also leads to pain when standing or moving on flat feet—which keeps high-heel aficionados from doing much of that (*Economist* 2010b).

In addition to helping to illuminate the symbolic and practical effects of what people do, the culture-as-practice framework also allows insight into culture change. For it is in the process of 'doing culture' that adaptations can happen. Something that works really well in one context may cease to work well in a new context, and so people may begin to do it in a novel, better-adapted way, introducing new variations into the biocultural mix.

Culture in Practice

No Body, No Culture

Culture cannot happen without bodies. Our very existence as social beings is body-dependent. To make and maintain social relationships, we look at each other, touch each other, and talk to each other. Through talk—the specialized use of teeth, breath, tongue, lips, and larynx—we teach each other things: cultural things, like how to do a certain type of dance or how to eat or dress in a way that is culturally acceptable. Culture is embodied in a very literal way when we pay attention to the shape and the size of our body and try and sculpt it into the shape or size that our culture idealizes.

We even embody culture when we sleep: culture helps determine whom we sleep with, when, and where. In some cultures whole families sleep in the same room or immediate vicinity; in others, mothers sleep with young children and fathers sleep with other village men elsewhere; in still others, where each nuclear family shares a home, only the conjugal duo shares sleeping quarters. Offspring here may share a room due to lack of funds, but the cultural ideal for children is a room of one's own.

Culture and Emotion

Emotional expression comprises another arena for culture's embodiment. For example, Japanese culture, like many others, favors emotional control as part of an overarching **ethos** (worldview; fundamental value set) that places the group's well-being ahead of self-centered priorities. This is in contrast with mainstream America's ethos, which prioritizes self over society in many circumstances. Not surprisingly, then, when researchers showed Japanese students (in Japan) and American students (in the United States) pictures of faces of people either hiding or expressing their emotions, the Americans were less successful at accurately gauging how each photographed

TABLE 11.1 Comparable proto-emojis

	Happy	*Sad*
Eye-focused	(^_^)	(;_;)
Mouth-focused	:)	: (

person really felt. The Americans did not grow up needing to develop skill in deciphering hidden emotions and so they were not as good at it as the Japanese (Yuki et al. 2007; see also Nisbett and Masuda 2003).

As the researchers explain, the trick to accurately reading masked emotions hinges on looking at eyes rather than mouths. Eye muscles reflecting emotional states are hard to voluntarily control and so are more likely to give away our real emotions. Our lip and cheek muscles, however, are easy to override when we want to hide emotion. That is, because of the physiological and anatomical legacy of language, we can quite freely manipulate smiles, frowns, grimaces, and so forth if we want to. Japanese people are enculturated to look at the eyes, because they know mouths might be lying; this culturally engendered capacity exists as an adaptive response to culturally fostered emotional inexpressivity. Japanese people are thereby better able to discern hidden feelings than mainstream Americans who, lacking similar cultural training, naively focus on the mouth.

Variation in the way populations embody their emotions has implications not only for interpersonal relationships, which may in fact be enhanced by the ability to see what people really are feeling, but also for the unlikely arena of digital communications. When emojis first became popular, they were not cartoons but typed combinations of letters and symbols. Japanese-born people and cultural groups with similar approaches to emotion typed combinations focused on the eyes; others typed forms that focused on the mouth (see Table 11.1).

Culture and Cognition

I wish we had room for a full chapter on culture and emotion and one on culture and cognition. The variation we see in how people feel and think across cultures is eye-opening. Much human thought is of course universally determined (for example, binary contrasts such as up and down, right and left, good and bad are found across all cultures, and children in all populations seem primed to develop particular cognitive capacities in more or less the same order). However, again, as with our more corporeal (physical) selves, different environmental conditions lead to between-population diversity—not just in content, as might be expected, but in form as well.

Take, for example, the cognitive patterns found among US youth today, bombarded as they are with diverse digital stimuli. I teach online, and many students tell me they listen to and view online lectures while simultaneously texting and being texted by friends, checking and posting to social media, and so on. Some argue that current use of information technology strengthens the brain, much like going to the gym and lifting weights strengthens the body. People who use a lot of technology, or so the argument goes, are much better at, for example, multitasking than our electronically bereft great-great-grandparents ever could have been.

Evidence, however, shows clearly that multitaskers are, in reality, so distracted that any net gains relating to doing lots at once are offset by reductions in how well each task is completed. The more experience a population has with multitasking, the less distracted they will be—but in the end, there is no free lunch: overall performance suffers for all digital multitaskers. In the words of Clifford Nass, whose research team was in fact "absolutely shocked" at their findings:

multitaskers are "terrible at ignoring irrelevant information; they're terrible at keeping information in their head nicely and neatly organized; and they're terrible at switching from one task to another"—although they (we) my believe the opposite (Dretzin 2010).

Leaving aside other ramifications of screens (see Supplement Box 11 for one example), the collateral damage a population incurs from multitasking may be quite far-reaching. Habitually flitting from one thing to the next seems to correlate with an inability to follow through in various areas and what some call a 'soundbite' mentality. Some experts explain the dramatic increase we've seen in varieties of attention deficit disorder (ADD) in the United States as stemming from our constant over-stimulation.

SUPPLEMENT BOX 11: EMBODIMENT OF CULTURAL NORMS FOR YOUNG PEOPLE'S EDUCATION AND RECREATION

Over the past 100 or so years, Western peoples have habituated themselves to taking in information not so much by doing as by viewing. For this we use screens. We have large screens in movie theaters, chalk and now whiteboard as well as projection screens in classrooms, television or entertainment screens at home, computer displays, and the screens on mobile devices. Even book pages serve as a kind of screen for the visual display of abstract information.

As a society we have become habituated to two-dimensional visual media—and to staying inside to use it. This is a key cultural practice for us (that is, for financially richer Western populations)—and it makes our bodies different from those of populations whose daily lives, particularly when children, are spent active and outside.

Take for instance skeletal alignment. In days past, a stoop indicated older age while today it can be read as a sign that someone regularly uses a smartphone or works hunched over a portable computer. More than that, it can be used to predict disability: who hasn't been warned about the long-term debilitation that 'text neck' and 'computer slouch' can lead to? Whether or not such injuries prove to be epidemiologically significant rather than simply part of a moral panic regarding so-called screen time, it is a fact that populations that walk and stand less and look at screens more—particularly screens on mobile devices—differ posturally from those that walk and stand more and view screens less, such as foragers.

High-screen-use groups also differ from low-use groups in terms of eyesight, particularly when it comes to nearsightedness or myopia. Myopia occurs when the eyeball grows too long horizontally. This places the eye's focal point in front of, versus right on, the retina (the light-sensitive structure at the back of the eye that cues the brain to create images). The problem can be corrected, in the near term at least, with optical lenses that refocus where light lands in the eye. In the long term, however, myopia places people at heightened risk for glaucoma and other degenerative eye diseases.

Myopia is not a foreordained human condition. We know this because of rate changes over time and across cultures. Myopia in young US adults has doubled in the past fifty years (Dolgin 2015). The problem is particularly notable among children. A recent study in the Kaiser Permanente Southern California health system found that nearly 60 percent of pediatric patients (those eighteen years of age or younger) are myopic. In contrast, only 0.8 percent of Laotians in this age group and 10 percent of Africans measured at age fifteen have myopia (Theophanous et al. 2018). A study done by Francis Young and colleagues among the Inuit in Northern Alaska found that only 2 of 131 adults who had grown up in the traditional way had myopia, while more than half of their offspring were myopic (as cited in Dolgin 2015).

In developed nations within East and Southeast Asia, myopia now affects between 80 and 90 percent of young people (Theophanous et al. 2018). In Seoul, South Korea, 96.5 percent of nineteen-year-old men are nearsighted (Dolgin 2015).

Genetics certainly play a role in one's vulnerability to myopia but population differences seem to have much more to do with screen use and, more directly, with the number of hours spent in the classroom or study hall or indoors at home. Why? Time spent inside displaces time out-of-doors and the less time children spend outside the higher their risk for myopia.

For one thing, viewing distances indoors are shorter than those we experience outside. For another, bright light from the sun seems to trigger the retina to put out chemicals that, among other things, slow eye growth. In this, sunlight is a regulator of sorts. The less exposed one's eyeballs are to sunlight, the longer they grow—and the more likely one is to thereby develop myopia (Dolgin 2015). This certainly puts the old practice of sending children outside to play after school in new perspective, and increases the relevance of today's pro-recess movement and related calls to delay academics until age seven (as Finland's state schools and independent Steiner or Waldorf schools worldwide do already).

It also forces us to rethink assumptions about conditions we have come to take for granted as 'normal.' Population-level variation in visual acuity and in habitual skeletal alignment (posture) reflects cultural diversity in not only access to screens but also the priority given to time spent indoors. Whether our children stay indoors due to a focus on academics or a penchant for video games or YouTube shorts, the results are the same. These include visual and musculoskeletal issues that do not plague people growing up in less Westernized or alternative community settings where more time is spent outdoors.

Culture and Carriage

Culture is perhaps most obviously embodied in our carriage—in how we move and hold our bodies. This becomes clear when we interact with members of another culture and see people carry themselves in different ways. Often how they move looks funny to us outsiders. Alternative ways of moving may look even funnier when we try them—at least before we gain mastery over movements we are unused to, such as sleeping in a hammock (without falling out), moving quickly and quietly up a steep rocky path (without tripping), walking on a moving hand-made river raft (without going overboard), or carrying a load atop the head.

Expected modes of movement—including not just how people sit, stand, or walk, but how they gesture and even how they express emotions—differ across cultures, and they also differ between cultural subgroups. Take gestures, for instance. While a nod of the head means "yes" to most Americans, Greek and some other Mediterranean people traditionally have said "no" by a similarly vertical movement of the head (Gross 1992, 109). A mainstream American might beckon a person forward by wiggling the index finger out and in with the palm up and other fingers and thumb closed or bent in toward the palm. This could be insulting elsewhere, however; for instance, this gesture is used in the Philippines to call animals (Galanti 2008, 47).

We also train bodies about what kind of personal distance from others to attempt or expect. For instance, while many Americans, Canadians, and British people prefer about four feet between themselves and other people, Latin American, Japanese, and Arab people prefer about two (Gross 1992, 109).

More total differences in bodily carriage also exist. Japanese *geisha* for instance—female entertainers in the classical tradition—are trained from an early age to hold their bodies (including their heads and eyes) in a certain way and to walk in a particular manner (see Figure 11.1). The geisha walk is also to some degree constrained by the beautiful kimonos and shoes that geisha must wear (see Figure 11.2). The kimono, the shoes, the walk—these are cultural artifacts.

It is true that geisha belong to a very special subculture whose practitioners undergo intensive physical training—not just for walking but also for playing musical instruments, dancing, serving tea or other beverages, and so on. Yet even without specific training or openly acknowledged rules for posture and movement, from day one a person's carriage is influenced by culture. It is influenced by cultural expectations for particular roles, genders, or professions. All movement is culturally influenced. Even the way that people interact with one another sexually differs across cultures. Sure, there are some universal similarities, but what is so interesting here are the things that are not universal—the diverse patterns expressed by people as members of particular cultures.

One way we might test this would be to play charades, in which people are instructed to act like members of particular cultural groups. Someone given, for instance, a card that says "cowboy," will bow the legs slightly and saunter forward, thumbs hooked in belt loops, miming a tip of the hat to some 'ma'am' in the audience. The next player, given, say, "rapper," might move their torso, arms, and hands in a particular way to show this identity. The body is trained how to move according to culture or subculture.

Another place to look for carriage diversity is along gender lines. In most cultures, different genders entail different ideals for how bodies should be carried. For example, in the old days Anglo-European US boys bowed and girls curtsied when greeting an adult or someone of a higher social status than themselves. Beyond different rules for gendered greetings, there are different rules for how men and women should walk, get into a car or onto a horse, and even sit down. For instance, even today in mainstream US culture women do not sit with legs spread wide—that posture is reserved for men.

FIGURE 11.1 Traditionally, geisha training entailed an apprenticeship. "Geishas" (n.d.). George Grantham Bain Collection, Library of Congress, LCCN LC-B2-5534-5A, LC-DIG-ggbain-32973.

FIGURE 11.2 Geisha in uniform (n.d.). George Grantham Bain Collection, Library of Congress, LCCN LC-B2-5189-7, LC-DIG-ggbain-30458.

Embodiment

While other groups might find either mode of sitting strange, the point here is not so much the exact recommendation but that there are recommendations at all—that there is a 'right' way to sit and a 'wrong' way to sit. Of course, the well enculturated person will sit as taught without thinking about it. Culture is habitual and often out-of-mind this way. Nurtured, cultural modes of carriage thus become thought of and talked about as if 'natural' or biologically given. The damage such determinism can lead to notwithstanding, the point here is that modes of sitting (and so on) can become so ingrained or deeply embodied that we do not even realize how cultural our actions are. This, in fact, is one definition of **embodiment**: the literal 'making physical' of culture.

Continuity and Change

Watching someone sitting 'like a lady' (or gentleman) also drives home the way in which culture is created in the practice or performance of it. Every time that a 'lady' sits properly upon a chair or as she enters a car she performs as per her culture. If she has paid attention to what 'ladies' around her have been doing—if she has studied her culture—her performance will go very well.

At the same time, if enough people refuse a cultural recommendation—if enough people choose to sit in a different fashion—and this could happen when, say, the recommended way of sitting becomes dysfunctional or it stops fitting well with other components of the culture—cultural evolution can occur. For instance, in regard to our sitting example, this might happen when the economy changes and women start to do different kinds of jobs.

Biological Predispositions

It is certainly true that some of how we move is determined by the makeup of our bodies. We have hinges in certain places, and we are bipedal. If we were not, we might not need to sit, let alone to learn how to do it. Still, given what we have—given the bodies that are born with—culture plays a key role in the diversity of carriage recommendations within and across societies.

Gendered Bodies

As seen for sitting, gender is a prime arena for embodiment: it is a crucial biocultural arena, with serious ramifications in the majority of social settings. For example, in highland Papua New Guinea, people grow yams. Just as Americans have Thanksgiving every November, people in Highland Papua New Guinea often have yam harvest festivals. Why bring this up here? It turns out that yams can be male, or yams can be female. Gender is such an important organizing principle that it can even be applied to vegetables.

Some cultures see gender as so important that it is part of the language; each word may be gendered so that even a lamp or a table may be feminine or masculine. Likewise, in many cultures, someone overhearing a phone call may well be able to discern the gender of the person on the other end of the line because gender determines forms of address and other modes of speech. Some cultures segregate men's worlds and women's worlds, guarding carefully against cross-gender contact. Even in our own society, to this day, there are certain arenas left to the women and certain arenas left to the men.

Indeed, gender is such an important organizing principle, even for us, that often the first thing we do when hearing that someone has a new baby is to ask if it is a boy or a girl. This information helps us to know how to properly comport ourselves, culturally speaking, in relation to said infant. We know what kind of gift to get, and what to say when the baby cries. We have a whole repertoire of cultural reactions to draw on in responding to the child based on a perception of its gender. Parents today often want to know if they will have a boy or girl even before birth so that they can decorate the baby's room in a so-called gender-appropriate fashion, using blue if the baby will be a boy, or pink if she's a girl.

These choices are, of course, culture-bound—and time bound as well. One hundred years ago, the color choices of Americans would have been the reverse. Pink was for boys—pink being an attenuated or child-friendly version of red, the manly color of power, heat, and action (accordingly, later it was the color of superman's cape). Baby blue, being a version of the color of the Madonna's mantle, was for girls. Things have changed, but in any case, and whatever color scheme might be *au currant*, most people's comfort level goes up if they know which sex—and so (they may believe) which gender—to expect.

Gender Is Not Sex

We have established that gender serves as a key organizing principle for social life. Although gender is what many of us think we learn about when we ask a baby's sex, it is not. **Sex**, simply and technically put, is a biologically differentiated status related to one's chromosomes, reproductive apparatus, and genital format. Indeed, sex generally has been determined by looking and by seeing and ascribing meaning to whether a person was born with a penis or a vagina. Today, we also can use genetic tests. Note that sex is not the same as sexuality, which relates one's preferred way of expressing sexual feelings. While *sexuality* is what one does or would prefer to do sexually with one's body (including one's mind), *sex* labels the body one has—generally as male or female.

While sex is a material fact, **gender** is a **cultural construction**. That is, like race, gender comprises cultural ideas about how humans can and should conduct themselves (in this case to be seen as masculine or feminine)—ideas that often have nothing to do with biology or other features of the material world. It is not that biology never affects gender; indeed, in human systems connections abound so, for instance, the sex hormones to which a fetus is exposed will have some impact on brain development; and different combinations foster different behavioral proclivities, which in turn are interpreted through a gendered lens. However, culture helps determine how much, or how little, we make of such differences.

As a cultural construction, gender variously entails cultural expectations of how to move, dress, modify the body (for example, by shaving certain parts of it), and so on. Depending on the culture, gender entails various ideas about what a person should eat or drink (steak or salad, beer or wine), appropriate labor or professions, and so on.

Sometimes gender even entails assumptions regarding to which sex a person should be attracted. Mainstream Americans generally believe that men are attracted to the female sex, women, to males. Historians such as Jonathan Katz (2007) relate the focus on heterosexuality per se to falling eighteenth-century middle-class birth rates in the West and related demographic worries. Prior to that, reproduction and lust were, in theory at least, quite separate matters. As the heterosexual imperative caught on, and in similar contexts in other cultural settings, homosexuality threatened to undermine gender claims just as heterosexual exploits fortified them. Gender is not always so tightly linked to sexual expression, however; this differs cross-culturally and, it seems, with demographic trends. The looser the link, the less notice people take of one's sexual partners' sexes.

Likewise, gender (the cultural construct) and sex (the biological entity) can exist to a large degree independently. In such a context, if they become uncoupled, it does not cause such a fuss. Accordingly, we must define sex and gender as separate entities if we are to use them to make intellectual progress in unpacking the hows, whats, and whys of gender diversity.

Opposing Examples

The idea that sex and gender are conceptually different things was at one point in time a revolutionary proposition in the United States. Not until Margaret Mead published *Sex and Temperament* (2001 [1935]) and *Male and Female* (2001 [1949]) did people really start to come to grips with this idea. Prior to that, many clung to the notion that gender expressions stem from the natural propensities of male or female bodies to act in certain ways.

Margaret Mead was a student of Franz Boas who, as shown in Chapter 4, stood very much against racism and the ethnocentrism it represented. He countered biological determinism with cultural determinism, which puts causality on the side of culture rather than nature. In this, cultural determinism is of course wrong: both nature and nurture are important and in fact they are conjoined. Nonetheless, Boas provided an alternative to the biological determinism rife in the late 1800s and early 1900s.

Margaret Mead applied a culturally deterministic framework in her research, too, as she worked to counter the biologically deterministic understandings that most people in the West had about gender. To do so she asked if the same beliefs that US people had about how males and females should and should not act, and practices thereof, were found across all cultures. If they were found universally, then perhaps there was some basis to the notion that sex begets gender, or that gender naturally arises out of sex. On the other hand, if beliefs and practices differed cross-culturally, then the hypothesis that sex determines gender would have been falsified or disproven.

To test the hypothesis, Mead went to New Guinea to do ethnographic research. The resulting book was widely read by the literate American public. It reported that, in the three cultural groups that Mead visited, there were some very different gender configurations. Mead found that among one group, the Tchambuli, men acted in a way that most folks in the United States at that time would consider feminine. In other words, Tchambuli men took care of children, were concerned with social relations, and so forth. Women there acted in a managerial and undomestic way that we would consider masculine. In a second group, the Arapesh, everybody—male and female—acted in what we would call a feminine manner. In a third group, the Mundugamor, everybody acted in what we would call a masculine way. Mead highlighted these differences to demonstrate the culturally constructed nature of gender. Mead's findings would have been even more interesting had she not imposed her US-derived comparative framework; she might have seen, then, that other cultures can have wholly different expectations—not just ones that reverse our own. Nonetheless, her work did make clear that gender conformity is culturally relative. A woman who wanted to stay home with the children among the Tchambuli would be bending gender expectations; she would be an anomaly in the cultural context of her world.

Another example that turned expectations upside down came from among the Wodaabe, a nomadic cattle-herding group that traditionally has lived in the African Sahel. Among the Wodaabe a time comes when all nubile marriage-eligible men parade themselves in front of the women, showing off their beauty in a way that might remind us of the Miss Universe competitions. The men preen for hours prior to their performances; they take great care with their hairdos and apply cosmetics to ensure that the women's highest male beauty expectations are met (see Figure 11.3).

FIGURE 11.3 A Wodaabe man preparing for a Yaake demonstration; Niger, 1997. Photo by Dan Lundberg via Wikimedia Commons, CC-BY-SA-2.0.

The women, for their part, get to choose in the evening which men they will take home with them that night.

Despite this, Wodaabe men are all much more physically fit than many of men in the West today. This is due to the active Wodaabe lifestyle and the lack of labor-saving devices in their world, as well as due to the nature of their masculine labor—labor with which they, therefore, shape their bodies as they live. So in many ways, and on average, Wodaabe men meet the West's stereotype of masculinity much better than average Western men in sedentary professions can do.

Biological determinism in regard to gender holds little water. Sure, there are some features of gender in most cultures that do link back to biology—that can be tracked in terms of sex-linked genital, mammary, and hair growth pattern differences, for instance. Yet there are large swaths of gender that do not overlap with sex. Cross-cultural differences in the ways people enact masculinity and femininity demonstrate this. Some readers may not be immediately comfortable with the fact that Woodabe men wear makeup, or that Tchambuli women so firmly occupy the public and not the domestic arena. Nevertheless, seen relativistically, this is just part and parcel of human biocultural diversity.

Anomalies

Unsettling Sex and Gender as Organizing Principles

When it comes to subverted gender expectations in mainstream American culture, even staunch relativists may not be so sanguine. One reason for this goes right back to those babies, and the question "Boy or girl?" In mainstream America, without insight into someone's gender role, we find it difficult to organize our social interactions.

That is exactly why some people become unnerved by gender-fluid individuals—those whose gender shifts—and by undifferentiated or by **androgynous** people—those who seem neither clearly male nor clearly female (the term is derived from Greek: *andr* means man and *gyne* means woman; the suffix *ous* means possessing or full of). Similarly unsettling to the mainstream may be people who cross or combine gender traits together or express gender in ways that are not, on the face of it, suited to or concordant with their body's apparent sex (male or female). Some examples here are the female body-builder or the skirt-wearing male who maintains a beard. Mainstream Americans had gotten so used to using an oppositional or dual version of gender as a cue to social interaction that androgyny or nonconcordance let alone gender neutrality can leave us fumbling, confused, and embarrassed.

In anthropological parlance, the gender-fluid, gender-neutral, androgynous, or gender-non-concordant person is an **anomaly**—such a person will not fit into preconceived, categorically separate (binary) gender ideals. Note here that our culture's traditional insistence on static gender dualism is so deep that we lack agreed-upon neutral or nonbinary singular pronouns.

We may generate all kinds of moralizing rhetoric in trying to explain any anxiety that the dress-wearing male or female body builder might provoke for us. Such effort really is not necessary, however, because the explanation is quite simple: such individuals subvert or challenge the existing theoretical social order. They cross or bring together what we have been taught, culturally, to think of as separate.

Our body ideals (and by extension our ideals for body decoration and apparel) exist in part because of the way that they reinforce the culturally preferred social order. Another way to say this is that a culture's preferences for social order are recapitulated in that culture's preferences for bodies. So, for example, if a person comes from a cultural world that prefers to keep male and female or masculinity and femininity separate, a physical or embodied performance that overtly brings them together, for instance expressing the opposite sex's gender, fails or refuses to support the illusion (or culturally agreed-upon convention) of their categorical separateness. In this, the androgynous or trans- or cross-gendered person makes members of that culture very nervous indeed.

Here, a word about terminology is in order. *Trans* is Latin for across from or on the other side of; a transatlantic flight brings people from Manila, say, to New York. Similarly, the **transgender** individual expresses or performs (or simply identifies with) the 'opposite' sex's gender. In contrast, the **cisgender** individual performs gender in a way that both meets cultural expectations for expression of that gender and—this bit is crucial—matches his or her apparent sex (*cis* is Latin for on this side of). Following from this, **cisnormative** gender expression entails performing gender in a way that appears concordant with one's sex—or enables one to 'pass' without strangers knowing that one's gender and sex are not in accord.

All this said, the idea that gender and sex have to line up—for instance that a male who wears dresses necessarily wants to or should be female—and that any gender performance must be cisnormative is culturally constructed. Furthermore, the underlying assumptions of stable binarism that fuel such propositions—that there are only two genders and only two sexes, and that one's gender and sex are permanent—do not square with the evidence.

More than Two

Sex used to be ascertained by a quick genital inspection but in recent years chromosomes have taken center stage. Males generally have both X and Y chromosomes in the 23rd pair; females generally have two Xs. However simple that sounds, sex is a highly complex and unstable outcome that unfolds genetically, hormonally, cellularly, and anatomically as we develop. One often ends up male or female enough to be categorized that way, but not always. Key here is that sex entails numerous traits, many of which gain expression along a (multidimensional) continuum; so we always are generalizing when we think of someone as if 100 percent male or female. Also, a frankly intersex position possible. An **intersex** individual may be, for instance, female chromosomally but male in all other regards; or that individual may be, to the naked eye or not, 'in the middle' (Henig 2017).

All embryos start out essentially the same sex-wise: generally, male sex traits begin to develop only when certain Y-linked genes turn on. Sometimes mutations occur, or unusual alleles are inherited, so a person could, for instance, have a mixed phenotype, or the phenotype might seem, outwardly at least, 'opposite' to the genotype. For instance, a genotypic male (XY) lacking ovaries or a womb might be endowed with female genitals. This person could live her whole life in the feminine mode; or reproductive irregularities or any 'masculine tendencies' expressed could become problematic, depending on cultural context (more of which later).

A special form of this situation hinges on the enzyme 5-alpha reductase, which fosters the development of male traits during gestation. Individuals with low levels of this enzyme often have female genitals—until puberty. At about twelve years of age, they grow penises and can gain body and facial hair, a lower voice, and male musculature, although these secondary traits may not be as pronounced as they typically are in males.

Developing a penis at puberty can be disconcerting when the likelihood of it has never been raised, or in the context of a culture that does not acknowledge intersex possibilities. Yet, because 5-alpha reductase deficiency is caused by a recessive X-linked allele, some communities see it at high enough rates to have developed familiarity. For example, families in a region in the Dominican Republic know it so well—it happens to one in ninety children there—that they have a third sex category: *guevedoce* (from "huevos a los doce" or testicles at twelve). Although some *guevedoce* prefer to continue living as if females, many adopt the masculine role when their genitals change. Indeed, many who do so report having identified as boys earlier, when their parents still thought of them as girls. Although culture does have a heavy hand in gender ideals and preferences, 'male' hormones circulate in and through a *guevedoce* fetus. Although gender is much too complex to hinge only on sex hormones, this exposure may predispose *guevedoces* toward a masculine gender identity, particularly where other circumstances in the sociocultural landscape support it (Mosley 2015).

Intersex conditions are on the upswing in the West, as is gender dysphoria—a condition occurring when one feels forced into the 'wrong' gender (Rich et al. 2016; Henig 2017). Although increased social media use and having one or more peers who have recently come out can be predisposing factors, another reason for the uptick in both arenas is that our head counts are becoming more accurate as sex and gender are better understood and their diversity less stigmatized.

In addition, niche construction may be at work: endocrine-disrupting chemicals (EDCs) introduced by humans into the environment—certain pesticides for instance—are likely associated with the recent increase. Absorbed into our bodies, EDCs can mimic hormones, increase or decrease their production, and block or alter our hormone receptors. Bisphenol A (BPA), used in plastics, is an EDC; so are some polychlorinated biphenyls (PCBs).

Recall from Chapter 10 the review of the Akwesasne Mohawk nation's experience with toxicant exposures, PCBs included. Although sex and gender diversity have not yet been investigated in relation to PCBs, we do know that they have caused reproductive difficulties, and studies of adolescents have shown that they hasten menarche in girls and reduce testosterone in boys (Schell 2012). We also know from animal studies that EDCs can affect tissue differentiation during gestation. Studies undertaken with humans after occupational exposures and industrial accidents confirm this, demonstrating for example nonconforming genitalia (France) and altered sex ratios (Italy). Broader findings regarding the dysregulation of hormonal pathways in humans even suggest that the ways people associate with their sex physiology and the ways they enact gender may be affected (Rich et al. 2016). Such causality notwithstanding, our reception of sex and gender differences is highly culture-dependent.

Anomalies Are Powerful

Any anomaly draws its disruptive power from the simple fact that it does not fit into or stay put within the given categories that make up a particular cultural world. For instance, steak-flavored ice cream would be an anomaly for the mainstream American. As such it would challenge the human penchant for order.

We approach the world with a preconceived notion of what an orderly life should consist of, with everything in the right place. A thing not where it is supposed to be—"matter out of place," to quote Mary Douglas, a social anthropologist who has thought a lot about these issues (1966)—has power because it disturbs, or represents a disturbance in, our very sense of order. In Douglas's view, the key model for order in life comes from the social structure.

Whatever the ultimate source, the proximate or immediate power of the anomaly inheres in its anomalous nature. The mandrake root, featured in Harry Potter's story and often used in magic or sorcery and in traditional medicine, is powerful because of its form (see Figure 11.4).

Mandrake roots look human, as if they have legs and arms. In this, unlike the simple carrot, the mandrake seems to cross or bring together the world of the human and the world of the plant. It draws its power from that simple fact—from its anomalous, category-crossing nature. Many plants used in magic or medicine have crossover or transcategorical characteristics, whether in color, smell, or shape.

Anomalies have power. What that power can do, and whether it is considered sacred and pure, or profane and polluting, depends on the cultural context within which we interpret and how we therefore handle it. Take, for instance, a chicken with two heads. In some cultures, that chicken would be revered as a saint, or seen as a harbinger of good luck. In others, it might be considered polluting, and people would sooner not see it: it might harm them, for example, by causing their own children to have two heads. Either way, contact with the chicken—let alone eating it—would be **taboo** or strictly controlled or prohibited.

While we are considering anomalies that might be eaten, here are a few more examples. In some cultures the dolphin's nature as an air-breathing mammal that lives in the water when mammals 'should' live on land makes it taboo. An egg with an embryo in it rather than just yolk and white, and which therefore brings together death and life, may be taboo similarly—it may be considered a delicacy because of this. Importantly, in cultures where such an egg is eaten,

FIGURE 11.4 Mandrake (*Mandragora officinarum*) from Tacuinum Sanitatis manuscript, c. 1390.

certain ritual precautions will likely need to be taken to ensure that its power is productively directed. For instance, the egg could be eaten as part of a fertility rite. If eaten without proper precautions, it might beget octuplets—or lifetime infertility.

Handling Anomalies

An anomaly's power must be handled carefully. Oftentimes, those entrusted to deal with anomalies are specially trained priests or shamans. In the *Harry Potter* series, students in the first few years are not proficient in executing magical spells or controlling magical items or substances because they have not yet received enough needed training. Concurrently, at Hogwarts, or anywhere else that magic is practiced, anomalous and powerful items or persons often are carefully stored or otherwise kept contained. They and their supernatural powers may be the focus of worship and supplication or subservience.

The concept of anomaly and the related issue of power (whether holy and purifying, or polluting) make it easier to understand why some people find disturbing a body that mixes some characteristics defined as masculine and some defined as feminine. It also explains how and why, in some cultures, such bodies, as well as other anomalous bodies (the deaf and blind, those with extra fingers), are treated respectfully. Such individuals may function as a link between the supernatural and natural worlds. Accordingly, many cultures greet gender diverse or intersex people with awe and respect.

Sex and Gender, Unlinked

A Zuni Example

In many cultures, the overlay of one gender on the 'opposite sex' does not raise eyebrows because sex and gender are not tightly linked. Let us take, for instance, the Zuni people, who live in the Southwest United States and have been the subject of a good deal of work on gender because of the existence of a third gender role, that of the *lhamana*—generally occupied by a male who dresses and works as a woman, or a 'two spirit' person (see Roscoe 1992). The lhamana-type role is not only available but also valued and, in fact, necessary in cultures that have it. In fact, in some cultures, if the lhamana role (however that culture terms it) is not filled voluntarily for a village or group, someone will be recruited.

In our Zuni example, the fact that sex and gender are conceptually decoupled is seen in the way that, traditionally, a child is not addressed as a he or she until the child chooses a gender to live by and express, for example, in dress or in the labor that he or she chooses to do. For instance, a Zuni male child who likes to do the labor of weaving—feminine labor—would be choosing to live as a she (see Figure 11.5).

Conceptual Possibility

The powerful hold that dualism has on the Western mind has limited inquiry into gender roles that cannot be read in terms of what we think of as masculine and feminine, or our binary format. The possibility of taking on a gender role that transcends duality is greater in cultures with cosmologies or religious systems in which the material and spiritual worlds are vitally conjoined or where transformation and ambiguity are central themes. In cultures where, for instance, gods or spirits interact with humans, transforming themselves into various living things with various sexes, the crossover role often emerges as a culturally sanctioned choice.

FIGURE 11.5 We'wha, a Zuni lhamana, weaving. Photo by John Hillers, c. 1871–1907. Smithsonian Institution, Bureau of American Ethnology, National Archives no. 3028457.

Functions of Gender Crossover Roles

Gender crossover roles have a number of concrete, generally conservative functions. These include economic, spiritual, socially mediating, and role modeling functions.

If the sex ratio is somehow skewed in a society, and it benefits one sex to take on the labor of the other, cultural notions of masculinity or femininity could change (adapt) to include those types of labor. Another way to get that work done is to use a crossover role. Samoan tradition, for instance, includes a feminine-like gender role for biological males termed *fa'afafine*. This third gender role is open to any male who prefers it. When a household has no girl children to do 'girl' work, and one of the boy children hasn't already taken up the fa'afafine role, a boy child could be assigned to it—that is, to labor and dress as a girl—ensuring that the household's home-economy needs are met. Somewhat similarly, in Albanian tradition, a female with no brothers may take on a male gender role to support the lineage.

Yet, females acting as men happens more rarely cross-culturally than males acting as women. This may be because in many cultures masculine roles carry more value, in which case it may seem unremarkable that females wish to fill them. Another theory is that when all children are raised by women and have little access to men, they only really know the feminine role. In such conditions, a masculine identity can be hard to internalize. Female-only childrearing is common across cultures and therefore more males world-wide are interested in acting feminine than vice versa, or so the theory goes. Whatever the reason, recognizable, respectfully labeled, culturally sanctioned, culturally available roles for feminine males are more common than those for masculine females cross culturally.

Beyond economics, a second and perhaps more important group-wide function of the gender-diverse individual is as a spiritual functionary. For instance, gender-diverse people can work as high priests or shamans to keep the supernatural and natural worlds in balance, or to ensure that rain falls and crops grow and so on, aiding in the continued survival of their people. Alternately, someone having trouble conceiving might call on a two-spirit individual to conduct certain rituals to help with conception. Other rituals can be performed to secure success with a love affair or even for good luck in exams or a bet on the horses.

A third function of someone in a crossover role is to serve as a physical go-between or a mediator for men and women here on earth. A two-spirit person could broker heterosexual marriages or carry messages between boyfriends and girlfriends.

Last but not least, gender diversity provides role models. Commonly, the individual who chooses to take on sex-nonconcordant labor and dress practices does so in a way that puts to shame those who are 'naturally' of the gender enacted. For example, take We'wha, the famous Zuni lhamana pictured in Figure 11.5. We'wha's weaving simply was the best: it was weaving that any girl or woman could learn from. Likewise, an Albanian female who took on a masculine identity to become the family heir could shoot a rifle and hold liquor better than any man near or far. Or take the 'drag queen' role. As RuPaul noted in the 1990s, "When people say 'you dress like a woman', I say 'I don't think I do because women really don't dress like this.' I dress like our cultural made up version of what femininity is, which isn't real" (Harron 1995).

Cultural Critique

We internalize many images and ideals for gender whole hog, and transgender behavior based on such ideals can and does reinforce them, supporting cultural conservation. Still, some forms of gender diversity openly question or even critique cultural norms.

Take, for example, the motion picture comedy *Some Like It Hot*, which concerned two men (played by Tony Curtis and Jack Lemmon) who got themselves in trouble and so hid out by traveling with an "all-girl band" (see Figure 11.6). Compared to the 'real' women in the band, Curtis and Lemmon (or anyhow their characters) simply could not pull womanhood off, and that was the joke of the movie. In this case, the cross-gendered men served as anti-role models, reinforcing the need for men to stay manly (Wilder 1959).

On that basis, you might think I should have included this example in the earlier 'cultural conservation' section above. Some theorists do focus on the pro-patriarchal, misogynistic (anti-woman) sentiment exhibited in this kind of crossover act. Certainly, there is some misogyny in

FIGURE 11.6 Curtis and Lemmon dressed as women in *Some Like It Hot*, released in 1959.

the film. But there also is something else going on. In some ways, the gender diversity of *Some Like It Hot* was subversive: it questioned mainstream norms.

In the film, a man (Osgood, played by Joe Brown) falls in love with a man dressed as a woman ("Daphne") but a man nonetheless (as played by Jack Lemmon). That in itself was no big deal. There are plays from Shakespeare's time and before, and in non-Western cultures, too, in which this kind of thing happens, exposing the constructed nature of gender.

What is different here is that for Daphne the 'misunderstanding' does not get resolved in the way that it does in a culture-conserving comedy of errors. Instead, at the end of the film, Daphne and Osgood get into a motor boat, with Osgood intending to speed his little lady away into the night and, eventually, to wedded bliss. Osgood says something like, "I called Mama. She was so happy she cried. She wants you to have her wedding dress." Daphne replies, in falsetto, "Oh, Osgood! I can't get married in your mother's dress." Osgood offers, "Oh, that's all right, we can have it altered." Daphne says, "Osgood I have to level with you. We can't get married." "Why not?" Osgood asks. Daphne explains, "I'm not a natural blonde." "Doesn't matter," says Osgood. "I smoke, I smoke all the time." "I don't care." "I can never have children." Osgood consoles, "We can adopt some." Finally, completely exasperated, Daphne says, "But you don't understand." Ripping off her wig, dropping the falsetto, Daphne says, "I'm a man." Osgood remains smiling, driving the motor boat like nothing happened: "Nobody's perfect," he says with love's contentment (Wilder 1959).

The film thus not only calls attention to the constructed nature of gender, but also threatens to undo it. This was why it was considered so subversive on release that its posters were stamped "not suitable for children."

In sum, gender diversity is not always conservative in its function; it does not always act to preserve or reinforce existing expectations. It can also be very subversive, depending on how it is deployed. While in some contexts flouting norms may be punished, in others it can lead to culture change.

Rites of Passage

Constructive Function

Until this point, we have focused on differing cultural expectations, not only for what is masculine and what is feminine, but also for how tightly or how loosely sex and gender are coupled to start. We've explored how people shape, train, and carry their bodies as per gender expectations, which differ cross-culturally. The argument that culture constructs gender, and in doing so impacts the body, gains further reinforcement when we consider rites of passage. In many cultures, throughout time and around the world, only a very specific rite of passage can turn a person from a child into a gendered adult.

What Is a Rite of Passage?

A **rite of passage** is simply a set of ritual acts intended to move a person or a group of people from one social status to the next. Take for example the first haircut: in many cultures this brings a baby from being a nonhuman organism to being human, or from being a gender-neutral baby to being a boy or girl. The point of the rite is to help the individual make the transition and to mark, for others too, that it has been made.

In Orthodox Judaism, for instance, a boy's first haircut comes at about the age of three. This rite of passage is actually one in a series of rites of passage that altogether create a person into a man.

The little boy has already undergone a circumcision, as an infant. Now, as he grows, he will every few years undergo a new rite of passage moving him, through a series of social statuses, further and further toward manhood. In each rite of passage, other community members participate in helping to transition the lad from one status to the next.

Social Birth

Rites of passage are necessary for our social birth or our full incorporation into a cultural group, including for gender membership. Generally, a boy can be created into a man only through the concerted efforts of men; likewise for girls and women.

To continue with the example of ultra-Orthodox Jews, a boy's birth includes him getting circumcised one week after his mother delivers him. Technically, his father must do that, and, technically, it is not just a clip of the foreskin that is entailed. Circumcision is part of a long process of uncovering the real man within. When a fetus is in the womb, it is in the amniotic sac and sometimes, when babies are born, bits of that sac actually can adhere to them and must be pulled off. The circumcision is a continuation of this kind of unsheathing. So, in effect, the father who circumcises the boy is the one who 'really' births him, socially speaking.

Recently, as described by Yoram Bilu (2003), ultra-Orthodox Jews have brought together the first haircut and school initiation. This cultural adaptation happened partly because of increased pressure to assimilate. One idea behind doing school initiation earlier and combining it with the first haircut was to get the little boys into their private schools sooner, ostensibly to better protect them from the influences of the broader world. Of course, the cultures of other Jewish populations have evolved in other directions. The ultra-Orthodox example is just one of many.

In this community, these two rites of passage have come together so that each little boy is simultaneously separated from infancy and brought into the male world of learning, and removed from the female domestic domain. Mother's breast milk, rabbis say, is now replaced by the sweet nourishment provided by learning. Accordingly, during the rite of passage, the rabbi who replaces the mother holds the young one on his lap. He puts honey onto a piece of paper that has all the letters of the alphabet listed, or writes the letters out himself using honey for 'ink.' He then offers the page of sweet letters to the boy, who licks them up in lieu of breast milk. In the ultra-Orthodox series of rites, this introduction to letters and words—to the world of knowledge, or of the book—further separates the boy from his mother.

This rite of passage is orchestrated and witnessed by men and boys in the new initiate's community. It is a sweet occasion, and the boy's release from his coverlet of hair as well as the opening of his eyes to learning are commemorated by sharing (what else?) a honey cake. All the boys and men get a bit of the cake and thereby not only share the sweet experience but also incorporate it into their bodies through the process of ingestion. The act of shared eating, called **commensality**, is done in many ritual contexts; it is done whenever people ceremonially break bread together. Two people who have eaten from the same loaf cement, through that shared substance, their social bond to one another. This kind of sharing will happen again for this little boy at his Bar Mitzvah and then at his wedding and of course when he circumcises his own sons; family building is a key goal of manhood in the ultra-Orthodox Jewish tradition.

Becoming an Adult

Across cultures, there are more rites of passage bringing boys into manhood than girls into womanhood. Some theorists explain this as due to the female body's capacity to menstruate, which provides a natural marker of the onset of physical womanhood. Other theorists, particularly those with psychoanalytical training, believe that the journey to manhood is just harder, because in most cultures women raise children.

In cultures that have constructed highly distinct and segregated male and female spheres, in which men occupy a completely separate world from women and vice versa, it is much harder for little boys to become men because it is a longer jump for them to make. They are not simply moving from one sub-status to another sub-status in a gender-integrated world—one in which they have male role models to interact with and study all the time. Rather, in gender-segregated cultures, little boy children stay with female kin. Because men often stay away, living separately, these boys have little idea what men do. While girls see women all the time, boys growing up in a gender-segregated context may find masculine identity formation truly difficult, at least initially.

Elaborate rites of passage for boys in late childhood and early adolescence help sort this kind of thing out (see Figure 11.7). Indeed, male rites of passage in more gender-segregated societies often are intense and long. The series of trials they involve can include feats of physical strength and endurance or resolve. Some cultures circumcise boys as teens. Sometimes, more intense genital surgeries are required (the hazards of which may be seen in the epidemiological profile of the group implicated). Certain groups in Melanesia, Australia, and elsewhere, particularly those that are extremely gender segregated, practice subincision, in which the penis is split on the underside from the urinary opening. Other groups require tattooing or the like during a rite of passage. Still others require a boy to kill a large mammal or to experience a vision or to demonstrate his worth by other means.

Because the body is central to the biocultural approach, let us explore in a bit more depth some of the cultural understandings about the body that are worked through in male rites of passage (also see Burton 2001). In some cultures, boys raised by women are understood as thereby filled with female qualities, in part due to drinking mother's milk. To create them as men, that milk must be replaced, as it was with the sweetness of learning in the ultra-Orthodox Jewish school initiation rite. Other people replace it with other things, including sometimes male semen, a masculine life force.

FIGURE 11.7 Nine to ten-year-old boys of the Yao tribe in Malawi participating in circumcision and initiation rites. Photo by Steve Evans from Citizen of the World, CC-BY-2.0.

This is perhaps most famously done among the Sambia, whose initiation rites Gilbert Herdt studied (1994 [1981]). In this culture, and some others, boys perform oral sex and thus feed on their male guardians as part of their rite of passage. Ingestion of semen and its biological incorporation into their bodies helps make them men.

Among the Sambia, as in many other societies, the boy-to-man transition is a group affair. This highlights boys' importance to the community and helps reinforce community bonds—not only among the elders conducting the ceremonies but also among the initiates going through them. Children who undergo their rites together have a very strong bond with one another for life.

Group rites also have practical importance materially. These ceremonies are not only intense in terms of content—they also are time and resource intensive. It is more economical—in terms of time, effort, and resources—to ferry a group of children through a rite of passage together.

Structure and Process

Every rite of passage has three main parts, as represented in Figure 11.8. It always begins with an act of separation separating individuals from an initial social role. This is represented at the junction of the first horizontal and vertical lines, with the number "1." The lower horizontal axis represents the mundane or everyday world, and the higher one represents sacred time and space, during and in which the second, transitional (liminal) phase happens. Rebirth (reincorporation) is phase three, and the end of the rite of passage.

Separation

To begin a rite of passage, initiates must be separated from their old social position. Essentially, a switch must be thrown. While this may happen as gently as being awakened with a sweet song not sung on any other day, sometimes this happens in a most frightening way for the children or child involved. It might happen at night, when slippage between the everyday world and the supernatural realm is most likely. A child could be sleeping peacefully with his mother one moment and kidnapped, so to speak, the next by male elders or the spiritual leaders of the group—people with whom he often has not spent much time and who may be disguised anyhow (I describe rites of passage from the male perspective in keeping with the points made in the previous section; but of course females go through them, too).

Ripped from mother's bosom, the child may be stripped of any regular clothing, perhaps even of his hair, and taken to a spot far from the village. Only when an individual is expelled, dismissed, or separated from a previous status can he assume the next. Death symbolism is prevalent during the separation process.

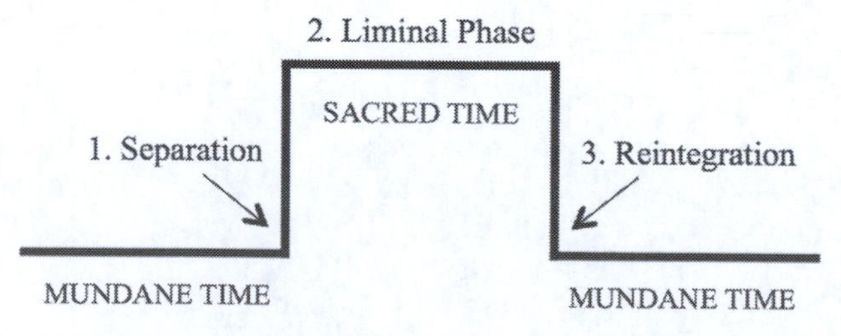

FIGURE 11.8 Three-part structure of rites of passage: 1 = Separation or pre-liminal phase; 2 = Transition or liminal phase; 3 = Reincorporation or post-liminal phase (see Leach 1979 [1961]; Van Gennep 1960 [1908]; Turner 1979 [1964]).

Transition

The second, transitional stage is famously termed the **liminal** period. This label we have inherited from Arnold Van-Gennep who, along with Victor Turner, greatly enhanced our understanding of rites of passage (Turner 1979 [1964]; Van Gennep 1960 [1908]). The label comes originally from the Latin word for threshold, or for the small area dividing one room from the next, or the inside of a house from the outside. A person going through a rite of passage is, in effect, standing in the doorway between social statuses.

Once our boy's old status is gone and he has been, in effect, removed from the secular, mundane world and its everyday social structure and trappings, he comes to occupy 'inter-structural' time and space. In Figure 11.8, this liminal arena is represented by the higher register on which this phase is depicted. The person in transition here is "betwixt and between" social statuses (as Turner famously put it). As such, he occupies a powerfully anomalous, sacred space.

Recall that the power of those who cross categories or do not fit into them must be handled with care, as taboo. Oftentimes, initiates are therefore sequestered: the mere sight of them might harm regular village members—or vice versa. If this sounds farfetched, consider that in mainstream American marriages it has been considered bad luck for the bride and groom to see each other on the day of the wedding prior to taking vows.

Just as the bride, when getting dressed for her wedding, might be given special instructions for how to be a success as a wife by her mother or other female relatives who already have undergone that particular rite of passage, so, during their confinement, are boy initiates taught what they need to know to be men. While being treated as liminal already heightens awareness or arousal, many cultures induce altered states in initiates to help their minds grab onto key information or to open them up to the change that must be undertaken. They may be kept up all night for several days; they may be forced to fast; they may be asked to sing for hours on end, or to participate in drumming. Pain and fear also can be used to reinforce the transformative effect of lessons learned in this phase of the passage (see Figure 11.9).

FIGURE 11.9 Platoon Sergeant Shawn D. Angell trains marine leader candidates at the Officer Candidate School, Marine Corps Base, Quantico, Virginia. Photo by John Kennicutt, US Marine Corps.

Symbols packed full of cultural meanings related to the transformation that is being orchestrated also are deployed to enhance and speed up learning. Often, such symbols are monstrous combinations and they can be very scary. The combinations may emphasize qualities that come together in the new status. For instance, a statue of a man's body with a lion's head may be used to emphasize the union of brute physical strength and the power of the intellect in the grown man. This head and body combination represents an inversion, because lions are stronger than men and men smarter than lions—but the message lodges in this very topsy-turvy fact. Symbolism in rites of passage is often turned around, as fits the crazy, mixed-up, upside-down world of liminal, intra-structural time and space.

Also and accordingly, often in these ceremonies, death and life images and male and female ones are combined. Further, male or female images will have exaggerated features, such as impossibly huge genitals or dozens of breasts. While being shown the various symbols, for instance via ritual objects such as statuettes, or in costume decorations, the initiates generally hear stories explaining, explicitly or implicitly, what is and is not proper behavior for someone of the status that they are about to earn. The initiates' heightened arousal gives messages extended resonance and staying power.

Largely due to the intensity of rites of passage, people going through them together form a bond of community with one another, called **communitas**. One might be the son of the king and another the son of a sharecropper, but in a rite of passage, every individual is of equal status—because nobody has any status whatsoever. Initiates have been stripped of that (see Figure 11.10).

The loyalty of communitas can persist to old age. Some cultures formalize rite of passage ties, giving each group of initiates or 'age mates' a name and assigning these groups certain ritual responsibilities. Socially, these groups can help bind a village or community, providing ties that cross-cut and so knit together disparate lineages.

FIGURE 11.10 New recruits in basic training ('boot camp') rally themselves with motivational chants while waiting to navigate the obstacle course at Lackland Air Force Base, Texas. US Air Force Public Affairs.

None of what I have described is unique to non-Western cultures. In the United States, we have fraternity and sorority rushes, complete with sometimes quite dangerous hazing rites. Boot camp entails the elements of a rite of passage. Through humiliation, degradation, pain, rituals that call for endurance, and the sheer experiential intensity of basic training, civilians can be transformed into military women and men.

Reincorporation

At the end of the transitional or liminal phase, when the elders have deemed the initiates sufficiently educated and sufficiently transformed to take on the rights and obligations of the status that awaits them, they are reincorporated into society. The symbolic imagery used here is frankly that of rebirth. In other words, some of the same rituals undertaken when a new baby is born also are performed when initiates emerge to occupy their new status. In one instance,

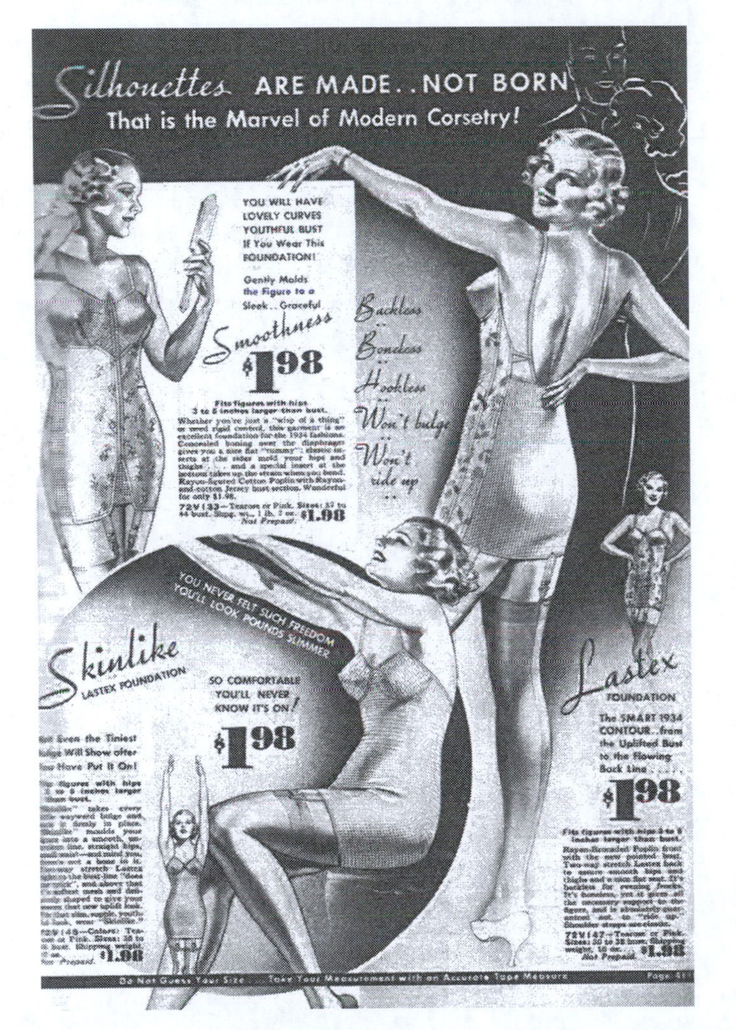

FIGURE 11.11 Corset advertisement from a 1934 catalog by National (a mail-order company). Courtesy of Katharine Zenke (anadaday.wordpress.com).

in a New Guinea group, freshly initiated young men being reincorporated actually crawl back out into their social world through a tunnel or birth canal created under the spread legs of their male elders. This is very much like the social birth that happens in Judaism with father-driven circumcision.

So men make men. Likewise, women make women in those female rites of passage that do exist. Long or short, individual or group-oriented, intense or fairly easy, men's and women's rites of passage work to create properly gendered beings. What is proper varies across cultures, of course.

Gender: A Key Site of Biocultural Interaction

Gender, far from being simply a natural fact, represents a major junction of biology and culture. Gender is created and put into play in diverse ways in different populations using bodies that are sculpted in particular ways through population-specific embodied cultural practices—sometimes quite knowingly so (see Figure 11.11). We will continue to examine the confluence of biology and culture in the next chapter, which looks more closely at how culture, itself a biological imperative for humankind, in turn affects our biological form and functio...

12

BODY IDEALS AND OUTCOMES

This chapter prepares you to:

- Recount examples of body decoration and modification cross-culturally, with an emphasis on their biological impact and an eye to the cultural contexts that support them
- Explore cross-cultural and historic differences in weight ideals in relation to food supply
- Examine weight-related health challenges in light of our evolved metabolic biology
- Identify and describe the functions of body modification and decoration cross-culturally, and discuss the main dimensions along which such practices vary

Ideals for the body, including its height, weight, and shape, and the ways in which it can be decorated, modified, and sculpted, have been around for a long, long time. Indeed, evidence of our interest in body symbolism dates back to culture's emergence 75,000 or so years ago. Among the artifacts found bearing witness to this transformative event are shell beads and pigments used for bodily adornment. Later, but not too much later, we begin to find tattoo kits, and evidence on human remains of tattooing.

We put culture into practice via our bodies. In part because we do this, our bodies can be treated as texts—as things to be read. When we see a person's body, we make assumptions about that person based not only on their apparent sex, gendered mode of dress, or habitual patterns of movement (see Chapter 11) but also their height and weight, insignias displayed, the way the hair is kept, and so on. We examined previously how social structure is written into or embedded in the body so that various populations' bodies differ categorically (see Section II and Chapter 10). Flawless skin and complete sets of teeth, on the one hand, or scars, missing fingers, and even internally inscribed indicators such as hypertension and diabetes or particular forms of bone and joint wear, on the other, can be read for insight into a group's social standing, or a whole society's priorities.

With knowledge of our culture's visual vocabulary, we may proactively inscribe ourselves or our loved ones. Body modifications such as circumcision undertaken in rites of passage are certainly one form of such self-inscription, but there are others. For instance, we may become bodybuilders, striving to alter our musculature, whether for aesthetic reasons, to amass strength, or to impress others with our dedication. We may undergo liposuction to conform more to our culture's weight ideals. We may paint our faces and color our hair or wear wigs as football fans to show the world which team we back. As fans, we want people to be able to read us—to know we support a specific team. Similarly, we may tattoo ourselves or simply label ourselves with a college sweatshirt so that anybody who walks past can read, on our very physical form, something about our group membership.

The embodied signaling we do also conveys information regarding whether or not we endorse the way our society is organized or the cultural values promoted there. For example, wearing a collegiate sweatshirt can convey that one values education, or that one is willing to play by the rules that one's society has regarding how to get ahead, for instance.

Invited to or not, people will gaze upon us. People will read whatever we offer visually. Meaning will be imputed in accord with established body ideals. These ideals vary from culture to culture, and within a given culture, they also vary from era to era. The biocultural context in which particular ideals are created and maintained as well as the varied cascade of biocultural effects they may have contribute to the range of human biocultural diversity.

Opening Examples

The Model Citizen

Take for example John Bull, an invented figure, who became a mascot for England in the 1800s, somewhat like the United States' Uncle Sam. He often was depicted out-of-doors, in the countryside, with a dog at his feet (see Figure 12.1). He was strong, but also fat, somewhat older, and decidedly middle class.

With these features, John Bull does not fit today's ideals. Nevertheless, in his time, his figure crystallized representationally all that was wonderful about the prototypically common English man, with his sensible outlook, emphasis on fairness and generosity, sporting nature, and willingness to take a principled stand. John Bull's inventor, John Arbuthnot, leveraged Bull's body, including its size and how it was dressed, to connote valued traits so that Bull's image stood in for the good social and cultural citizen of England. His body and, by extension, England's body, was fine indeed.

FIGURE 12.1 "Conversation across the Water"; John Bull and Napoleon, artist unknown, Vaisseau de Ligne, *Time Life,* 1979.

Today, of course, in England as in much of the West, stout bodies symbolize gluttony and greed as well as a lack of self-control or a willingness to sacrifice for the nation. Our model citizen today therefore has a different physique.

The Anti-Example

While the body can be used, as John Bull's was, to express pro-social and pro-cultural values, it also can be used to mark and so warn against nonconformity. Nonconformers often have, throughout history and around the world, been **stigmatized** or physically marked as targets for rejection. Their heads have been shaved, they have been tattooed, had their hands cut off, or been forced to wear clothing clearly labeled to let others know that they are not to be seen as upstanding or otherwise valued members of society. The scarlet letter, for instance, in the book of the same title by Nathaniel Hawthorne, was used to mark its bearer as adulterous at a time when such behavior, for a woman, was completely unacceptable. In the book, Hester Prynne had broken a very important rule, so her clothing—her social skin—was marked to show it and shame her, and to urge others who might do the same to get back in line.

Some stigmata are inherited. For a time in European history, a sunken-looking or small flat nose (for example, with a depressed bridge) was taken as a sign of personal and ancestral degeneracy, as nose deterioration problems were clinically associated with the sexually transmitted disease syphilis. Cosmetic nose surgeries focused on enlargement. When one saw a person with a small nose, one assumed that person was lecherous and corrupt; one wondered if the person's parents also were (a gap for a nose can result from congenital syphilis). Seeing someone with a small nose also served as a reminder to comport oneself in a way that would minimize chances of contracting the disease (Gilman 1999).

Preparing for a Deeper Dive

These examples give us some sense of the way that culture and biology come together in body ideals, with biology providing the raw material and culture the model for how to shape it. The shaping can, of course, have an impact, not only in terms of its superficial results but also its health outcomes, such as when eating too much red meat (which John Bull was wont to do) causes heart disease, or a commonly undertaken nose operation leads to nasal impairment, or a stigmatizing mark among members of one group is used by others to support that group's maltreatment. To better understand the biocultural significance of cross-population variation in body ideals, we will take several angles on the issue. After first focusing on overall body size and shape, we will then move on to consider how populations diversely adorn the body, whether through surface decoration or physical modification.

Body Build

As suggested above, each culture has a preferred body build. For reasons later reviewed, bigger has often been viewed as better, but not always across the board. For instance, among Azawagh Arabs, a semi-nomadic people who live in windswept western Niger (and who also have been known as Moors), men should be slender and women, fat. To this end, girls are urged and sometimes even forced to stuff themselves with fattening porridge and milk so that they can achieve great physical heft, thereby becoming women. This quickens the onset of menstruation and supports fertility. A swollen body is a demonstration of a proper feminine attitude (see Figure 12.2).

While men, in this culture, lead public lives, and are highly mobile, women are decidedly domestic and still, ideally. The body of a woman should be 'closed'; just as her stretched skin

FIGURE 12.2 "Mauresse [Moor woman], Dowiche" by P. D. Boilat, 1853 (from Esquisses Sénégalaises; physionomie du pays, peuplades, commerce, religions, passé et avenir, récits et légendes). New York Public Library Digital Library 1222721.

now contains fat, it too contains the blood and breast milk of kinship. It is crucial that women not diffuse this by sharing it outside of the community; female substance—and women's fealty—must be safeguarded. Big bottoms, stretch marks, and rolls of stomach fat therefore serve both to enhance female attractiveness and to show that a woman is doing her job of containment. It also helps women to do this job by grounding them, quite literally: the larger a person is, the harder it is to move about, and when a woman's role is as an anchor for the family this works out very well indeed, at least symbolically (Popenoe 2004).

There are other cultures that value fat female bodies too. In many of the West African societies to which African Americans trace much heritage, those who can afford to do so seclude their adolescent girls in special 'fattening rooms.' After a period of ritual education and heavy eating, the girls emerge fat, attractive, and nubile (see, for example, Simmons 1998). Richer families, of course, can afford to feed their girls well, and to do longer without their contribution to the household labor pool; thus, one's higher status is carried and displayed in one's belly, backside, and other big body parts.

The people of Fiji have traditionally liked everyone fat. Fatness signifies a wealth of social connections and financial resources and thus 'health' (Becker 1995). In fact, the social connections embodied in the fat individual are seen as literal, not symbolic; big bodies are co-created as people feed each other. The popular US idea that weight is a personal achievement is wrong for two reasons: first, social structural factors such as poverty and lack of neighborhood parks foster much overweight and obesity here; second (but no less importantly), weight is co-created by people in social relations.

Weight thus demonstrates the intertwined nature of our bodies. In this, it is not alone. We already have talked, in the context of the discussion in Chapter 6 of breastfeeding and co-sleeping, about various ways in which infants' and caretakers' (especially mothers') bodies are interdependent. Still other kinds of interdependencies and interconnections exist, as Fijian ideas about fatness suggest. Empathy is one (see Chapter 4). There also are interpersonal biochemical signaling systems, such as those involving others' scents, although a lack of space bars me from going into those here. I can, however, take time to note that, in some cultures, a conjugal or spousal couple becomes as one, physically speaking. For example, in some cultures men participate in pregnancy, experiencing morning sickness and other symptoms. The anthropological term for this is **couvade**. Just as fatness implicates more than one person, pregnancy, too, is not only a private bodily state but also a group one; in this case that group is the parents, together (and see Chapter 13).

Getting back to build, in Jamaican tradition, where a respected adult is called a "big man" or "big woman," good relations involve food sharing, and people on good terms with others are ideally large. I did my initial fieldwork in rural Jamaica where, when friends or participants in my research wanted to compliment me, they told me I'd put on weight (Sobo 1993).

Being large is not in and of itself a good thing, however. Jamaicans look and strive for 'good' fat—fat that is firm, not slack. The goal is a body that is plump with vitality. Slack fat signifies decline, such as when ripe fruit goes soft. The Jamaican preference is for fat that stands up, as do breasts plumped with breast milk. In this, Jamaican tradition, like most cultural traditions across time, favors bodies that measure in the United States as overweight—but not as obese. Although there are exceptions, such as with the Azawagh Arabs, extreme fatness (obesity) generally is considered worrisome all around the world.

The opposite of being too fat is being too slim. In Jamaican tradition, slenderness or weight loss signals social neglect. A Jamaican seeing someone grow thin might wonder about the sorts of life stresses that have caused the weight loss, rather than offering congratulations for it and attributing it to a 'good' diet, as many middle- and upper-class Americans do without reflection. US residents living in our inner cities, on the other hand, also worry about weight loss among friends or relatives. Thinness might be a symptom of disease, such as tuberculosis; it may indicate that a person is addicted to dangerous drugs; or it may simply signify poverty.

In her cross-cultural review on the topic, Claire Cassidy (1991) noted that bigness tends to ensure reproductive success and survival in times of scarcity. She also found that socially dominant individuals with sound relationships are usually large (relatively speaking). While in some cultures fatness is increasingly devalued due to the rise in diabetes and changes in diet and activity level that have meant individuals once merely plump are now obese, the long-standing equation between size and health has been harder to shake in areas where famine is still a likely hardship.

A Mismatch between Human Biology and Modern Life?

The recent Western increase in the pervasiveness of food (or things to eat and drink, anyway) represents a major environmental shift for human beings. For most of time, and in most places, food has been a sought after good not because of its deliciousness (although there always was some of that) but for sheer survival. Because food could be scarce, being able to store food energy as fat, and certain nutrients in fat, was selectively advantageous. In other words, our bodies adapted, genetically, to store at least some sustenance for lean times.

However, with industrialization and its correlate, increasing urbanization, came the "Western diet," which is high in fat, sugar, and processed or refined foods as well as calories, and low in polyunsaturated fatty acids and fiber. The changes in various populations' nutritional status after agriculture emerged were large; but the changes following industrialization and the related ease of access this gives us to (industrial) food were enough to usher a shift so big that it's been termed the **nutrition transition**.

The nutrition transition is not necessarily voluntary: for instance, it can follow migration, or it can be forced upon a population as a result of colonization or related environmental injustices that make living off the land too risky. This happened to some degree to the Akwesasne Mohawk community living downwind and downstream of factories disgorging toxicants (see Chapter 9), although they are making a concerted effort to resist it. This resistance comes in part because a Western diet is not physically beneficial (see below) but also because keeping food traditions alive is essential to cultural survival more broadly, including preservation of a group's cultural relationship with the land (Schell 2012).

Beyond not being wholly voluntary, the transition often is not faithful or exact: the ideal Western diet remains out of reach for many. That is, while colonialism and related changes in the global economy affect local food availability and price by diverting resources (land, crops, fisheries) into lucrative foreign markets, and by influencing local tastes, such changes do not necessarily afford a complete transition. In the Pacific Islands, for instance, traditional foods have been displaced by cheap imports, such as rice and mutton flaps (sheep bellies), which have become mainstays. Mutton flaps are an offcut or byproduct of sheep and lamb butchery. Rich (white) Australians and New Zealanders buy the preferred cuts, while these fatty remains are sent to Papua New Guinea, Samoa, Tonga, and the like, to be sold cheaply. Turkey tails are another modern food product exported to the Pacific Islands, this time by US corporations—when not sold for use as pet food (Snowdon 2013).

As a result of how the nutrition transition has played out in the Pacific Islands, average weights have gone up in keeping with dependence on low quality imported food; In Fiji, for instance, the number of overweight people doubled between 1993 and 2004—tripling among those under 18 years of age. Obesity among Pacific Islanders is not a fact of nature (Snowdon 2013).

While peoples affected by the nutrition transition are sometimes less prone to famine or hunger, in return they are more prone to the kinds of degenerative diseases linked to poor diet and lack of physical activity (e.g., diabetes, cardiovascular disease, stroke). The transition places our bodies in a compromised position: we not only inappropriately but also automatically leverage our evolved capacity for fat storage, packing on the pounds; our bodies also, again in relation to our evolutionary heritage, increase their metabolic capacity to do so. We seem in fact better able now than ever to take advantage of the food that is around us—in terms of stored fat.

Calories In, Calories Out

The upward trend in weights worldwide is linked to the nutrition transition. For one thing, even the poor in stable nations not faced with war and its fallout can afford masses of calories on a daily basis. Indeed, the cheaper the food, the more calories it usually contains; although one must often look past nutritional content. In other words, the poor (and others) are fed plenty, but they are not well-fed. The quality of the industrialized, mass-produced diet is vastly different, and deficient, when compared to the quality found in, say, the typical forager diet (see Chapter 6) or even an agricultural but pre-industrial one.

In addition, calories in are not balanced with calories out (see Supplement Box 12). Activity levels have dropped dramatically in the fatter nations. Whereas once people here undertook hard physical labor on a daily basis not only to get food but also to prepare it and to keep their homes in order, labor-saving devices abound now. Gone, for most readers, are the days when we churned butter by hand, beat the rugs to clean them, or walked to the river and back carrying heavy jugs of water, burning calories all day. Nor is paid labor physically laborious; many employees spend the day sitting—too much of which is, as it turns out, hazardous to our health.

SUPPLEMENT BOX 12: DIET OR DNA? THE NUTRITION TRANSITION IN ACTION

Although it may be nice to blame our genes for all that ails us, genes play a smaller role in rotundness than some like to think. This is clearly demonstrated when we compare the body mass indexes or BMIs (a measure of body fat based on height and weight) of the northern and southern Pima peoples. The Pima are a Native American cultural group that includes the Tohono O'odham described in Chapter 9. Historically, the northern and southern branches were part of the same population. With the birth of the United States as a political entity, those who lived in Arizona gained rights and obligations relative to the US government, while southern members came under the rule of Mexico.

The present diabetes and obesity problems of the Tohono O'odham stem from the unfair diversion of their river and the concomitant loss of their traditional way of life. Without their traditional farming and related foodways and physical activity patterns in play, the Tohono O'odham went from a fit population with no incidence of diabetes to one of the most diabetes-prone groups in the United States (California Newsreel 2008a). Their genome did not change; their lifestyle did. They began eating the lard, sugar, and white flour provided by the federal government as a replacement for farmed food. They began using less energy because they were not farming. The infamous Southwestern staple, fry bread, which is just that, is a legacy of this arrangement—not 'native tradition.' Concurrently, the Tohono O'odham experienced the chronic stress of displacement and marginalization (regarding some physical effects of which, see Chapter 10).

In contrast, the southern people, who live in the Sierra Madre Mountains in Mexico, have much lower rates of obesity, overweight, and diabetes. In one study, three of thirty-five had diabetes, in contrast to half (one in two) of their US cousins. A key reason for this is that, south of the US border, traditional subsistence practices remain in place. There was no nutrition transition. The southern group's lifestyle is physically rigorous, and their diet retains healthful ingredients (National Institute of Diabetes and Digestive and Kidney Diseases 2002, 5; see also Thomas 1998).

Nobody doubts that the Tohono O'odham who do not lead active lifestyles, and so forth, are vulnerable to what some have termed 'diabesity'; as human beings, of course they are. However, whether they are more genetically vulnerable than members of other populations remains a questionable point. Many populations have been subject to famine but have not gotten nearly as fat or sick. For example, in Ireland, there was the Potato Famine starting in 1845; Russia experienced famine in 1921; many nations embroiled in the world wars ran short of food. It seems likely, then, that (a genetic component notwithstanding) the present glut of fatty, sugary, calorie-laden foods in certain environments is the lever that sets the bodily fat storage machine in motion, given certain other conditions such as poverty, unemployment, and chronic stress such as in relation to minority social status.

Metabolic Issues

Food-rich environments—particularly those with industrially produced food—are mismatched to our evolved digestive and metabolic physiology and biochemistry. Many of the ingredients now contained in food are created through synthetic industrial processes, and our bodies often do not have metabolic pathways in place to deal with them. Some use or alter the course of pathways meant for real food components in ways that do not always turn out well.

When we expect to or are ready to eat, we start to salivate, and our metabolism or the rate at which we break down and use or assimilate nutriments picks up, at least under normal conditions. By normal here I mean when the food to be ingested is actual food and not an industrially created substitute. With those, anticipatory responses may differ.

Sugar substitutes, for instance, or foods containing them, may be anticipated as low in calories, which they indeed are, but the foods that often accompany them are not (consider the large fries and hamburger that your friend orders with a diet soda). Anticipating a low-calorie food meal based on its recognition of the sugar substitute in the soda, the body may not rev metabolic processes up to the standard food-digestion rate; a lower metabolic pace in times of shortage helps the body to stretch or make the most of a minimal intake. This is why people on low-calorie diets do, after a time, have lower metabolic rates. The body lowers its metabolic set points to try to regain homeostasis (this is the same kind of allostatic process described in Chapter 10, whereby a change meant to better regulate the body in the short run turns out to dysregulate it in the long run by throwing off other processes).

The problem with having a lowered metabolism when food actually is plentiful, and liberally eaten, is that with a lower metabolism more of the calories taken in are stored than usual (in contrast, the higher the metabolic rate, the more calories are burned). At least this is how things work in rats, whose metabolic processes in relation to sugar substitutes have been extensively studied. The rats also eat more overall when fed a diet laced with sugar substitutes. Scientists do not know exactly why overeating happens, but they suggest that the rats' bodies, anticipating calories from sugar and then not receiving them, send increased appetite signals, pushing the rats to seek out and take in those calories that seemed 'missing.' If these patterns hold true in humans, it may well be that drinking diet soda, for example, makes people fatter than drinking the real thing. Indeed, one study exploring the relationship between diet soda drinking and obesity found that for every can of diet soda consumed per day, the risk of being overweight increased 41 percent (see Park 2008).

It also may be that changes in prenatal exposures have triggered a rise in obesity in some populations via epigenetic mechanisms (see Chapter 3). There is some indication that exposure to certain toxins exuded by particular kinds of plastics predispose a fat-deposit regulating gene to remain activated or to over-express itself, at least in mice, who thereby can become very fat indeed (the toxin is BPA or bisphenol A; the gene, *agouti*; see Genetic Science Learning Center 2013 and Chapter 3).

Babies born to human mothers whose calories were restricted during pregnancy, whether due to dieting or famine, seem more likely to store fat more efficiently (that is, put on weight easily) as they grow, and they may be more prone to insulin resistance (this leaves more sugar in the blood, which is good if an individual is not receiving much sugar, but dysfunctional in our present, sugar-rich dietary environment, and may end in diabetes). In effect, the calorie-deprived environment that the fetus experiences in the womb of a mother whose diet is restricted is assumed by the fetus to forecast or be a predictor for the environment in which it will live as it grows up: if calories are restricted during gestation, the fetus is primed for a life of scarcity. This can have a negative impact if the actual environment is one in which food is abundant (Bateson et al. 2004). Cultural beliefs about how much to eat when pregnant, and structural conditions limiting access to food or the reverse, seem to play a role in underwriting a proportion of existing diversity in offspring body size from population to population.

Even when prenatal and postnatal environments match, and even when one eats food to which the digestive system is well-fitted, some aspects of appetite still are humanly created. I refer here not only to the foods on hand due to whichever cultural adaptations have been implemented and wherever in the world one may live, but also to flavor preferences determined even before birth (I mentioned in Chapter 6 how breast milk as well as foods eaten during pregnancy can lead to certain flavor preferences in infants and children). What we eat is bioculturally determined, and variable. When we eat, too, is a biocultural issue: we get hungry around the time that we normally eat due in part to habituation. We also can habituate to the hormone leptin which signals, in effect, fullness, instructing us to stop eating; the vast majority of obese people's bodies give this off, but because they have eaten through or over its signal so often they no longer 'hear' it (Kluger 2007).

Fat Phobia

Even in the United States, plumpness was long considered healthful and alluring. In the early twentieth century, the diet industry focused on helping people gain weight, not lose it. Those who were slender were caricatured as weak and unattractive. When and why did things change?

Journalists Kate Dailey and Abby Ellin (2009) trace the rise in fat hatred to World War I. At that time, some Americans believed that a fat population would not be able to compete. Moreover, food was in short supply due to the war; fat people were thought to either be hoarding, or to be taking more than their fair share of it. While in some societies where food is scarce fatness is simply taken as a sign of wealth and power, and thereby valued, this could not happen in the context of the war because it was contradicted by the cultural message that everyone would have to make sacrifices.

In short, then, emergent ideas about fat were related to ideas about self-control in the face of personal greed or weaknesses of the will. These ideas stuck around when the United States bounced back as a land of bounty. Other developments reinforced this view, such as the rise of sedentary desk jobs as well as labor-saving devices. The advent of ready-to-wear clothing, with its numerical sizes, provided people with a quick, systematic way to make size comparisons, too (Dailey and Ellin 2009). The easier it got to be fat, the more we began to favor thinness. Plumpness, which most cultures around the world and through time had sought after due to what it meant in the face of potential food scarcity, was turned by American prosperity into a fool's prize.

As fat stigmatization has risen in America, people whose bodies are plumper than the cultural ideal are caught in a health bind. On one hand, their risks for certain disease states may be heightened by their overweight status; on the other, if they have internalized the ideals of their culture (as good cultural citizens are wont to do), they can feel too ashamed of their bodies to get regular medical care, meaning that any potential problems may not be identified early enough for successful treatment.

Then again, while it is true that obesity carries certain risks, it also is the case that the ideal weights used by insurance companies to indicate health risks change periodically, and may be misestimated. Moreover, as we are coming to know, some people simply are heavier than—and just as healthy as—those whose weights meet the cultural ideal.

Body Decoration

Beyond size and associated body shape, ideals for which can vary cross-culturally, cultures have ideals for how and when bodies are cleaned, dressed, and decorated. They also have ideas about which body parts must be manipulated, such as through circumcision, piercing, adorning, covering up, and so on. Examples are provided below as we discuss first some of the dimensions along which body decoration and modification vary, and then the different functions they serve.

Dimensions of Variation

Body decoration and modification may vary along a number of not mutually exclusive dimensions. For instance, they may be voluntary or imposed (stigma); affiliative (normative) or disaffiliative (deviant); and flexible (temporary) or fixed (permanent).

Voluntary or Not

To begin, body decoration or modification can be voluntary or imposed. For example, while many people in the United States today elect to tattoo themselves, in other times and places tattoos were reserved to mark outcasts, who were forced to submit to them.

It is well known, for example, that the Nazis tattooed their prisoners (see Figure 12.3). They stigmatized many groups of people this way, not so much for breaking behavioral rules (although there was that), but for being of a certain heritage or biological lineage (the Nazis subscribed,

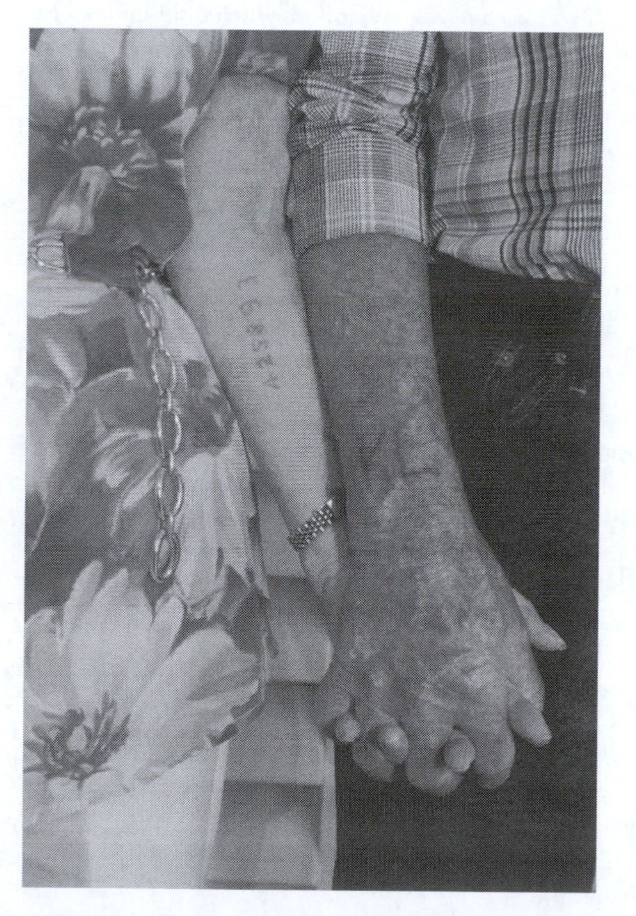

FIGURE 12.3 Rose and Max Schindler's concentration camp tattoos (2017), from *Two Who Survived, Keeping Hope Alive While Surviving the Holocaust* by Rose Schindler, Max Schindler, and M. Lee Connolly, 2019. The numeral tattoo was received in an Auschwitz sub-camp in occupied Czechoslovakia in October 1944. The alphabetical tattoo was received in Mielec, Poland in February 1943 (KL is an abbreviation for *konzentrationslager*, or concentration camp). Used with permission from MRS Publishing.

mistakenly of course, to an extreme and reductionist form of biological determinism). A catalog of stigmata was used by the Nazis during the Holocaust to mark individuals as specific kinds of outsiders to the so-called Aryan race. Conversely, the blonde, blue-eyed, Christian, heterosexual body, if robust and with all the expected parts, conformed to and supported Nazi social norms and expectations. Accordingly, people who would affiliate with the Nazi community voluntarily strove for such a look through, for instance, self-elected participation in physical fitness training initiatives.

Affiliative or Not

Another dimension of variation in body decoration or modification is whether or not it is affiliative or disaffiliative. If **affiliative**, it will symbolize one's affiliation or identification with a particular culture or subculture (see Figure 12.4). The military uniform worn properly, for instance, tells others that its wearers are team players, willing to conform, and interested in supporting their country.

A uniform—or other forms of body decoration or modification for that matter—may be 'disaffiliative,' too, signaling or placing distance between an individual and a particular group. A disaffiliative decoration or modification marks an individual as deviant, or as undesirous of affiliation with or of signaling membership with the norm or with the mainstream. The punk hairdo carried this meaning in the late 1970s. Tattoos, until recently, did too.

Here I should note that affiliative body modifications and decorations often entail an attempt to 'humanize' the body so that, in addition to affiliating with a particular group, we are, in conforming, affiliating with or demonstrating our very humanity. Conversely, to leave one's body in a 'natural' state, with hair uncut and unwashed, for instance, or to remain undressed is in effect disaffiliative. Many cultural myths regarding outsiders depict them in such a fashion, as wild women or men. Accordingly, in joining a group (or rejoining one as, for example, in the case of a man or woman who has been away at war), a person often will undergo some modification practices, such as hair cutting. In some contexts people must wash their faces and hands, and sometimes their feet as well. In mainstream US culture, people who have been away from

FIGURE 12.4 Youths show off their tattoos to photographer (Kenneth Paik), stating that everybody has them. Mulky Square, Kansas City, 1973. Environmental Protection Agency image 185/12/011052.

home sometimes become very preoccupied with a felt need to shower and shave, so as to 'feel human again.' We manipulate our bodies in ways meant not only to affiliate with our group or with humans, but also to disaffiliate with animals or outsiders.

We also do it for gender-affiliative purposes; we modify and decorate ourselves to feminize or masculinize our bodies. Mainstream US women, for example, shave their legs and armpits. This shaving is not a hygiene practice, although we will sometimes rationalize it as such. While other societies do not think about and sometimes even favor hair in those areas, such shaving is, in Western society, a means by which to differentiate men and women. Some theorists would go so far as to say that, in shaving her body, a woman seeks to turn herself back into a child (children do not have body hair), symbolically reinforcing the patriarchal social order through her doubled removal from power. Whether or not that is so, 'feminine' shaving definitely places a claim on a gender affiliation while disaffiliating a person from the masculine.

Here, let us bring in the Beatles, who ruled rock and roll in the 1960s (at a minimum, some would say). Even in their early years the Beatles made a certain statement with their chosen mode of body decoration and modification. That they did this was not wholly up to them; we all are available to other people freely as texts to be read, or read into. People do look; and people do talk.

The Beatles' hair was one of the main things people looked at and talked about (see Figure 12.5). Even at first, it was a little bit longer than the norm. Some people did see this as somewhat radical, yet because the Beatles performed in suits and with shaved faces most people could live with and indeed looked past it.

In contrast, any picture of the Beatles in their later years shows a real difference in grooming, hair length, and dress. Here, there is no question that their body decoration and modification choices reflected disaffiliation from the mainstream, and the so-called older generation. Disaffiliation was embodied in the ways that these musicians carried themselves, groomed themselves, dressed themselves, and marked themselves physically (see Figure 12.6).

FIGURE 12.5 The Beatles as they arrive in New York City in 1964. United Press International.

FIGURE 12.6 George Harrison in the Oval Office during the Ford administration, December 13, 1974.

Disaffiliation went beyond the body surface, too; the type of labor they did with their bodies, as musicians, and, moreover, their physical lifestyles in no way conformed to the type of labor or lifestyle that would have been considered respectable for young men in those days. Some in the mainstream were actually frightened by the Beatles' disaffiliative use of their bodies. They felt the same way, for similar reasons, about musicians with dreadlocks, or who wore leather, or who appeared as punks. Yet disaffiliation and rejection of the mainstream was not the only message that such musicians sent with their bodies. They also made an affiliative statement in regard to others like them; the Beatles, for instance, wished to be part of the pro-peace faction in the 1960s revolution.

That the Beatles' hair did change brings us to a third area of variation: the fixity or flexibility of a body decoration or modification form. Hair can be cut, dyed, braided, beaded; and it grows—changes to hair are generally flexible or impermanent. Flexible decorations and modifications are those we can take on and off much like judges or barristers in Commonwealth nations put on white wigs before performing judge work in court. Once a judge takes off that wig—and the robes and so on—other roles come to the forefront.

Permanent or Not

Importantly, the wig and robes, like other work uniforms, support the embodiment and enactment of an entire identity that is nevertheless partial or flexible. What we wear—our social skin, so to speak—does have a very real impact on how we comport ourselves. When people put on police officer uniforms, they feel like police officers and therefore act like police officers; that's as it should be.

Tattoos, in contrast, are generally fixed or permanent markers—particularly in times and places without access to removal services. Perhaps because those now exist, readers may find a better example of a fixed or permanent body modification in Sumatran teeth filing. In this tradition, people file their teeth to points as a way to create and maintain themselves as human. Other examples of fixed decorations or modifications include the use of metal neck rings or coils (which actually do not so much stretch the neck as push down the clavicles or collar bones) among the Kayan Lahwi, now of Burma; lip-stretching plates or plugs worn among, for

FIGURE 12.7 Māori woman with facial tattoo and baby. From *Mr. Oseba's Last Discovery* by Geo. W. Bell (1904), p.161.

example, Mursi and Surma or Suri women in Ethiopia and Suyá men in Brazil; ear lobe plugs; and scar patterns carved into the flesh. In contrast, face, body, and nail paint provides a temporary form of decoration, as does shaving your head.

Returning to tattoos, which have, after all, a very long history as a fixed form of body art, let us consider the Māori situation. These tattoos stand in contrast to decorations and modifications that are flexible or partial—that can be taken off or put on as one likes; or as is appropriate according to what one does for a living; or in keeping with cultural expectations as, say, for certain ritual events. Among the Māori, who are native to New Zealand, high-status adults traditionally are marked as high status with facial tattoos (see Figure 12.7). These are made not with percussive use of a needle but rather a chisel-like instrument. In any case, the extensivity and shapes of people's facial tattoos can be read as information regarding their ranks as well as their lineages. This lets others know how to behave in their presence—and what kind of behavior to expect of them. The Māori also have a tradition of extensive body tattoos and these too form part of the tattoo text. Although aesthetics are very important, traditionally, designs are chosen not for individual reasons but rather for what they connote regarding a person's social status, kinship ties, and accomplishments. They correspond, then, to things of social and not necessarily personal importance.

In some populations, rather than to tattoo such information onto the body, scarification techniques are used. In scarification, scars are made on the body purposefully, generally in geometric patterns. Small cuts are made over and over to generate particular designs, which cover broad swaths of the body, such as the belly, back, or forehead. The cuts are made with a clean and very sharp tool, ideally. They often are washed and covered initially with some kind of medicinal ointment or poultice to promote good healing (all cultures have some form of pre- and post-treatment process in place to insure against infection; a poorly healed scarification job is not only a health hazard but also a hindrance to the communications the scars are meant to send, so it must be protected against). Some Australian Aboriginal people have practiced scarification; many African groups do so. Scarification seems to work better on darker skin than tattoos for communicating vital information such as what tribe one belongs to, how many children one has had, whether one is married, or of high status or a chief. It also does something that tattoos cannot generally do: the raised lines or shapes made in the scarring process can be felt. There is a decidedly tactile aspect to this form of body modification.

The contrast between whether or not a given group is prone to fixed or flexible forms of body decoration or modification might seem, on the surface, like a contrast between modern or contemporary peoples and traditional peoples. It is true that they seem to be distributed that way, with more flexibility among the former and more fixed forms among the latter. Yet it would be much more accurate to say that the contrast is between closed societies on the one hand and open societies—or really, open class societies—on the other. **Closed societies** are those in which a person's social status is ascribed; often it is based on birth or anchored in a rite of passage from which there is no going back. **Open societies** are those with inbuilt class mobility (or at least the myth of such). Closed societies favor or have more forms of fixed body decoration and modification, which makes sense because little that is expressed in them will change; open societies have more flexible forms. This reflects the plasticity of one's social status in an open society, and the fact that one may go in and out of many social roles over even a short course of time there.

Functions of Body Decoration and Modification

In addition to the three dimensions of variation for body decoration and modification (voluntary–imposed, affiliative–disaffiliative, fixed–flexible), we can consider its functions. These are to communicate, protect, and transform.

Communicate

As will be clear by now, markings or decorations on or modifications to a person's body—indeed its very shape—can communicate to other people who that person is and what that person stands for. One thing the marks might communicate is that a person has adopted the correct attitude and supports the cultural values of the group. Assuming the same aesthetic or beauty ideals as the group communicates this message of affiliation.

However, more than simply being affiliative, modifications such as scarification, or Māori tattoos, and decorations such as medals or maybe a crown, can also communicate to other people important facts about, for instance, social position (who but the king or queen wears a crown?). Such symbols even can communicate information about achievements. If a person wears medals, for example, from fighting or otherwise in World War II, the person's peers and compatriots will know that much more about the person without ever having verbally been told. Some tattoos likewise mark achievements. All these are examples of the communicative function of body decoration and modification.

Protect

The protective function leverages the meaning response (described in Chapter 10) to provide a sense of safety. A symbol, embedded or emblazoned upon your body, evokes a meaning response in you. If you know that a particular shape or symbol is meant to protect, you will feel protected and go about your activities without being nervous and therefore not making nerves-related mistakes. This is one reason for wearing amulets or even particular perfumes or tattoos.

Clinton Sanders (1988) tells of a man who wanted a big, brutal bee tattoo; he thought it might scare away real bees, to which he was deathly allergic. If he encountered real bees when under the protection of a mean bee tattoo, he would feel safe and calm rather than experience an intense stress response. If he lived where there were many bees, this would surely cut back on his risk for chronic stress-related illness. The protective function of body decoration or modification can be directly physiological.

Transform

A third function of body decoration and modification is to transform a person from one status to another. We did discuss the transformative function already, albeit indirectly, when discussing rites of passage (Chapter 11). Undertaking a permanent form of body modification can be a very important **performative** part of a rite of passage, so it not only serves later to communicate that the person has been initiated into, say, manhood or womanhood, but actually helps to create this transformation. For instance, the pain from a particular modification procedure or simply the anticipation of it may create in the initiate an altered state or a sense of intensity, making the rite more meaningful and the lessons imparted more memorable. Further, through enduring it one may prove to others and oneself that one is worthy of the new status. The transformative function of body decoration or modification occurs in the liminal period of the rite.

Keeping a Relativistic Perspective

Sometimes, when outsiders consider the pain and so forth of a certain body modification process—particularly one not generally done in their culture and so not seen as crucial, or one done with children—they reject it out of hand. This begs the question of its culture-specific utility, and necessity. The importance of actually undergoing a body-modifying rite of passage is paramount for people if it is normative in their society. Remember, people without the proper markings have not identified themselves as members of the group. To fail to conform in that way would be to cut oneself off from the possibility of, for example, getting married and achieving other life goals held dear by one's culture. The stress generated this way would be high indeed to an insider with buy-in.

Moreover, if outsiders turn the mirror on themselves, they may see that their culture, too, entails practices that can appear brutal. Take, for instance, orthodontia in the United States; this long-term dental modification technique can be quite painful, physically and socially. While sometimes done for clinically legitimate reasons, it generally has cosmetic goals. Outsiders might see orthodontia as both vain and, when imposed on children, barbaric. They may be especially primed to see it as abusive in light of their knowledge of other mainstream US bodily practices imposed upon children. For instance, forced early weaning and forced solo sleeping for young infants are seen not only as sadistic and inhumane but also as incomprehensible in some cultural worlds: how could any parent be so cruel to their own child?

Appearances to others aside, we, too, are expected to present to the world a face of stoic forbearance during certain painful activities; we, too, have practices where enduring the pain is part of a transformational process. Take, for instance, US practices surrounding tattoos, face lifts, high-heel shoes that pinch, pectoral implants, vasectomies, wax-based hair removal, and so on. Even when undertaking fitness-focused exercise we often say "no pain, no gain." We have hopes and wishes for ourselves and our children just as do cultures that use body modification techniques in formalized and full-blown rites of passage. In sum, although there is notable cultural variation in how diverse populations will practice body decoration and modification and body shaping activities, and there are concomitant variations in bodily outcomes (including for health), we all do practice them. We do so with what are, ultimately, very similar and very human goals.

13

KINSHIP

So Relative

This chapter prepares you to:

- Describe the three most basic kinds of kinship, paying particular attention to cross-culturally variant dimensions of consubstantiality
- Explore diverse understandings of when, how, and if human life begins and ends; explain the consequences that various understandings of this have for biocultural practices such as abortion
- Explain how cultural ideas about kinship underwrite biocultural diversity in family-building strategies and so in reproductive and demographic outcomes

From a strictly biological standpoint, kinship regulates behavior in ways that promote the survival of genetic material equivalent to one's own, driving us to favor helping close kin over strangers. The physical mechanisms for recognizing kin could work somewhat like those in the immune system, whereby the body recognizes proteins that are self-generated and treats these differently from those that are not.

That said, a growing body of research suggests that culture is what triggers any biologically given predisposition to regulate our social behavior based on kin relations. Because culture has such a powerful influence over who we relate to and how, it shapes our kin groups, in time and space. It affects who we include as relatives, whether living, dead, or in between. It also affects our family building strategies. The demographic, relational, and next-generation outcomes that kinship entails are truly biocultural.

Kin Terms and Lineages

Kin terms make a good place to start our attempt to understand kinship. Kin terms name various kinds of kin. Although one might think that the explanation stops there, different cultures have different kin terms—and so different kin roles. For example, some cultures do not distinguish at all between siblings by gender; we do: we have 'brothers' and 'sisters.' Other cultures include information about relative age in their terms for siblings, and their kin terms include 'older sister' as well as a 'younger sister'—and 'older' and 'younger' brothers, too. Notice that I must resort to compound words in English to make that clear; in other languages, specific terms do that work; for instance, in Japanese one's own younger sister is *imouto* while one's older sister is *ane*.

Other cultures also may be more specific than our own when it comes to cousins. For example, if a culture prefers that cross-cousins marry, there will be words to distinguish them from parallel cousins (cross cousins are born to a parent's opposite-sexed sibling, such as one's mother's brother; parallel cousins are born to a parent's same-sexed sibling, such as one's mother's sister).

Although kin terms, and so kin roles, vary across cultures, what kin terms accomplish does not. What they do is to map out the rights and obligations that people have toward one another as kin. That explains why, in some contexts, it is crucial to know who your parallel or cross cousins are (you might have to marry one of them). Further, the roles specified for given labels (for example, father, mother, daughter, son) vary in different contexts. In the early days of the Israeli Kibbutz movement, for instance, the role of mother or father did not include an obligation for childrearing; children were raised communally by kibbutz members who specialized in the childcare role (see Figure 13.1).

We often use kin terms to call out the rights and obligations they index. For example, in mainstream American culture, when a parent sees his or her offspring leaving the house dressed in a way that contradicts parental values, he or she may say "No child of mine is leaving the house dressed like a [fill in the blank]." In doing so, the parent is calling on the child's kin-based obligation to honor or at least not to shame the parent and the family as well. Likewise, when a parent needs help with the kids, and so calls out to a partner "Mother [or: Father], could you please come put your children to bed [or: give them a bath, or discipline them]," the verbalizing parent is seeking to mobilize in another the kinship obligations that go along with co-parenting. Kin terms are used to remind people of their responsibilities as members of the family, as well as to remind them of the right others have to expect certain kinds of behavior from them.

Kin terms also map relatedness. When we hear somebody say 'uncle' or 'father,' we know how they are related—at least if we know their culture. If someone in mainstream US culture is related to a person as his or her father, he will have a very different kind of relationship to that

FIGURE 13.1 Communal child rearing was common in the Israeli Kibbutz movement's earlier years. Kibbutz Einat Archive via the PikiWiki—Israel free image collection project.

person than he would if he was that person's uncle. However, relatedness is culture-dependent. While in most Western cultures the father and son are considered more closely related, in some cultures it is the nephew and uncle who are the closer kin. This has to do with what kind of **lineage** a given culture prioritizes as 'real' or binding. In other words, it has to do with how they track heredity.

Anthropologists have identified many versions of lineality, or many ways to figure the family line. Cross-culturally, we have **matrilineages**, in which heredity is figured through the female line. We have **patrilineages**, in which heredity comes down through the male line. We also have bilateral lineages, in which heredity is tracked through both women and men (this is how many people in the United States figure heredity). There are other variations as well.

In most configurations those who have died can be included; much less frequently will those not yet born be thought of. One exception would be in a culture that practices transmigration of souls or reincarnation and whose population is stable. Yet, in most known cultures, family circles (or trees, or lines) only track backward, to ancestors—whose bodies may be gone but whose spirits can still be very much alive, either individually or as part of a collective spiritual mass.

In a patrilineage, children represent an extension of the father's lineage, and they generally take his name to show this. In a matrilineage the converse is true: children in matrilineal societies are part of their mothers' kin lines. Thus, while a father in a matrilineal society (let's call him Jack) may love his children dearly, and care for them as obligated to through the bonds of marriage and fatherhood, his real interest—kinwise—is in his sister's offspring. Jack's sister's children are the next generation of heirs for his lineage, and so he has great stakes in doing well by them. The relationship he has with them is termed an **avunculate** (the term uncle is embedded in there, making this an easy term to remember).

The avunculate is a wonderful invention. While a mother who gives birth to her children is concretely assured that she had some part in their creation, a father does not have such concrete proof of paternity. Short of inventing a way to claim it, as through naming and familial authority conventions such as we find in patriarchal societies today, a man may be better off focusing his energy on children he *knows* are in his lineage: his sister's offspring.

The avunculate is a very important relationship in many matrilineal cultures. When Jack's nephew marries, for instance, the boy and his new wife may assume avunculocal residence (they will come and live with Uncle Jack). This living arrangement contrasts with matrilocal residence (when the newlywed couple is meant to reside with the wife's kin), patrilocal residence (when they live with the husband's father), and neolocal residence (when they form a new household entirely).

Who Is Related, and How?

Kinship is about connectedness—embodied connectedness. It has a cultural aspect, sure; but it also is biological; furthermore, kinship culture affects biological outcomes.

In the past, anthropologists mapped out lineages using disciplinary conventions such as circles for women, triangles for men, equal signs for marriage, vertical lines for parent–child relations, and horizontal lines for sibling ties (see Figure 13.2). Those conventions are c.150 years old, and they reflect the family structure idealized by the scholars who invented them. They are therefore well-suited to depicting a nuclear family, with one mother, one father, and some descendants.

That said, they do not work very well when applied cross-culturally. It is hard, for instance, to demonstrate the avunculate with them (see Figure 13.3). They do not even work very well to

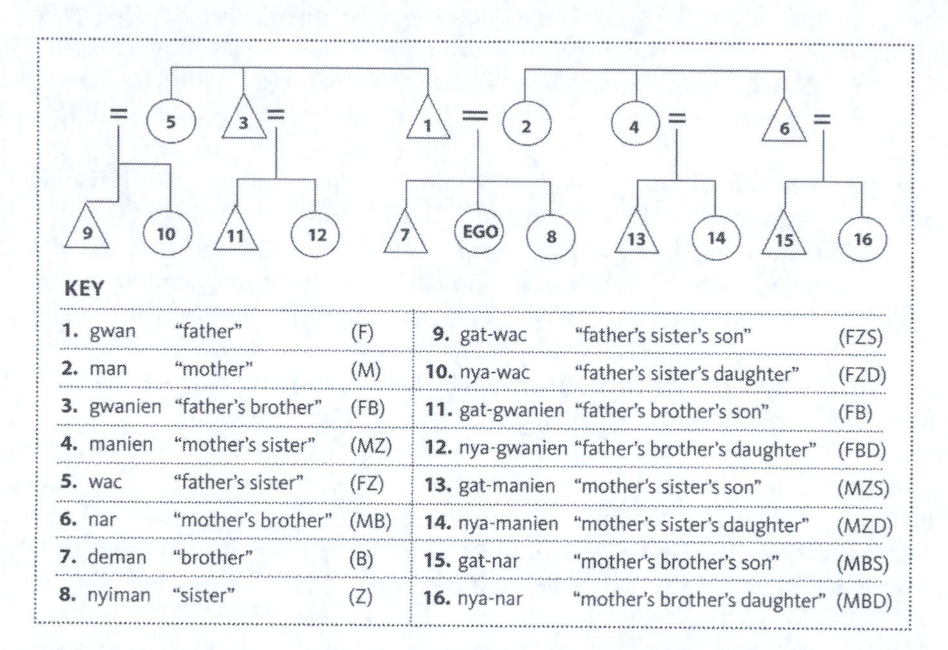

5
3
1
2
4
6
9
10
11
12
7
EGO
8
13
14
15
16
KEY
1. gwan "father" (F)
2. man "mother" (M)
3. gwanien "father's brother" (FB)
4. manien "mother's sister" (MZ)
5. wac "father's sister" (FZ)
6. nar "mother's brother" (MB)
7. deman "brother" (B)
8. nyiman "sister" (Z)
9. gat-wac "father's sister's son" (FZS)
10. nya-wac "father's sister's daughter" (FZD)
11. gat-gwanien "father's brother's son" (FB)
12. nya-gwanien "father's brother's daughter" (FBD)
13. gat-manien "mother's sister's son" (MZS)
14. nya-manien "mother's sister's daughter" (MZD)
15. gat-nar "mother's brother's son" (MBS)
16. nya-nar "mother's brother's daughter" (MBD)

FIGURE 13.2 Conventional kinship diagram for two generations of a Nuer (Sudanese) extended family; Nuer kin terms also are depicted. Courtesy of William Balée, *Inside Cultures*, 2012.

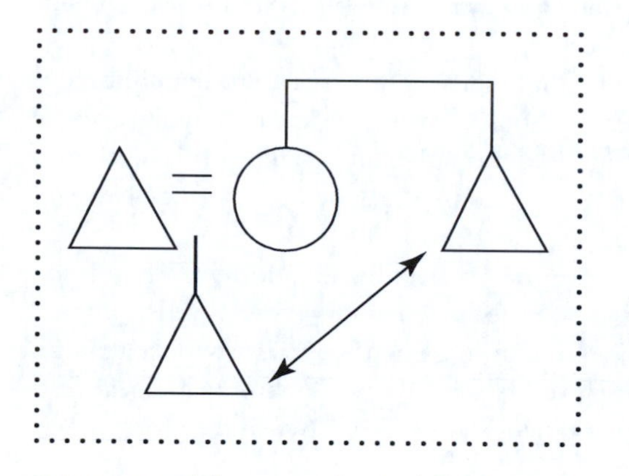

FIGURE 13.3 The avunculate is difficult to map using conventional kinship diagram symbols. Courtesy of William Balée, *Inside Cultures*, 2012.

describe the reality as it was on the ground 150 years ago, let alone today. One must add arrows, connecting lines, comments, and so forth to depict strong extra-nuclear relationships as in the avunculate. There is no inbuilt way to distinguish siblings by age, although sibling gender can be depicted—so long as it is stable. What if it changes over time? How might we depict a son who once was a daughter let alone of another gender? How do we show divorce and remarriage? There is no easy way: the conventions assume that gender, and marriage, will last forever.

Beyond not being able to capture or accommodate certain cultural realities, use of the kinship symbolism by anthropologists to describe lineages in other cultures resulted, in effect, in forcing square pegs into round holes. It described (or tried to describe) families with the anthropologist's kinship categories and from the anthropologist's point of view—not the native's. It certainly is important to describe kinship systems and to chart genealogies; often, information about who is related to whom, and how, illuminates otherwise puzzling behavioral patterns. In many cultures, nearly all behavior is guided by kinship rights and obligations. However, if kinship descriptions are ethnocentric to start with, crucial information will be missing from the ethnographic record. A swath of biocultural diversity will remain unaccounted for.

The question of viewpoint has been explored in relation to the experiences of immigrants to the United States who would bring their kin here to live with them. US courts of law depend on DNA evidence to define kin relationships. However, the DNA model does not match the kinship model used by many immigrant families. For example, remember the avunculate? If Uncle Jack tried to bring his nephew in, he would be denied—although he might be offered entry for his children-by-marriage even though to him they are nowhere near as related. Another problem with insistence on DNA testing is that it may bring to light extramarital affairs that had long been forgotten or which were outweighed by the parent–child relationship that mutual love and caring can create. Likewise, DNA samples may indicate that sons and daughters are not related genetically, and so they may be denied entry (Swarns 2007). The tragedy that such scenarios represent for many families cannot be overstated.

Now, one might say that DNA is the only true indicator of heredity, which derives from shared genes. That is a scientific fact: it is not in question. We are questioning here not the validity of hereditary genetic kinship but rather whether other kinds of kinship exist—and what kinds count. What kind of relatedness is important in or to a family (and therefore to biocultural outcomes)? This can be variously defined depending on what culture we are talking about. As we consider alternate definitions, bear in mind that parents' or caregivers' influence even on children who are not related genetically can be vast. That is, children can, so to speak, inherit traits even from nongenetic kin. This is due to the powerful developmental and epigenetic imprints that can be made on children by the environments in which they grow—by the environments that caregivers and parents, and whole cultures, create.

Kinship Types

The kinds of relatedness that social scientists have seen when looking around the world at different cultures, and when looking back through history to different times, boil down to three: classificatory, affinal, and consubstantial. The first category is in some ways a non-starter because **classificatory** kin are people who really are not kin but who are treated and addressed as such. A good example of this is when Grandma introduces her best friend to you as her sister or your aunt. I had an Auntie Doris; she was my granny's best friend. The two were not related, but they treated each other—relied on each other—as if kin. So Auntie Doris was a classificatory aunt of mine. Another example of classificatory kinship is when someone says that a person is like a daughter or son. Applying a kin term like this is tantamount to listing out the set of rights and obligations incumbent on that class of kin. A friend announcing "You are like a sister [brother] to me" accomplishes the same.

Some groups manipulate classificatory kinship to encourage altruism or selflessness and bonding in ways that mimic the promotion of what biologists term **inclusive fitness**: favoring the survival of one's close relatives or offspring equivalents (that is, of individual organisms with self-like genes), even at the cost of one's own reproductive success. In non-human animals this happens for example when one ground squirrel sounds an alarm in the face of approaching predators, aiding others' survival while itself attracting the predator's attention. This instinctual drive to promote the survival of, if not one's own genes, then genes most like one's own, may in effect be manipulated to various ends when we practice classificatory kinship.

The idea here is that we become willing to help others, and even to defend their lives, if we perceive kinship with them. Celibate nuns and monks provide classic examples. They not only refer to each other with sibling terms (brother, sister); they dress alike, emphasizing (familial) resemblance over difference; and they sometimes call those to whom they minister their children. Likewise, the parent–child relation often is replicated between church leader (father, mother) and congregant. The fraternity and sorority (Greek) system similarly leverages kin terms and capitalizes on the affiliative, familial imagery conveyed by conforming or uniform dress, as does the military. In this and similar ways, classificatory kinship regulates social behavior, mimicking to some degree what biological kinship may do, even between once-strangers.

Affinal kin also were once strangers or anyhow not from one's lineage, but they are (or have become) 'real' kin because they have been created as such by the law. In other words, and in Western terms, while we share no DNA with our affinal kin, they are kin nonetheless because we have entered into a contract that makes it so. Husbands and wives are affinal kin. Adopted sons or daughters are, too. Spouses and adopted children are treated by US law just like other lineage-linked kin; in fact, they sometimes even have more legal rights and responsibilities. Your spouse, if you have one, has more say over your medical care, for example, than does, for instance, your sister, or your mother for that matter (if your spouse is alive and you are of legal age, that is).

Last but, as we shall see, not least, are **consubstantial** kin: people who are related to us because we share some substance in common with them. We are conjoined by the substance we share. These kin generally belong to one's lineage or, as we used to say before the discovery of DNA, one's bloodline—and indeed, these kin used to be called **consanguineal**, in reference to that shared blood. 'Consanguineal' still is used by some, but it is more correct to treat blood kinship as the subcategory it is. 'Consubstantial' is the better word choice for the scholar, because this category of kin is not limited to those thought to share blood or even its Western replacement substance, DNA. There are many substances that, when shared, create the bodily bond—and experience—of kinship.

Forms of Consubstantiality

Blood Kin

The idea that all 'real' kin are consanguineal comes from the folk or popular logic espoused in many if not all cultures that babies are somehow grown from the blood of their parents. In some cases, the word 'blood' is generically used to mean any number of bodily fluids; it can be a term, then, somewhat like 'humors,' as in the humoral system of the ancient Greeks.

Jamaicans distinguish traditionally between two kinds of blood: red and white. Red blood is the same kind pumped by the heart; white blood includes semen, breast milk, and other light-colored fluids, including female sexual or reproductive discharge. Once we understand blood broadly, we can see how some cultures saw blood kinship as due to the mixing of male semen and female body fluids. While some cultures hold that a fetus, in the womb, takes in parental blood substances through its mouth, others teach that blood substances simply accrete or build up in layers once the fetus has been created (sometimes conception happens when parental bloods mingle). Until recently, Anglo-European culture shared this kind of belief.

The power of various bodily fluids to relate people can be used in love magic as well as in kinship creation. Love and kinship even can be thought of as one and the same: both entail enduring solidarity and mutual support. In some cultures, menstrual blood can be used to create a bond between two (or more) people. A menstruating woman might, for example, put a little menstrual blood into some food, perhaps soup, or some coffee; she then serves it to the person she would like to tie or bond to her—maybe someone to whom she would like to be married, or her partner if she has one who has strayed. Men, of course, also have substances with which they can do such magic. In incorporating some of someone's substance into your body, a relationship is created. Sharing blood has, historically, and generally speaking, provided the majority of cultures a key trope for this.

Genetic Kin

Nowadays in the West, the blood that begets kinship has been replaced in the popular mind with genetic material. Anyone who has passed a high-school biology class knows that **gametes** (eggs and sperm) contain chromosomes and that those chromosomes are what link a child's physical body back to its parents.

The idea of shared genes is in fact such a key part of the mainstream idea of kinship in the West, and kinship is still so important to us, that we often bring genes up when noting a family resemblance. In fact, we often search for such family resemblances, as if to prove to ourselves that genetics link the members of a given family (Becker et al. 2005). The mainstream idea that children should be borne in wedlock and other assumptions regarding the ideal family unit come into play in such searches. If one child in a family has, say, red hair, in a family where everybody else's hair is black, people look for a genetic explanation. One helpful relative may offer, "Well, it was probably from your great grandmother. Did you know she had red hair?" Another may confirm it. Then another, who has not heard the first conversation, might offer that grandpa's brother (on the other side) was a redhead. Whichever story the family agrees upon in effect does not matter; the important thing is that a connection can be claimed. If not, then the child's position within the family might be called into question, at least according to longstanding ideals regarding American family structure and the place of (genetic) kinship in it.

Indeed, children's parents may participate in such family history-making even if, or in fact because, they know differently. They may know, for example, that the child was adopted or otherwise does not share genetic material with the side of the family that relatives are pointing to. They may remain silent about this because of understandings regarding the importance of genetic connection. In this way, they become co-conspirators in supporting mainstream kinship and family building ideals that actually are in some sense oppressive to them.

Nurturance Kin

In other cultures, and in some subcultures in the United States, there are other options for kinship creation or confirmation besides those offered by genes. In many cultures, the main substance that connects people, creating kinship, is love or nurturance. This entails the effort put into raising a child, and the emotions invested. Feeding a child, providing guidance, paying school fees, creating a loving home—all this is enough, in some contexts, to create an actual kinship link with that child. The love and nurturance given is substantive; it creates a consubstantial link. It may not have a material presence, and so may seem non-substantial to outsiders, but like electricity or sound waves it is indeed there; it has real, substantive, measurable effects and is considered as if embodied by insiders.

Milk Kin

Breast milk is often considered a valuable substance for solidifying kin ties that might already exist or for creating kin ties where there are no prior ones. The Azawagh Arabs, whom we met in the previous chapter, practice milk-based or lactational kinship. That is, they share breastmilk intentionally to create specific kin ties (Popenoe 2004).

Why would people do this? For one thing, milk kin are kin just like blood kin, and so must welcome each other into their homes as family. The more kin one can claim, the larger one's social support network will be. For another thing, a woman might see a benefit in manipulating her children's possibilities of marrying certain people. Thus, she might strategically breastfeed a little girl she wants as a sibling to her own children. All of the children of these siblings (her grandchildren) will be cousins to each other—and cousins are preferred as marriage partners for Azawagh Arabs (this is part of the 'containment' approach that this culture subscribes to; see Chapter 12). So a breastfeeding session with the child next door might in fact be part of a long-term plan for one's grandchildren's marriages.

Commensality

The Importance of Co-Ingestion

The baby that becomes milk kin does so by ingesting the bodily product of its milk mother. Other nutriments also can support consubstantiality. This is seen in how commensality—sharing a meal or breaking bread together—not only signifies but also creates and helps maintain relationships.

Rituals that reinforce group solidarity often involve food sharing. Commonly, the food shared is a staple item in that culture—something that members feel they cannot live without. For example, in Japan, there are many rites involving sharing rice. In a traditional Japanese wedding ceremony, sharing a bowl of rice becomes crucial to solidifying and demonstrating the kinship bond being created by husband and wife, or between the husband's family and the wife's family.

In other weddings, such as Macedonian Orthodox ones, bread is shared. In Macedonian tradition, the bread should be sweet, so that the bride and groom will share a sweet life together. Besides the couple, their families all eat some of the bread. Thereby, these two separate families and two separate individuals join together. The social distance between them is eclipsed by the fact that they all, within their bodies, carry a piece of the same shared substance. They are now, in effect, one people. The same symbolism holds for shared wedding cakes.

Without food sharing, many of our social relations would lack glue. Just as crucial, however, are rules about with whom food should *not* be shared. Sharing food with them would create relationships that are too intimate to be acceptable—relationships that are taboo.

The Hua

The Hua of Papua New Guinea have food sharing rules that concretely relate to kinship; these are described by Anna Meigs (1987, 1997), whose work we explore here. The Hua live in the highlands, where they raise pigs and grow vegetables (see Figure 13.4).

Hua food rules take into account that whatever a person produces, or cooks, contains some of his or her *nu* or vital essence. Nu inhabits all of our body parts, including our flesh and blood as well as things extruded, such as breath, hair, nails, skin flakes, pus, feces, urine, and reproductive tract fluids. Anything that has come in contact with something from one's body, then, will absorb some of one's nu. This is how food sharing or commensality among the Hua creates consubstantiality—through the vital essence that food carries: its preparer or producer's nu. For this reason, if your sweat gets onto another person's skin, where it moves through the pores into that person's body, a kinship connection can be created.

FIGURE 13.4 Washing yams in the Highlands of New Guinea, 2010. CC-BY-SA-2.0.

SUPPLEMENT BOX 13: KINSHIP AND CANNIBALISM

Like all people, the Hua of Papua New Guinea have food sharing rules. As Anna Meigs explains, these have a lot to do with kinship (1987, 1997).

To begin, some Hua food rules stem from the absolute properties of or qualities in given foods and the understanding that eating them would recreate those properties within the eater, via contagion. Women, for instance, do not eat men's foods; and vice versa. We do the same, at least minimally; take, for example, how there are gendered patterns to what we drink at bars, and how bartenders decorate or in what kind of glasses or containers they place 'lady' and 'guy' drinks when serving; or consider how eating certain 'ethnic' foods—injera, pancit, posole, kimchi, Marmite, grits—can help members of an ethnic group to reinforce that ethnic aspect of their identity.

Other rules concern not what foods to eat or not, but from whom one can or should accept foods. These rules focus on the essential properties or qualities of those who offer the food rather than the food itself. Meigs calls these 'relative rules', and while the first kind of rules ('absolute rules') are about contagion, relative rules are, in a sense, about cannibalism—or really, how to avoid it. The idea behind relative rules is that whatever a person cooks or produces in his or her garden thereby contains some of his or her *nu* or vital essence.

When eating food, one also is eating the nu of the food preparer and sometimes even of the food producer. When Hua people work in their gardens they may breathe on the plants. Sweat from their bodies goes into growing the food. Fingernails may break and fall into the soil that nurtures the gardened items. Bits and pieces of a person therefore literally enter the food produced. The same thing happens to us when preparing food—when making a sandwich or baking cookies or roasting turnips. Our hands touch that food and, as they do, a little bit of our essence is transferred. A little bit of our breath, for example, falls on the cookie batter as we mix it. A hair might fall onto a sandwich or some skin flakes may come off onto the bread. When someone else eats that food, that someone else eats a little bit of us, too.

The Hua find more significance in this than we do, and so, for example, Hua parents should not eat food from their own children: to take and ingest food that one's own child has prepared or produced is almost like eating that child for dinner. Likewise, people should not eat their own best produce because that would be like eating themselves. Food sharing or commensality among the Hua creates consubstantiality—but it is not the food matter itself that creates this. It is the vital essence that the food carries: its preparer or producer's nu.

Importantly, nu conveys not only the substance of kinship but also feelings. Nu from an enemy thus can be destructive and must be avoided, unlike the nu of family members who, in theory, wish to support their kin. Many Westerners subscribe to some version of this theory, eating freely from kin but maintaining a wariness of food offered by strangers and surely refusing food prepared by people with certain moral failings, no matter how well their hands are washed or kitchens cleaned.

Building up kinship among the Hua takes time. People can purposefully add bits of their bodies to others' food to speed up the process. Mothers solidify ties with their children through breast milk as well as by rubbing babies with their own sweat and body oils. Parents cut their skin and feed the let blood to their children. In the past, sometimes even the flesh of deceased individuals was fed to others to ensure lineal ties. If this sounds strange, consider how in some Christian traditions congregants eat the flesh of Christ in the form of a communion wafer, or how in the United States people who have received organs from unknown donors (strangers) and those organ donors' survivors often report feeling an after-the-fact kinship connection (Sharp 2006).

Anyway, the Hua method for creating relatedness makes affinal or legal kinship irrelevant. Spouses may start out as mere affines but become conjoined consubstantially after a time. It does take a while—maybe fifteen years—because to marry, a man and woman must not be directly related, meaning there is no notable baseline of kinship to begin with for them. Still, after enough time, and enough consubstantial exchange, spouses are linked bodily. The family will hold a rite of passage in which a woman gives her husband, in public, some leafy greens from her garden (in Hua culture, greens are loaded with their producer's nu). The husband's elder brother proclaims that the husband can eat freely from his wife now—as freely as from his own mother (these are the terms expressed). The man and now, too, his family (which until then could not eat at all from the wife because she was unrelated in their system) can share food. The wife is thereby transitioned from a mere in-law (affine) to a consubstantial relative.

When Does Kinship Begin?

Another interesting thing about the Hua—and there are many similar-thinking groups—is that the currency of nu exchange matters: real kin are people with whom you are right now exchanging nu (Meigs 1987). That is, Hua kinship does not last forever. In the West, most hold that lineal kin always are kin: parent–child links cannot be denied (although they may be ignored). Among the Hua, however, and a number of other New Guinea peoples, unless individuals are constantly sharing substances, filling up bodily with each other's nu, a kinship bond can dissipate; it can fade away. This is one reason why some do not like to go away to college or to migrate for work far away from their kin. By doing so—unless they come back periodically for a corporeal reinjection of kinship-building substances—they may cease to be family members.

The larger question this brings up concerns when kinship bonds can be established. In Western, scientific thought they are established prenatally or before birth, through conception.

TABLE 13.1 Features of prenatal and postnatal kinship

	Prenatal or birth kinship	*Postnatal kinship**
Onset	Instantly given (e.g., when egg and sperm meet)	Additively achieved (e.g., with breast milk)
Flexibility	Static—closed to change	Dynamic—open to change
Sustainability	Endures under all conditions once created	Needs boosters for long-term endurance

* Note that in societies where postnatal kinship creation is possible, prenatal ties also can be made, generally by similar means (e.g., accretion).

In this system, kinship is a static thing. Western kinship is a closed system, and immutable or impervious to change. Once sperm and egg have come together, an individual's relationship to his or her mother and father has been created and sealed. It cannot be changed.

Instead of having this kind of static, closed, prenatal kinship system, the Hua—and many others, too—have a dynamic, open system. In such systems, kinship can be changed. Open systems can accommodate postnatal kinship creation (after birth) because kinship, rather than being instantly created, is built up over time by accretion. Indeed, in some cultures, even prenatal kinship depends on that: continued sexual intercourse is necessary to feed and grow the fetus. Every act of intercourse adds substance to the bourgeoning clump of matter that over time becomes a baby. In this kind of additive as opposed to instant system, a baby can have more than one father (see Table 13.1). The technical term for this kind of co-creation, in which each father provides part of the matter that becomes a child, is **partible paternity**. While semen is generally implicated, partible paternity could also be a post-natal achievement, depending on the rules of consubstantiality to which a given culture adheres.

The flexibility of open and additive systems increases one's kinship options, for instance in terms of having multiple parents. However, the demand for continued maintenance that this kind of kinship generally calls for has costs, too, one of which is in the sheer effort of upkeep. It also has ramifications for how adoption is experienced.

A person might have come from a particular parent's body, but when that parent stops doing what parents need to do to solidify parent–child kinship, the child can become the son or daughter of another parent (or parents) completely. An adopted child among the Hua, for example, has no claim on a genetic parent simply because that parent is genetically related. Parents give up the mother/father title and any related claim on a child when they fail to share nu. In such a context, any feelings of linkage are culturally minimized; they have no cultural basis. Conversely, the profound sense of kinship often expressed by US adoptees toward their genetic parents and vice versa—even without knowing them—is catalyzed by US cultural values.

When Does Life Begin?

Related to the question of when kinship starts and how it is established is the question of when human life begins. The way that a given culture answers this query—or whether it even asks it—has ramifications for decisions that are concurrently biological and cultural and, in turn, have concrete biocultural significance.

Quickening

In many cultures, pregnancy is confirmed when a mother feels a fetus moving in her womb. This can begin to happen in about the fourth month of pregnancy, or at the beginning of the second trimester (biomedicine divides pregnancy into three three-month phases). We call this point in pregnancy the **quickening**. The Catholic Church has called it 'ensoulment.' In Catholicism, people traditionally believed that a soul would not enter a fetus until after it had formed up enough to have a humanlike look to its organism—which generally occurs when the first trimester has ended.

However labeled and interpreted, the quickening is a universal experience, and many cultures interpret it as the point in time when the fetus becomes alive. From this point of view, an early-term abortion is not a cultural and moral problem because what is being aborted is not 'human.' In fact, it was not until the late 1800s that the Catholic Church became sure itself that abortion before the quickening could be culturally and morally problematic. At that time in European history, birth rates were declining among certain classes. Church leaders decided, after some wavering on the point, that ensoulment happened earlier: simultaneously with conception. Abortion at any time during a pregnancy then became nearly impossible to justify. Spiritual truths aside, as seen within the political and economic (and so demographic) context of the times, this decision by church leaders made sense.

Other Beginnings

Many cultures today believe that human life does not begin until the quickening. Others put its commencement at the point of birth. Still others put it a certain number of days after the baby has come into the world and proven itself viable. In fact, many cultures do not even give a baby a name until the dangerous phase just after birth has passed, if the baby lives through it. Some cultures at that point will name the baby, considering it now a human being—or making it one via the performative act of baby naming. Others wait until the teeth have begun to come in or for other milestones to pass before they call a baby 'human.'

Here, we might consider the interesting and very biocultural question of how babies turn into (or are turned into) real people. In some cultures, a vast amount of biological and cultural work must be done in support of this process. We saw in Chapter 7 that, in babyhood, this work might involve early weaning to promote independence. It also could entail favoring breastfeeding; in Anthroposophy (an Anglo-European spiritual subculture descended from Theosophy), it is thought that breast milk awakens the child's spirit; it contains 'will' forces necessary to jump-start proper child growth and development (Steiner 2007, 153–154).

As informative as exploring the details of healthy child development in this and other worldviews might be, here we should get back to the question of when life begins. It is a question with great salience in the United States today. However, it makes no sense in the context of cultures for which life never begins or ends.

Buddhist Perspective

Take, for instance, traditional Japanese Buddhism which, as William LaFleur describes it (1994), considers life as a wheel (see Figure 13.5). The soul is not on a linear journey from some starting point that finishes with death and perhaps an afterlife. Rather, its journey is circular; it is continually reincarnated. It moves back and forth between the realm of the sacred and the secular, 'real' world.

Souls stay in the river-filled, liquidy realm of the spirits until the time comes for them to be born in human form. At that point, they make a journey from the other world into this one. According to traditional Japanese Buddhism, as children grow and age, they become more and more solid. Becoming human, then, is somewhat of a thickening or materializing process.

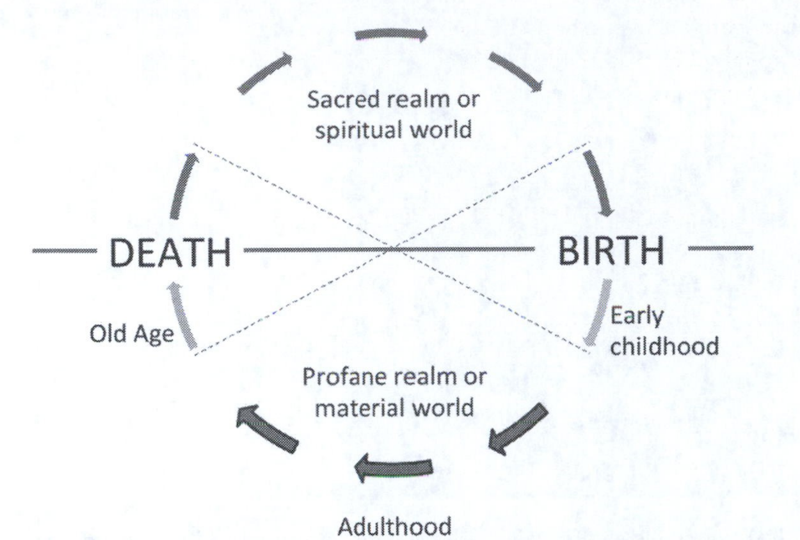

FIGURE 13.5 Traditional Japanese Buddhist conception of life, depicted as a repetitive, circular process (see LaFleur 1994, 35).

Children eventually lose their otherworldly nature, but it is not really until we are teenaged that our humanity solidifies to the point that the spirit realm is forgotten.

Adult life is mundane compared to the life of the child because adults are so solid. They have lost that quick connection to the other world. Nevertheless, later in life, our human solidity will slowly melt away. Over time, assuming that we do live a long life, we will begin the journey back into the other world, slowly rejoining it. Dementia thus can be taken merely as a sign of an inevitable progression. We come back in touch with the sacred world and eventually, upon human death, re-enter it completely, sooner or later to be born again. In this way, life is everlasting.

As LaFleur describes it, the sacred world is very liquidy. Rivers, streams, and springs are seen as conduits to that watery world. Often, if a woman miscarries, has an abortion, or has a stillborn child, that fetus or infant will be buried under the house. Even if there is not a real river under the house, the idea here is that water flowing below the floorboards, real or symbolic, will carry its soul back to the sacred realm safely. Moreover, there are certain rituals to ensure this, and to protect the soul of that unborn child so that when the time is right for it to actually enter the human world, it will come back, and it will have a good life.

That ritual is called the *mizuko kuyō* ritual. *Mizuko* means 'water child' and refers directly to the unborn (the fetus, the aborted, the miscarried, the stillborn), whose soul is sent back to the watery spirit world. *Kuyō* means 'to offer' or 'to nourish,' so the mizuko kuyō involves an offering meant to nourish or keep the mizuko. In other words, a mizuko kuyō is a memorial service for a water child. In it, an offering is made to the Mizuko *Jizō,* the guardian deity, sometimes spoken of as the womb of the earth, who protects not only pregnant women and water children but also travelers and pilgrims—anyone in the midst of making a passage.

The shrine areas where these offerings are made are full of little doll-like statues (see Figure 13.6). Those represent the miscarried, stillborn, or aborted souls. Some have little red bibs or knit caps, just like a living infant would. Some have pinwheels in their hands as a living child might hold a toy. Placing one of these at a mizuko kuyō shrine and making associated

FIGURE 13.6 Zōjō-ji (mini Jizō) in temple garden, Minato district, Tokyo, Japan. Used in memorial to the unborn. CC-BY-SA-3.0.

offerings gives people a very concrete way to handle the grief that they feel on losing a potential child and family member.

Mizuko kuyō does not exist just to make parents feel better. It is crucial for the protection of the water child; it helps protect the released soul, and ensures as well that it will come back when the time is right. The thinking here is that bad timing underlies miscarriage and abortion. Within the family it simply is not the right time to bring another soul into the world. Maybe there is a very sick sibling who needs a lot of attention and help; the parents therefore would not have the time necessary to give a new baby the love and care that it needs. Maybe there is not enough money available, and so sending this soul back is a way of making sure that it can instead be born when the time is right and family finances are back in order.

In other words, abortion is not a final life-ending act but rather a simple delay. Abortion is, then, from the traditional Buddhist point of view, the right thing to do if a family or woman wishes to bring a child into this world properly. In this tradition and others where life is not linear—where it does not have a certain start point and a certain end point but moves in a circle—abortion promotes family values and serves as a very pro-social act (LaFleur 1994).

Family Building

Infertility

While abortion is, by definition, something done on purpose for various reasons, miscarriage (also sometimes called 'spontaneous' abortion) is not. Many times, when a conception miscarries—when the body ejects or rejects an unviable fetus—the woman whose body is involved does not even know. About half of all fertilized eggs miscarry. This frequently happens very quickly, before a woman can know that she is or has been pregnant. Still, about 10 to 20 percent of all known pregnancies also end in miscarriage. One or two out of ten is a notable amount.

Of course, providing figures like this, as if all populations are exactly the same, is misleading. Miscarriage happens more in populations that are undernourished or otherwise stressed. Miscarriage rates also tie in with infertility rates, which again are higher when health is taxed.

Malnourished bodies, for instance, often cannot physically support a pregnancy. In addition, they are less likely to be ovulating or producing viable sperm. We saw, too, in Chapter 9, that toxicants such as those to which the Akwesasne Mohawk community has been exposed cause reproductive irregularities. These are just some examples of how infertility can be linked bio-culturally with social structural issues. Infertility ties in with poverty. It ties in with a lack of access to medical care as well, because certain diseases, if they go untreated, can lead to infertility. Today, although again this figure is generic, about 15 percent of all couples experience infertility.

Adoption

Because infertility is and has been so common, most cultures have certain strategies for how to deal with it. In many cases, these entail adoption or **fosterage** (fosterage is social but not legal adoption, sometimes temporary, yet often long term). Adoption, legal or not, is in many cultures a viable, respectable, normal mode of family building. Remember that in many cultures people can adopt children who have other parents, too, and adoption is not limited to people without biological children: recall the Azawagh Arabs' milk kin customs.

Beyond adoption, people can contract or bring in a **surrogate** or stand-in biological parent to help build their families. There is a story about surrogacy in the Bible, in the Book of Genesis; this story also features obliquely in the Quran. It is a story about a couple, Abraham and Sarah, who could not conceive a child. Sarah was thought to be barren (later she does bear a child, but that is another story). To solve the immediate dilemma—to generate for Abraham an heir (theirs was a patriarchal world)—Sarah's maidservant Hagar was called upon to have sexual intercourse with Abraham and bear a child for the couple. Hagar was Sarah's surrogate. She bore a son for Abraham and so for Sarah too: Ishmael, who became the patriarch of the Ishmaelites, or Arabs. So surrogacy is important in all of the Abrahamic faiths.

Assistive Reproductive Technology

Today, another option exists in the form of ART, or assistive reproductive technology. One kind of ART is *in vitro* fertilization (IVF). Eggs from the genetic mother and sperm from the genetic father are physically extracted and mixed together in a Petri dish to promote fertilization. Any resulting embryos are implanted into a womb, either that of the genetic mother, that of the intended birth and adoptive mother, or that of a gestational surrogate. A gestational surrogate is a stand in; with gestational surrogacy as with old fashioned surrogacy, such as Hagar practiced, the baby is handed over to the family that requested or contracted for it.

The new technology is awesome. It also is expensive. Moreover, the process is time consuming and risky. Whosoever's body the eggs come from must ingest or inject hormones for a while to stimulate ovulation. Instead of the one or at most two eggs that normally ripen, if any do, in a typical menstrual cycle, a woman undergoing IVF might produce from five to twenty eggs. This in itself puts extra stress on the body and overstimulated, enlarged ovaries may result, leading to concurrent health problems. The operation to withdraw the eggs involves general anesthesia, which always entails a degree of risk. Emotionally, IVF also is a risky, stressful business, due to uncertain outcomes and the strength of the (culturally relative) desire for a genetically related baby that drives people to undertake the process.

Then, there are the risks to the person who will carry the fetus to term. For example, clinically assisted implantation entails a risk for infection and medical error that might lead to harm. Carrying the child to term puts certain stresses on the pregnant person's body, and there is no guarantee for the health of the expected child or the gestational mother, although surrogates generally are well monitored by clinicians and in some cases receive supplements to ensure a healthy diet and so forth. The scheduled cesarean section, often used with gestational surrogacy so that the receiving parent or parents can be present at the birth, is major surgery that has its own risk for complications list.

In discussing risks I do not wish to downplay the benefits of IVF and surrogacy. For example, a child is conceived and born, helping an infertile person or couple to fulfill a culturally underwritten dream of parenthood. Surrogates benefit, too, financially for instance. Still, the risks mentioned and other costs exist. They inscribe themselves into national statistics regarding reproductive outcomes, morbidity and mortality included. Read superficially, comparative figures may be taken to confirm biology-based population diversity (the people from *x* suffer more from *y*) when the outcomes really are biocultural. Because culture shapes them, these outcomes are something over which we do have some control.

Culture's Role

Normative cultural ideas about kinship affect our strategies for building families and help to generate our desires for particular kinds of families to begin with. They also affect legal regulation of family building strategies, including not only adoption but also regarding ARTs.

When a new ART comes on the scene, heated debate follows about who has the right to use it, and under what conditions. For instance, people ask if there should be age cut-offs or behavioral standards imposed on applicants, whereby, for example, people over forty or with experience in sex work might be denied access. Experts and lay people alike debate whether we should limit the use of ARTs to the normative familial context (often a legally married male–female couple) or whether we should allow male–male or female–female or even multiply partnered spousal units and single people to use them. These decisions are made on a cultural basis, not a medical one.

Normative cultural ideas about kinship and the family affect the choices we make in the face of infertility or when we want to have more children than we can without assistance. For instance, today, most Westernized people will opt to use ARTs when they can afford them because of cultural understandings about 'real' kinship as anchored in shared DNA.

These understandings are, in fact, a key driver in the expanding market for medical travel, in which people cross borders to seek out medical or dental treatments that they cannot access at home, whether because such treatments are too expensive or not yet done there or simply not legal in their home country. Reproductive services using ARTs account for a notable share of medical travel, and they often involve purchased eggs and gestational surrogacy.

In terms of the former, people raised to believe that shared DNA is crucial to kinship often seek sellers (euphemistically, 'donors') who look like themselves, to help disguise the fact that the resulting baby will not have the mother's genes (they do likewise with sperm, although that is more easily available domestically). White working-class North Americans, for instance, often turn to the Czech Republic for eggs (Speier 2016).

Here we begin to understand why some prefer the term SRT to ART, with S standing for selective. By choosing which traits we wish to see in our offspring, we are in effect adding another tool to our selective reproduction kit, affecting the frequency of genes in the next generation's gene pool. Of course, beyond the coarse-grained practice of mate or sexual selection (driven purely by biology before culture came along), people have long practiced selective infanticide, child abandonment, and neglect, thereby culling offspring with 'undesirable' qualities. Such practices have affected not only the prevalence of certain hereditary challenges or diseases in a given population but sex ratios as well. Such biocultural outcomes also are worked into the SRT equation when, for instance, gamete donors are chosen for traits that, expressed in resultant offspring, will imply that the receiving parent or parents were themselves fertile, bring status to the family, or help ensure a child's success given the socio-cultural context into which that child will be born. Importantly, then, SRT decisions are not purely individual choices: they reflect internalized cultural beliefs regarding the ideal shape and character of the population (Gammeltoft and Wahlberg 2014).

Whatever the source of gametes may be, for gestational surrogacy many would-be parents have turned to India, which established itself early on as a prime, low-priced, customer-oriented market (Kumar 2008). From one perspective this has been win–win. People who otherwise could not afford the process, say in the United States, get it for a fraction of their homeland cost. Moreover, international gestational surrogacy provides people overseas who otherwise may be struggling with a source of income. The women in India who hire their wombs out can use the money to put their children through school, provide food for the family, or even buy a house.

Yet there are lots of risks. In addition to the health risks to the surrogate already mentioned, such as with cesarean operations, there may be increased risk of multiple pregnancies and all that they entail, because in order to get one embryo to stick, clinicians may be more likely to gamble in a surrogate's womb and deposit too many, as a form of insurance that at least one will survive. In some nations, regulations limit this.

Plus there may be social risks to the surrogate. In many cultures, this kind of service—or being in a situation in which one is forced to provide it to survive—is very much frowned upon. The fact that a woman is so ill-provided for that she must rent out her womb says something to others about her social network's (or her own) 'failings.' Surrogacy can thus be a self-stigmatizing act. Psychic risks include this plus also the cost of coming to see and experience one's body as partible, or more as if a machine with interchangeable pieces than a holistically integrated human organism.

Population-wide, the practice has ramifications for demographic patterns. As ART access rises, the demand for children to adopt goes down. An increase may be noted in orphan or abandoned child populations.

In response to some of these issues, in 2016 India passed a law limiting surrogacy services to married Indian citizens. The bill was controversial, in part because the market had been established. Surrogacy agencies already are bypassing it by moving their gestational surrogates across national borders, such as into Nepal. Another scheme brings Kenyans in for implantation and, at the end of the second trimester, sends them to Nairobi to complete gestation and delivery. Needless to say, such arrangements make surrogates even more dependent on organizing agencies than before, increasing their vulnerability and the risks to which they are exposed (Rudrappa 2017).

Still, for mainstream Americans, use of gestational surrogacy or other ARTs is often the first choice that comes to mind when seeking to build a family and sexual intercourse does not suffice. Despite the risks, it is the choice that makes cultural sense. Our belief in the importance of DNA to kinship is so powerful that it drives people toward ARTs and away from adoption on their family building journeys.

In non-Western contexts or where DNA is not the main substance of kinship, adoption is much easier. It really fits well for family building in open, flexible kinship systems—the kind that allow people to construct kinship additively postnatally from substances such as breast milk, food, nurturance, and so forth. In that case, a resort to ART would seem ridiculous. People will not take those risks if they can just adopt and create consubstantiality at home, cheaply, safely, and easily. Indeed, the Hua would say that those who favor ARTs over adoption are out of their minds.

It is true that a tradition of adoption exists in the West: kinship can be created affinally. Nevertheless, culturally speaking, and as historical records demonstrate, this kind of kinship has been seen as second rate. In past generations, adoptions often were kept secret because of the potential stigma and for fear that the parent–child relation would somehow be compromised if the adopted child knew, or that sibling relationships might suffer. Adoptive children and parents were put on the defensive by the normative standard that lineal kin be genetically related (the same was true in earlier historical periods when bloodline standards reigned).

Although secrecy regarding adoptions does seem to be on the wane, we still see culture-induced defensiveness in the fact that adoption is remarked upon at all. We see it when adoptive parents and offspring justify their relationships as if 'real,' saying things like, "My mom and dad are like real parents to me. I think of them as my real parents. They are my real parents." The fact that such statements are even made is culturally revealing.

Human Communion

American defenses of the relatedness of adoptees and adopters, like the feeling of connection reported after organ transfers, reveal something else also: even while we overtly prioritize genetic kin ties, we hold additional, contrastive understandings regarding kinship that have much in common with other cultures—cultures that prioritize diverse sources for kinship, including nurturance. Although some may explain this pejoratively as merely atavistic (as if a regressive holdover from our so-called primitive days) there is a better explanation: our universal, evolved human preference for community.

As Chapter 4 taught, a capacity for empathy, shared intentionality, and cooperation were essential ingredients in the emergence of behaviorally modern human beings. In this light, our acknowledgment of the importance to kinship of things well beyond DNA bears witness to one of the key characteristics of being human: our need for, and our ability and impulse to build, social connections. It is the case that there are numerous and diverse ways to build and explain these connections and myriad possible sources for consubstantiality when the entire ethnographic record is taken into account. Ideas about when connections can be or are built, or how long they might endure, also vary. The biocultural diversity entailed in and by these differences notwithstanding, all of humanity shares at least this: a universal and inherent drive for communion.

CONCLUSION

Respecting Connections

In Part I of the book, our focus was on how we adapt to the geographic or natural environment, with its predators, plants, and so forth. We paid special attention to co-evolution, the related benefits of biological diversity, and the role of niche construction. We saw that many early adaptations fed into each other in our evolution, synergistically shaping our species. In Part II, we began to think about biocultural diversity spurred by the many changes brought about by our ancestors after they spread across the globe—changes to both the geographic environment and the social one—particularly once we settled down. These changes came about because of how we decided to manipulate the environment in our quest for more food, and how we decided to organize ourselves when we intensified our subsistence strategies. These changes had benefits, but they had certain problems, too. In Part III, our focus shifted to how we adapt to culture itself—to the beliefs, values, and ideals we hold, which in many ways are imprinted upon and in the body. In other words, this book has focused on human adaptation to geographically, socially, and culturally generated selective ('environmental') pressures and the biocultural diversities that result.

After establishing the value of a holistic, systems approach, the book approached genetic inheritance as a portal of entry into a sphere of inquiry that is simultaneously biological—physiological, anatomical, genetic, and so forth—and at the same time cultural—social, political, economic, historical, ideological, and so on. We started on the biological side because some of the key concepts used today in the social sciences were developed first in that arena (for example, the theory of natural selection). By the last chapter, however, we had in effect come full circle: we ended by investigating inheritance and the importance of kin or relatives again, this time in much more cultural terms.

We might, of course, have begun with this cultural end. In European anthropology, the comparative study of kinship came first; it fostered anthropology's emergence as a discipline in its own right. In the United States anthropology as an academic discipline emerged in response to racism, for instance via Franz Boas, and so had a very physical, biological focus initially, but in Europe, origin stories hark back to William Halse R. Rivers and his genealogical method, used perhaps most famously on the 1898 Torres Straits Expedition. I bring this up to reinforce the argument that multifaceted issues can and should be approached from multiple entryways. The placement of the kinship chapter at the end, and therefore near to Chapter 1 in the circle the chapters create, represents a final request for holistic thinking.

Having come to this point—having been through all three sections—four key propositions should be obvious to all:

- Biology and culture co-create each other and the human experience
- Holism and systems thinking are useful
- Variation is good
- We are all connected

Before closing, then, I shall review these propositions one by one. In doing so, I shall also remind readers of some areas that need attention, both scientifically and humanistically.

Biology and Culture Co-Create Each Other and the Human Experience

A key question asked throughout the text is: How do culture and biology interact? *Biology and culture interact synergistically, creating together something that each cannot create on its own.* Biology and culture (once culture emerged, that is) reinforce one another; they co-create one another; they produce what we know of as the human experience.

While some of the things we have studied, such as tattooing or the distribution of skin pigmentation between populations, imply that biocultural interaction occurs only where we can see it, much of the interaction between biology and culture takes place inside of the body. A lot of the important effects of biology–culture interaction happen out of sight, at the cellular and even biochemical level deep inside human beings. Think here, for instance, about some of what we learned when studying the allostatic load and epigenetics. We should always stay aware of the invisible, which sometimes is more important than the seen. We should also stay aware of the importance of regional differences; what we learned about geographic clines highlighted the error of lumping groups of people together too broadly. Significant differences often occur on a much smaller scale (as well as across different lines) than those who are wedded to the old-time conception of race would believe.

Holism and Systems Thinking Are Useful

The book's focus on the synergy between biology and culture directly reflects anthropology's celebrated holistic perspective. It also demonstrates anthropology's systems perspective, which proposes that unless one understands how a system works—and in this case our interest is human systems—one really cannot understand in any intelligent fashion what's going on in that system. It makes no sense, from anthropology's holistic, systems-oriented viewpoint, to examine the parts of our existence on their own, or stilled from action. Rather, we should examine the parts working together, dynamically, in movement. We should be interested in the relationships between parts of human systems, and in what emerges out of the interactions between those parts. Whatever outcome may be achieved is achieved by the system as a whole. Applying this insight to investigations of such varied phenomena as educational achievement gaps or differences in disease rates or athletic and other such capacities between defined groups of people can fuel discoveries regarding such diversity.

Not only are human societies systems, they're complex ones; we used the concept of the complex adaptive system to gain a toehold in our understanding of how human systems work. Understanding how systems work is key to constructively changing them. In a complex adaptive system, the parts of the system can adapt—they can learn, altering the system's direction. A system that can adapt is better able to survive than a mechanical (nonadaptive) system when its environment changes.

Our whole investigation of the dynamics of biocultural diversity was informed by an interest in adaptation. For much of the book, adaptation was measured with the index variable of health. The health of a population is a good indicator of how well or how poorly that population is adapted to the context in which it lives.

With that in mind, the book's key lesson is basically this: neither biology nor culture can stand alone. *Culture and biology are related in a dynamic, synergistic way, co-creating one another.* Our biology would not be what it is today had culture not influenced it, and our culture would not be what it is if biology had not had influence over culture as well. Despite the implications of the fact that English only gives me separate words here—biology, culture—they and their influences are conjoined.

Long ago, that fact was obvious. But over the last few hundred years, as universities grew and multiplied, disciplinary walls were created. Scholars carved off and laid claim to different aspects of existence, building ring fences around them. As Rudolf Steiner explains, the common approach in science has been to pull the fabric of life apart in search of abstractions. The legacy of conventionally bounded disciplines reflects this on a general level. However, as Steiner further notes—and here the emphasis is his—"*To explain a thing, to make it intelligible*, means nothing else than to place it into the context from which it has been torn .... A thing cut off from the world-whole does not exist" (2008 [1894], 74).

The fallacy of reductionism aside, habitually viewing things—for instance, facets of human diversity—as artificially bounded, abstractable phenomena existing in isolation can leave us less able to draw back and take a wide-angled view of things. It can blind us to the connections that do exist, even to the point of supporting an extreme form of materialist, capitalist logic that says anything goes when profit is at stake. A more holistic way of thinking—a unified point of view—may help us to realize a more humanitarian future in the long run as well as promoting, in the nearer term, a more realistic and comprehensive understanding of human diversity.

Variation Is Good

We know much more today than we used to about the range of variations between human populations. For instance, as examples included in this book demonstrate, we understand now how many variations occur—and we know now the value of holism for increasing the richness of this understanding. But there is still much we do not know, including regarding the full range of human plasticity, and the enablers of and barriers to (as well as the limits on) various forms of this plasticity. Another area among many ripe for investigation is the transgenerational transfer of epigenetically acquired traits. Such knowledge gaps make biocultural studies a fertile field for future research.

Relevant here is the axiom that variation is good. For example, genetic variation helps ensure that an ample number of able individuals reproduce, helping the group as a whole to survive into the next generation. Social and cultural variations, too, are advantageous in terms of our population's survival; they provide us with a wide range of ways to be—some of which of course are more adaptive (or, better fitted to our environment) than others. Tamping out variation, through whatever means, would be to put all our eggs in one basket. As experienced investors and innovators know, this is not a good idea. It lessens our resilience and creativity in many ways. Seen in this light, biocultural diversity's usefulness comes clear.

We've learned throughout the book about diverse ways of organizing social groups and of embodied cultural practices that may not make sense to outsiders but that are perfectly logical when viewed from the inside perspective. We have put aside ethnocentrism in favor of cultural relativism so that we, too, might understand—and benefit from viewing—foreign practices in their own light.

Of course, just because something 'makes sense' in cultural context does not make it right. Some practices may be better than others in terms of allowing all people to optimize their potential. One force holding people back from this in some areas is the global political economy, which affords some people more benefits than others—benefits that come at a cost to those who do not have them. Oftentimes, the costs are embedded in people's bodies, as through job-related injuries or diseases stemming from the chronic stress brought on by the conditions of low-wage labor, unsafe housing, lack of educational and health care access, and the like.

We Are All Connected

Lest I end this book on a melancholy note, I must restate that the global system is global: we are part of it. The lives of the global poor and even of the poor in one's own nation, not to mention one's neighborhood, often seem as if they are lived in completely different worlds from the lives of the middle class, let alone the rich. These worlds are not separate, however; they are intimately connected. They are part of one system and exist in dynamic relation to one another. An action in one part of the system can have effects—even large ones—in another of its parts. We need to better leverage this fact.

As a final reminder of our participation in complex adaptive systems and an example of how small changes can lead to big ones, let us reconsider the food movements gaining traction today (see Chapter 7). By fostering, on small farms, the kinds of multistranded relationships that were prevalent, prior to industrial agriculture's onset, between biotic populations and the earth, air, water, and sun, and by striving to bridge the contemporary divide between consumers and producers, these movements can help offset some of the ills we humans have created over recent centuries as a downside of factory logic.

It is the case that organic food produced by individual or small-holding polycultural farmers can be hard to find and expensive to buy (finding plain 'organic' food is much less a problem these days due to its industrialization). However, research suggests that as this subsistence style catches on, the benefits to society will include fewer toxic pollutants, less use of fossil fuels, and increased floral and faunal biodiversity. Health expenditures may be lowered, too, as health outcomes improve.

Potential health gains may extend beyond those brought about by the better nutrient profiles that many nonindustrial crops offer. For instance, without the irrigation ditches that large-scale monocropping can demand, the risk for malaria (in regions that have it) may go down and with that, over the long term, so, too, may the frequency of the sickle cell allele. Further, if consumers become more involved in food production, even if only through contributing a few hours of labor each week as part of a cooperative agreement, risks related to physical inactivity could be lowered. Gains in people's sense of connectedness and overall well-being also can occur, with concurrent positive effects on populations. Right now gains are local, but members of most branches of the new food movements—biodynamic farmers, for instance—demand a social justice approach that targets the globe (Biodynamic Farming and Gardening Association 2012).

A good case in point involves *Navdanya*, a movement for biodiversity conservation and farmers' rights based in India. The name, Navdanya, refers to nine seeds that are part of India's agricultural legacy. Seeds are life sources; properly fostered they end up as whole plants, which naturally produce their own seeds for self-regeneration (see Figure 14.1).

This is not so, of course, for industrial hybrids, which is one problem that Navdanya was established to stand against: after large agribusiness interests gained prominence, farmers once able to supply their own seeds (from last year's produce) now had to buy them. They also had to buy expensive pesticides, to safeguard the nonindigenous crops. The hazards entailed went

FIGURE 14.1 Cecilia Joaquin gathering seeds with traditional Coast Pomo seed beater and burden basket. Photo by Edward S. Curtis, c. 1924. Library of Congress Prints and Photographs Division LC-USZ62-116525, cph 3c16525.

well beyond exposing the soil, water, flora, fauna, farmers, and, by extension, farmers' families to toxins. They included indebting farmers, too, and inducing stress-, malnutrition-, and disease burden-related impairments in farmers' children—impairments that reduce cognitive capacities and so reinforce a cycle of poverty (see Chapters 9 and 10). The pesticides also have been linked to a titanic wave of farmer suicides, which itself has had numerous down-the-line effects on households and communities. Navdanya would turn farmers who have grown dependent on and been impoverished by commercially purchased seeds and chemicals back toward self-sustainable organic methods.

That said, when Navdanya was conceived, previously longstanding farming techniques as well as local crops had come under threat of extinction, having been cast aside as farmers turned toward modern methods. To promote food sovereignty, Navdanya had to work to keep old traditions alive, for example by training seed keepers and establishing community seed banks where heritage seed varieties could be cultivated, stored, and distributed. Within twenty years, there were fifty-five such seed banks (seed banks have enjoyed rising prominence in other nations too; see Figure 14.2). India has also passed some legislation protecting small farmers. Navdanya and its instigator, ecologist Vandana Shiva, have gained the world's attention. They serve as models for others who seek similar ends (Navdanya 2009).

The links between human biocultural diversity and small-scale organic farming take sophistication to see. However, readers of this book by now can recognize that subsistence arrangements and the larger political economies that they are part of underwrite many of the embodied differences seen between populations today. Incorporating into our subsistence tool-kit polyculture such as Navdanya or the biodynamic approach recommended could offset negative, health-sapping differences as well as promote a more humanistic social order. Because this kind

FIGURE 14.2 Sculpture by Peter Randall-Page, "Inner Compulsion"; commemorates the seed collection of the Millennium Seed Bank (in background), Wakehurst Place, West Sussex. Photo by Jim Linwood, 2010, CC-BY-2.0.

of farming requires regional sensitivity, so that indigenous or at least geographically appropriate crops and livestock are emphasized, helpful biocultural differences also can, in theory, be maintained.

Some may assume that the kinds of broad-based changes necessary to support the widespread adoption of models like Navdanya's can never happen because the current way of life is too well entrenched. However, more extraordinary changes have been witnessed, such as those entailed in keystone paradigm shifts in science. Comparably, shifts in how people understood biocultural differences related to geographic ancestry supported the passage of the Thirteenth Amendment (outlawing slavery), the military's integration after World War II, school desegregation in 1954, and the civil rights protections enacted in the mid-1960s. These shifts required remodeling large-scale and heavily institutionalized political-economic systems of inequality. Changing such systems seemed as far-fetched to some people back then as the idea of someone walking on the moon—but changes did happen.

As history has shown repeatedly, and as we know from studying some of the mechanisms of human biocultural adaptation, change is part of life. Given the right conditions, even the smallest of variations can turn out to have huge ramifications over the course of time. It behooves us to pay attention to this—and to do as much anticipatory work as is possible to help ensure that whatever emerges through change is to our world's benefit. Even changes such as those discussed above regarding food could have negative consequences if all angles are not proactively considered and risks properly managed.

My point here is that our future actions, and insights gained through the application of the unified approach promoted in this book, can have just as much if not more impact than the actions and discoveries of those who went before us. We can strive to set right past mistakes, and we can work proactively toward the future we desire. We can scientifically investigate the puzzles of human difference from a more holistic perspective that, because of its comprehensive emphasis, enhances and adds to the body of knowledge that precedes us.

A unified approach to the magnificent array of human variation reveals that our experience as human beings comes out of the biology–culture relationship. Nature and nurture together create us as we are. Diverse as human populations may be, we all live in the world together. Let us leverage what we have learned about biocultural diversity's causes and effects in a concerted effort to optimize humankind's potential and thereby the potential of the larger systems we are part of.

GLOSSARY

A

Abiotic: nonliving chemical or physical factors in an environment, such as soil, atmospheric gasses, sunshine, and water

Acclimatization or acclimation: short-term, reversible changes in an organism's biology, such as when a summertime beachgoer develops a seasonal tan; contrast with 'developmental adjustment'

Acute disease: a sudden onset ailment; comes and goes (or kills) quickly

Adaptation: a responsive, survival-enhancing change in a system brought about in reaction to some kind of atypical or new and stress-producing change in the environment

Adaptive immunity: immunity brought on through vaccination or inoculation

Affiliative: an act or mark (e.g., a body decoration or modification) that symbolizes one's affiliation or identification with a particular culture or subculture; antonym is 'disaffiliative'

Affinal kin: kin related by law, such as through adoption or marriage in the United States

Agency: the ability to impose one's will or make significant choices

Agent: pathogen, substance, or process that causes morbidity and/or mortality

Agricultural Revolution: the emergence of settled agriculture starting about 10,000 years ago and its concomitant biocultural effects, including those stemming from new forms of niche construction following from our domestication of plants and animals

Agriculture: settled farming, practiced with plows, irrigation, and/or fertilizers

Alleles: variants of a single gene; see also 'polymorphic'

Allocare: the provision of childcare services to children who are not one's biological offspring; literally, 'other care'

Allostasis: the process of re-creating homeostasis by changing the body's initial set points or 'factory settings' to accommodate chronic stress

Allostatic load: cumulative multisystem physiological dysregulation resulting from chronic stress; a higher allostatic load places an organism at a higher risk for poor health

Androgynous: expressing both masculine and feminine characteristics

Anomaly: something that does not fit into preconceived cultural categories, is unclassifiable, or combines traits from two or more categorical types (the latter also is sometimes termed a 'monster'); an anomaly has power because it disturbs, or represents a disturbance in, our culturally influenced sense of order

Anthropocene: the current geological epoch, in which human activity has become the dominant force shaping Earth's geology and ecosystems; may have begun with the Agricultural Revolution

Anthropogenic: humanly created or generated; often applied to describe diseases or problems related to human-made changes to the physical environment, such as via agricultural practices

Anthropology: study of humankind; holistic in perspective

Antibodies: special proteins that can lock, selectively, onto the invading cells' antigens and effectively disarm them

Antigens: biochemical substances that mark cells as belonging to ourselves or not

Archaeological record: all material culture or artifacts and other remains of historic and prehistoric societies

Artificial selection: purposeful breeding of plants or animals, by humans, with the intention of producing in their offspring particular traits

Asexual reproduction: self-replication, for instance by splitting, so that each generation is a duplicate of the last; reproduction involving no sex; see also 'cloning'

Assimilation hypothesis: holds that populations of *Homo sapiens* who left Africa in earlier waves were absorbed or assimilated into newer populations

Attenuation: when a pathogen evolves, via natural selection, to have less of an impact on its host, and its host, as a result, can carry or spread it for longer

Avunculate: matrilineal relationship between a man and his sister's children

B

Balanced polymorphism: exists when selective pressure for one form of an allele is balanced or offset by selective pressure against that form, so that both alleles persist in a population, having achieved some kind of balance

Band: egalitarian, family-based socio-political group; leadership is achievement based

Bioavailability: quality of a food item affecting the degree to which an eater can extract and make use of nutrients; bioavailability is affected by such things as food storage conditions, cooking processes, and juxtapositions with other food items

Biocultural diversity: all population-based human variation generated in or reflecting the dynamic, synergistic communion of biology and culture

Biological determinism: the belief that biology alone determines one's capacities and characteristics (often used in opposition to 'cultural determinism')

Biotic: living factor; a living component of an ecosystem, such as a plant or animal species, that affects other aspects of that ecosystem

C

Carrying capacity: the total number of a given species that a habitat or niche can support without collapsing in whole or part

Chiefdom: hierarchical socio-political group run by a hereditary bureaucracy, in which people inherit their positions

Cholera: potentially fatal, frequently epidemic acute and often waterborne bacterial infection of the small intestine; symptoms include copious amounts of watery (i.e., often fatally dehydrating) diarrhea

Chromosomes: organized, structured packages of DNA found in the center or nucleus of each cell and passed along to offspring during reproduction

Chronic disease: lasts over a long time span

Cisgender: when gender is expressed or performed in a way that both meets cultural expectations for expression of that gender and appears concordant with sex (*cis* is Latin for on this side of); see also 'cisnormative'

Cisnormative: generally used in regard to a gender performance that enables one to 'pass' without strangers knowing that one's gender and sex are not in accord; see also 'cisgender'

Classificatory kin: fictive kin; people who are referred to with kin terms and said to be related but who are known by everyone not to be

Cline: see 'geographic cline'

Cloning: self-replication, for instance by splitting, so that each generation is a duplicate of the last; reproduction involving no sex; see also 'asexual reproduction'

Closed societies: societies in which a person's social status is ascribed; often it is based on birth

Colostrum: antibody-packed liquid that comes out of breasts prior to breast milk in a new mother

Commensality: eating together to create and cement a bond; sometimes this bond is literally one of shared substance

Communitas: sense of bondedness or community loyalty that people who go through a rite of passage together have toward one another

Community immunity: see 'herd immunity'

Comparative method: entails viewing each culture in comparison with others or from a comparative perspective so that universals can be derived

Complex adaptive system: system or network of dynamically interrelated parts between which information can flow; this system can change itself adaptively in response to changes in its environment (compare to 'mechanical system')

Complex society: chiefdom or state; heterogeneous in structure (refers to social organization, not culture)

Concordance: when features or traits co-occur with predictable regularity

Consanguineal kin: kin who share blood or are from the same bloodline; no other substance but blood is implicated in consanguineality, which is therefore a subset of consubstantiality (see 'consubstantial kin')

Consubstantial kin: kin related through shared substance, whether immaterial (e.g., nurturance, love) or material (e.g., bodily fluids such as breast milk; DNA, blood, but see 'consanguineal kin')

Co-sleeping: purposefully sleeping in close proximity, for instance in the same room (but not necessarily the same bed) as per cultural recommendations

Couvade: male participation in pregnancy, sometimes as demonstrated through the male experience of food cravings or morning sickness

Cultural consonance: the degree to which one's lifestyle fits with the lifestyle that one's culture recommends and that one thereby aspires to; includes (so see also) 'role incongruity'

Cultural determinism: the belief that culture alone determines one's capacities and characteristics (often used in opposition to 'biological determinism')

Cultural Revolution: dramatic shift in the archeological record occurring about 75,000 years ago indicating the onset of true behavioral modernity in *Homo sapiens sapiens*, including full-fledged capacity for culturally supported cooperative social life; see also 'upper Paleolithic Revolution'

Culturally constructed: created and maintained by culture; ideas about a category (e.g., success, proper greetings, headache) that a culture provides

Culture: a uniquely human adaptive strategy; the totality of each human group's shared, learned heritage; this includes social, political, economic, religious, cosmological, linguistic, health promotion, and other systems

Culture-bound syndromes: culturally unique, culturally named conditions that bring together, as syndromes, a variety of symptoms; often linked to stress

Cystic fibrosis: recessive genetic disease that disrupts the body's ability to absorb nutrients, breathe, and sweat; generally fatal in childhood if untreated

Cystic fibrosis transmembrane conductance regulators (CFTRs): see 'transmembrane regulators'

D

Deliberate practice: systematic, focused, practice engaged in purposefully with the conscious aim of improving particular skills or skill segments

Developmental: related to or occurs during the time that an individual organism is growing or maturing

Developmental adjustment: a responsive, irreversible change in growth or biochemical processes during development that occurs in reaction to environmental conditions

Diabetes: noninfectious chronic disease in which the body's ability to regulate glucose or blood sugar via insulin is compromised; can be controlled through diet, exercise, and medication; found more frequently in populations with higher levels of obesity and a more sedentary lifestyle

Dietary staple: food (usually plant-based) that a group eats a lot of; examples for various cultures are bread, tortillas, mongongo nuts, and rice

Disease ecology: theoretical framework that focuses on the germ, the immediate or proximal environment in which it lives, and the context in which germs are spread

Division of labor: extra-familial economic specialization, in which different groups within a society do different jobs and are therefore interdependent

DNA (deoxyribonucleic acid): biological molecule containing genetic information; each strand of DNA is essentially a string of genes

Dominant alleles: alleles (variants of a gene) that make their own instructions take precedence over recessive alleles so that only they are expressed as a trait; a dominant trait is a trait expressed in this way

Dynamic: in motion; characterized by change or activity; not static

E

Ecological selection: an increase in a given, pre-existing trait's frequency in the next generation's gene pool that happens when a change in the environment makes that trait advantageous; natural selection's workhorse

Ecosystem Services: system-benefitting maintenance functions of various parts of the ecosystem

Ecosystems: balanced systems comprising multiple abiotic or nonliving materials and biotic or living populations, each occupying a particular niche, coexisting in a balanced way, via energy flows, so that species' population numbers and volumes of abiotic materials hold steady even as time passes

Embodiment: the literal 'making physical' of culture

Emergence: a novel property or pattern that comes about as the result of interactions between parts of a system that aim to keep the system working and, in doing so, lead to something entirely new and otherwise unpredictable

Enamel hypoplasia: a dental condition; a sign of nutritional stress; there is less enamel in some places on the teeth than there would be on healthy, evenly nourished teeth

Endemic: disease that is local to or well-established in a particular region and exists in balance with the population that hosts it

Environmental pressure: anything in the environment that reduces a population's ability to function or puts a damper on its potential; includes disease, deprivation (through drought, famine, and so forth), disaster (flooding, earthquake), and an increase in predators

Environmental reservoir: abiotic substance that can hold agents so that hosts come in contact with them (e.g., water, soil); biotic contrast is 'reservoir host'

Epidemic: disease that spreads over a population with great speed

Epidemiological polarization: when the epidemiological profile of the poor entails high levels of mortality from infectious disease and lots of death at young ages, while the rich live longer and die later from noninfectious diseases

Epidemiological profile: a profile or picture of a given group describing what diseases and other health challenges they are experiencing or have experienced, and to what degree

Epidemiological transition: shift from an epidemiological profile highlighting infectious diseases with high mortality rates to one in which noninfectious or chronic diseases with low mortality but high morbidity rates feature centrally

Epidemiological triangle: triangle representing the relationship between the host, environment, and agent

Epidemiology: study of disease distribution and its determinants

Epigenetic change: a phenotypically relevant modification in how a gene is expressed that is engineered by an epigenetic event or events; may be maladaptive and may be reversible

Epigenetic events: biochemical processes in the epigenome that turn genes on or off or otherwise significantly affect their expression

Epigenetics: the study of interactions between genes and their environment that bring the phenotype into being (as per Waddington 1942)

Epigenome: a layer of biochemical processes overlying the genes and affecting their expression

Ethnicity: facet of identity related to shared national or regional origins and shared culture

Ethnocentrism: putting one's own culture at the center of any interpretation; viewing other cultures through one's own culture's lens

Ethnographic record: all ethnographic accounts, old and new, taken together

Ethnography: written account of one particular culture, traditionally in book format

Ethnology: comparative study of cultures

Ethos: a culture's overall worldview or fundamental values

Evolution: change, in any direction for any reason

Exposure: being in close proximity to a chemical, pathogen (germ), radioactivity, or extremes of weather.

F

Fauna: animals and other nonplant life forms (often used in the phrase 'flora and fauna')

Fight or flight reaction: when a stressed individual is immediately prepared via internal changes (a stress response) for a fight or for fleeing the scene; thought to aid species survival; see also 'stress'

Fitness: a measure of adaptation related to mortality and fertility; when comparing two groups, lower mortality with higher fertility is an indicator that one group is better fitted or adapted to an environment than the other

Flora: plants (often used in the phrase 'flora and fauna')

Foraging: gathering vegetable and animal or other foods directly, also known as hunter-gatherer lifestyle; often, foragers are nomadic, at least seasonally; about two-thirds of the forager diet is plant based

Fosterage: social but not legal adoption, sometimes temporary but often long-term

G

Gametes: reproductive cells (eggs, sperm)

Gender: culturally recommended scripts for enacting masculinity or femininity; cultural ideas about what tangible sex differences mean

Gene: a discrete sequence of deoxyribonucleic acid (DNA) that contains (stores or codes) the recipe for a particular protein, or regulates the expression of protein-coding genes; genes are passed along from parent to offspring through the process of reproduction

Gene flow: genetic evolution due to migration of genes from one gene pool into another

Gene pool: the entire population's sum total of genes

Genetic adaptation: an adaptive change in the frequency of a given gene or genes in the gene pool from generation to generation caused by natural selection; a specialized form of genetic evolution

Genetic drift: genetic evolution due to random chance

Genetic evolution: a change in the frequency of a given gene or genes in the gene pool from generation to generation caused by a variety of mechanisms, including mutation, genetic drift, gene flow, and natural selection (genetic adaptation is a particular form of genetic evolution)

Genome: the entire stock of genetic information carried by a given species

Genotype: the set of genes an individual carries (as opposed to their manifestation)

Genus: a broad subfamily of organisms; each genus includes or is made up of various species

Geographic cline: incrementally or gradually changing distributions of traits over geographic regions, which are related to incrementally changing environmental pressures or challenges

Geographic luck: living, fortuitously, in a location that about 10,000 years ago had great floral and faunal candidates for domestication (Diamond's term; 1997); see also 'Agricultural Revolution'

H

Health–wealth gradient: the positive correlation between socioeconomic status and health and well-being

Hemoglobin: oxygen-binding or carrying molecules found in red blood cells

Herd immunity: immunity to a vaccine-preventable disease measured at the population level; achieved once a threshold proportion of the group has been immunized, decreasing the likelihood of transmission, even to unimmunized individuals; see also 'community immunity'

Heterozygous: refers to inherited gene pairs; occurs when a person has two different alleles for a pair

High-demand, low-control roles: social roles (typically, jobs) whose occupants are ordered around by bosses or superiors and who have little control over what they do (e.g., assembly line workers)

Historical particularism: a viewpoint in which, because different societies have different histories, their cultures are therefore unique and particular (compare to 'unilineal evolution')

HIV/AIDS (human immunodeficiency virus/acquired immune deficiency syndrome): AIDS is a chronic, potentially fatal viral infection of the immune system; HIV, the virus that causes it, is spread via direct contact with bodily fluids; death generally comes through opportunistic infections such as by pneumonia or tuberculosis

Holism: a perspective that views parts of a system within the context provided by the system as a whole; holds that single parts of a system cannot be understood in isolation and that the whole is more than the sum of its parts

Homeostasis: literally, steady state; a balance achieved when small changes are made that do not notably alter the system but instead allow it to run as it has been running

Homo erectus/ergaster: literally, 'erect man'; ancestral human-like species that stood erect and walked habitually on two feet; emerged about 1.5 million years ago

Homo habilis: literally, 'handy man'; ancestral human-like species that systematically made and used tools; emerged just over 2 million years ago

Homo sapiens: literally, 'wise' or 'knowing man'; fully anatomically modern human beings; the earliest *Homo sapiens* fossils date back to about 300,000 years ago

Homo sapiens sapiens: behaviorally and anatomically modern human subspecies to which contemporary human beings belong; became prevalent perhaps 100,000 years ago; signs of the full emergence of this subspecies as culture-carrying beings enter the archaeological record about 75,000 years ago (earlier some places; later in others)

Homozygous: refers to inherited gene pairs; occurs when a person has two identical alleles for a pair

Horticulture: gardening lifestyle (no irrigation, no plows, no fertilizer, no permanent settlements)

Host: person or individual organism sickened by an agent's active presence

Hunter-gatherer: see 'foraging'

Hygiene hypothesis: holds that the immune systems of children exposed to more microbes have greater tolerance for the irritants that trigger asthma and allergies; see also 'old friends hypothesis'

I

Inclusive fitness: genetic outcome of altruistically favoring the survival of one's close relatives or offspring equivalents (i.e., individual organisms with self-like genes) even at the cost of one's own reproductive success

Index trait: a trait that indexes a potential mating partner's fitness relative to others in the group; in sexual selection, organisms rely on index traits (albeit generally not consciously) to help insure the viability of the next generation

Industrial melanism: when industrial processes darken the environment, leading to increased fitness for members of a species in which more melanic pigment is expressed and so supporting the natural selection of the melanic trait

Industrialized agriculture: farming (often of one crop only) for profit, rather than food, using highly mechanized means

Infectious disease: a disease resulting from the presence and activity of a pathogenic microbial agent that can be communicated from one person to another (contrast with non-infectious diseases, which involve no agents or germs)

Innate immunity: genetically inherited immunity

Institutionalized racism: differential access to society's goods, services, and opportunities that is produced through organizational or institutional practices, such rules regarding who can own private property

Intersectionality: convergence of various dimensions of identity such as gender, race or ethnicity, age, and class; overlapping, interlocking systems of discrimination or oppression based on such dimensions that support the status quo

Intersex: non-binary sex development; when sex characteristics including chromosomes, gonads, genitals, and sex hormones are nonconcordant or do not follow typical 'male' or 'female' patterning

Intensification: the process of doing more to get food; more intensified subsistence strategies (e.g., agriculture) manipulate or interfere with the environment more than less intensive strategies (e.g., scavenging) do

L

Lactose tolerance: the ability to digest milk sugars (lactose) until late in life

Lamarckism: the idea that acquired characteristics can be inherited (named after naturalist Jean-Baptiste Lamarck)

Leukocytes: white blood cells; key to good immune system functioning

Liminal: occupying a threshold; standing in limbo between social statuses; the middle phase of a rite of passage

Lineage: descent group with shared ancestry

Lymphocytes: those leukocytes or white blood cells that help our bodies remember and recognize previous invaders and support our bodies in destroying them

M

Macroevolution: genetic adaptation that brings about whole new species

Macronutrients: the three key nutrients: proteins, carbohydrates, and fats

Malaria: a disease caused proximally by parasites belonging to the genus *Plasmodium* and carried by the *Anopheles* mosquito; entails cyclical high fevers, headaches, and often death in humans; one of the top ten causes of death worldwide

Market exchange: impersonal exchanges, often involving money; goods have no links to their producers; once a good is paid for no further obligation ensues

Matrilineage: descent group in which heredity is figured through the female lineage.

Meaning response: self-healing attributable to knowledge or belief (to the cultural meaning) that a practice or process carries; see also 'placebo effect'

Mechanical system: system of dynamically interrelated parts that cannot change itself or adapt to changes in its environment (compare to 'complex adaptive system')

Melanin–folate–vitamin D triangle: relationship between melanin, folate, and vitamin D that is naturally selected for in relation to UVB exposure rates in a region; must be balanced so that the skin has just enough melanin (pigmentation, darkness) to enable vitamin D production and so calcium absorption while still protecting folate stores from degradation

Metabolism: the biochemical process of breaking down and repurposing food components for bodily use

Metapopulation: a group of separately located populations of the same species that interact in some way and so maintain their identification as one species

Microbiome: the pooled genetic material of the microbial organisms living on and in our bodies

Microevolution: genetic adaptation that is not major enough to lead to speciation (the emergence of a new species)

Micronutrients: vitamins and minerals without which the biochemical processes entailed in fueling human life cannot happen

Monogenic: traits traceable to just one gene; most traits, in contrast, are 'polygenic'

Morbidity: sickness

Mortality: death

Multiregional metapopulation model: holds that *Homo sapiens sapiens* evolved not in one geographic spot or as a single lineage but in many spots through many intertwining lineages concurrently, as genes flowed back and forth between ancestral populations and across various regions of the earth; with enough such gene flow, regional populations never diverged enough genotypically from the population as a whole for speciation to occur; see also 'metapopulation'

Mutation: genetic miscopying, often due to some kind of exposure, for instance to a toxic chemical, or radiation, or a particular virus

N

Natural selection: the key mechanism of genetic evolution; a process by which the genes for expressed traits that happen to give an organism an adaptive advantage for survival under given environmental conditions are therefore more likely to be found in the next generation's gene pool, if the same environmental conditions persist; descent with modification

Neolithic Revolution: see 'Agricultural Revolution'

Niche: species-specific way of making a living or subsisting

Niche construction: when organisms, through biological and behavioral processes, act upon (construct or modify) the environmental niches that they (and other organisms) occupy just as those niches act upon them

Nonconcordance: when things (features, traits) do not co-occur with predictable regularity; often referenced to dismiss claims that biological race exists

Nonlinear: when ramifications of an event do not follow from the event in a predictable manner but instead occur as surprises that lead to surprises of their own

Nonrandom mating: selective mating, driven by conscious rules or unconscious impulses; happens in human populations for instance when kinship or class-related rules or ethnic boundaries limit who can build a family with whom; see also 'sexual selection'

Normal science: scientific practice that is driven by a shared, paradigmatic theory about the world that has long been agreed upon

Nutrition transition: switch from a traditional diet to the "Western diet," which is high in fat, sugar, and processed foods and low in fiber; increases a group's risk for diet-linked diseases such as diabetes and heart disease

O

Old friends hypothesis: updated version of 'hygiene hypothesis' focusing on particular microbes with which we have co-evolved, terming them 'old friends' because of the services they provide to our bodies; see also 'microbiome'

One Health: theoretical model emphasizing connections between animal and human health whereby the health (and health-related actions) of one influences the health of the other

Open societies: societies with inbuilt class mobility (or at least the myth of such)

Organic analogy: society is like an organism, with differentiated systems held together by mutual interdependence; it will evolve over time from being simple and homogeneous (like an amoeba) to being heterogeneous and complex (also sometimes called 'organismic analogy')

Oxytocin: hormone involved in labor and lactation that can also promote trust and empathy and dampen the stress response

P

Paleolithic diet: diet said to mimic that of our pre-agricultural forager forebears, generally without taking into account how geographic diversity and various other complicating factors actually shaped said diet

Paleolithic Revolution: see 'Cultural Revolution'

Pandemic: disease that spreads around the world with great speed

Paradigm: a theoretical framework that guides scientists in deciding what questions to ask, and forms a lens through which scientists interpret data

Paradigm shift: a radical transition in thinking; see also 'paradigm'

Partible paternity: occurs in additive kinship systems when more than one man provides part of the matter or substance (generally via semen) that becomes a child, and so a child has more than one father; see also 'consubstantial kin'

Participant-observation: a hands-on approach to ethnographic data collection that involves living among members of the culture under study, speaking their language, and participating in as much of their daily lives as is possible

Passive immunity: immunity acquired by infants from breastfeeding; antibodies are passed along via breast milk

Pastoralism: herding lifestyle; nomadic or semi-nomadic.

Pathogen: germ or infectious agent

Pathogenesis: the creation or genesis of damage or pathology in the body

Patrilineage: descent group in which heredity is figured through the male lineage

Peppered moths: species of moth found most famously in the forests of England; its coloration can evolve from lighter to darker and back again when the environment changes so that one coloration is more adaptive than the other

Performative: a communicative act that does not just describe a situation but creates it, as when pronouncing a couple married makes them married

Phagocytes: those leukocytes or white blood cells that eat or otherwise get rid of pathogens

Phenotype: the manifestation or measurable expression of the genotype

Phytochemicals: components or chemical compounds in a plant; can provide ingrown or natural protection from predators and disease

Placebo: an inert substance or an act that is not biomedical and does not effect a clinical cure

Placebo effect: a measurable, observable, or felt improvement in health that is not directly attributable to biomedical treatment; see also 'meaning response'

Plasticity: malleability, particularly in relation to an organism's developmental range

Political economy: a theoretical perspective that considers phenomena or processes in political and economic context; rather than to focus on proximate factors, the focus is on ultimate causes

Polyculture: old-fashioned or traditional (pre-industrial) agriculture, involving multiple species and thus multiple circles of flowing energy and matter

Polygenic: traits that develop through the interaction of a collection of numerous genes (poly means many)

Polymorphic: having various forms (poly means many); see also 'alleles'

Polyphasic sleep: sleep taken in several segments versus in one long, consolidated overnight bout

Porotic hyperostosis: bones that are more porous than they should be; a sign of iron deficiency, which (whether caused by malnutrition or infection) triggers the expansion of bone tissues that form red blood cells

Primary health care: a public health movement that focuses on preventing problems before they happen.

Proximate: immediate; very near; proximate causes are direct causes

Punnett square: table used for the graphical representation and calculation of possible allele combinations resulting when two individual organisms breed.

Q

Quickening: the point in gestation at which fetal movement can be felt by the birth mother

R

Race: a clearly differentiated subspecies or subgroup within a species; races have features that regularly co-occur; human subspecies do not exist

Racialization: the simple classification of people according to so-called race; nonevaluative (contrast with 'racism')

Racism: an evaluative position in which people are classed by so-called race and the races are rated and ranked (contrast with 'racialization')

Recessive allele: gene form that can be expressed only if paired with a self-same allele; can be dominated or overridden by a dominant allele; a recessive trait is a trait overridden in this way

Reciprocal relationship: an exchange relationship infused with emotional value, and meant to be of long-standing duration

Red Queen Hypothesis: the idea that a population in which genetic variation is not maintained is at a distinct disadvantage in a world that is constantly producing new environmental pressures

Redistribution: goods are taken from one subgroup for use with another, or for use society-wide (e.g., to build roads)

Reductionism: the idea that an entire system can be explained by a single aspect of that system

Reflexivity: the act of deeply examining one's biases and motives to grasp how they influence one's perceptions and conclusions

Relativism: evaluates the ideas and practices enacted by members of a given culture by that culture's own standards

Replacement hypothesis: holds that populations of *Homo sapiens* who left Africa in earlier waves were decimated by diseases newer population waves brought, or were driven out or otherwise replaced by the newcomers

Reservoir host: living organism that carries a pathogen as a host and serves as a vector or pathogen-carrying organism

Resilience: strength in the face of otherwise stressful situations

Resistance: ability of an organism to withstand a particular pathogen or stressor; in pathogens, an evolved capacity to remain unaffected by an antibiotic due to vulnerable organisms dying off prior to reproducing themselves

Rite of passage: a set of ritual acts intended to move a person or a group of people from one social status to the next

Ritual: set of actions performed for their culturally relevant symbolic value; a multimedia event that, when enacted, increases group cohesion

RNA (ribonucleic acid): biological molecule known for its role in protein synthesis; also found to regulate gene expression

Role incongruity: when one is not living up to the expectations entailed in the role one has been placed in or has elected to take on; a specific manifestation of low cultural consonance

S

Scavenging: minimum-impact lifestyle in which subsistence depends on picking up dead organisms and fruit (etc.) that has dropped off the plant

Seasonality: when changing seasons lead to shifts in diet quality and quantity

Sedentism: when a population lives in a permanent settlement, as agriculturalists do

Sex: biologically differentiated status of male or female, related to genital and chromosomal endowment; regarding non-binary sex differentiation, see 'intersex'

Sexual reproduction: reproduction or production of offspring in which more than one organism contributes genes, and so genes are shuffled around, generally with half coming from one parent and half from one other; ensures variation, as every offspring can receive a slightly different combination of genes

Sexual selection: selective mating, driven by unconscious impulses or conscious rules; happens in human populations for instance when kinship or class-related rules or ethnic boundaries limit who can build a family with whom; see also 'nonrandom mating'

Shared intentionality: having the motivation and ability to collaborate with others (as a "we") with a joint objective in mind

Sickle cell anemia: gene-linked disease-causing red blood cells to carry less oxygen and to form an abnormal crescent shape that can clog circulation; generally incurable

Simple society: band or tribe organization; homogeneous in structure (refers to social organization, not culture)

Smallpox: potentially fatal, frequently epidemic acute viral infection accompanied by an extensive pustular or blister-like rash and high fever; spread via contact with pustular fluid (pustules can be near or in mouth or nose, so that droplets in an infected person's breath also may lead to contagion)

Social cohesion: when a society holds itself together

Social condensation: when larger groups break into smaller factions, much as water vapor will condense into droplets on a cold drink bottle

Social death: social treatment of an individual as dead, or as a non-person; removal of persons from the category of those deemed worthy of normal social interaction

Social justice: the equitable distribution of basic human rights such as the right to healthful living conditions, and equal opportunities for equal outcomes among all social groups

Social soundness: when aid groups work with community leaders and members to create a good fit between programs and cultures

Social stratification: hierarchical social organization

Social structure: refers to the way in which a society is structured or built from the relationships between the classes or groups of people within it

Somatization: the projection of mental attitudes or concerns onto the body so that they are expressed as physical symptoms; often this provides an outlet or channel for stress

Specialization: when different groups within a society do different jobs and are therefore interdependent

Species: a discrete organism type that can reproduce itself, for instance by interbreeding and giving birth to fertile offspring; reproduction of fertile offspring is impossible across species boundaries

State: hierarchical socio-political entity in which bureaucracy is centralized, with multiple parts, including a formal legal system

Stigmatized: marked for rejection, generally with a visual sign (e.g., a tattoo on a visible part of one's body, a sign on one's clothing)

Stress: the body's immediate response to environmental pressures; includes reactions of the nervous, hormonal, and immune systems; protective in the short run; key biochemicals entailed are cortisol and catecholamines, including epinephrine; see also 'fight or flight response' (regarding chronic stress, see 'allostasis'; contrast with 'environmental pressure')

Structural violence: occurs when the shape of a given social structure harms or is harmful to the people who occupy certain positions within that social structure

Subsistence strategy: approach to or means of making a living or extracting food from the environment, such as by foraging or industrial agriculture

Surplus: extra food or material items that can be stored

Surrogate: stand in mother or father for procreative purposes; surrogate mothers (gestational surrogates) can carry (gestate) children for their legal mothers.

Syndemic: a cluster of health problems that work together synergystically, reinforcing and often exacerbating each other

Synergy: when the combined effect of two or more substances or agents is altogether distinct from the separate effect of each; see also 'complex adaptive system'

System: two or more elements whose organization serves a common purpose or outcome

Systems thinking: a point of view that highlights relationships and what emerges from them; properties of a system can be neither explained nor determined by examining its parts alone

T

Taboo: rule against coming in contact with something, often due to it being a well of power; see also 'anomaly'

Theory of mind: the ability to attribute beliefs, intentions, desires, emotional states, etc., to the minds of others, and to grasp that these may be different from one's own and may affect others' behaviors accordingly

Toxicants: humanly introduced poisons (as opposed to toxins, which occur naturally); found, for instance, in industrial pollution

Transgender: when a person's gender identity does not correspond with their birth sex as seen from the two-sex, two-gender perspective; in some cultures what mainstream Westerners see as transgender roles leave a two-gender system behind

Transgenerational epigenetic change: epigenetic change that resists stripping during embryonic development and persists into the fourth generation in the absence of its initial stimulus; see 'epigenetic events,' 'epigenetics,' 'epigenome'

Transmembrane conductance regulators (TRs): biological structures that move (conduct, transfer) things across membranes, such as the intestinal membrane, regulating their movement or flow across these membranes; certain TRs are implicated in resistance and susceptibility to cholera and in cystic fibrosis

Tribe: egalitarian socio-political group made up of a number of bands that have cross-cutting inter-band ties; leadership is achievement based

Tuberculosis: potentially fatal bacterial infection of the lungs; can spread to other organs and the skeleton; long-lasting; spread via droplets released when a sick person coughs or sneezes

U

Ultimate: in the end; ultimate causes are where the buck stops

Unilineal evolution: the idea that culture (or anything) evolves in one direction over time, progressively improving (nonimprovement would be 'devolution')

Upper Paleolithic Revolution: dramatic shift in the archeological record occurring about 75,000 years ago indicating onset of true behavioral modernity in *Homo sapiens sapiens*, including full-fledged capacity for culturally supported cooperative social life; see also 'Cultural Revolution'

V

Vector: organism that transmits a pathogen to another organism (e.g., mosquito)

Z

Zoonotic: a disease that originated in an animal species and then jumped to humans

REFERENCES

American Anthropological Association. 2007a. "Global Census." In *Race: Are We so Different?* Retrieved from http://understandingrace.org/GlobalCensus.

American Anthropological Association. 2007b. "Only Skin Deep." In *Race: Are We so Different?* Retrieved from http://understandingrace.org/OnlySkinDeep

Anonymous. "Hunters and Gatherers: The Search for Survival - An Introduction." *Cultural Survival Quarterly Magazine* 8, no. 3 (September 1984). Retrieved from https://www.culturalsurvival.org/publications/cultural-survival-quarterly/hunters-and-gatherers-search-survival-introduction.

Antón, Susan. "The Many Faces of Early *Homo*." *Bulletin of the General Anthropology Division* 25, no. 1 (2018): 1–5.

Archibold, Randal C. "Indians' Water Rights Give Hope for Better Health." *New York Times*, August 30, 2008, A20.

Baer, Roberta D., Susan C. Weller, Javier Garcia de Alba, and Ana L. Salcedo. "Ethnomedical and Biomedical Realities: Is There an Epidemiological Relationship between Stress-Related Folk Illnesses and Type 2 Diabetes?" *Human Organization* 71, no. 4 (2012): 339–47.

Balée, William. *Inside Cultures: A New Introduction to Cultural Anthropology*. Walnut Creek, CA: Left Coast Press, Inc., 2012.

Balter, Michael. "Human Language May Have Evolved to Help Our Ancestors Make Tools." *Science*, 2015. Retrieved from http://www.sciencemag.org/news/2015/01/human-language-may-have-evolved-help-our-ancestors-make-tools.

Barr, D. *Health Disparities in the United States: Social Class, Race, Ethnicity, and Health*. Baltimore, MD: Johns Hopkins University Press, 2008.

Bart, P. "Why Women's Status Changes with Middle Age." In *Sociological Symposium*. Blacksburg, VA: Virginia Polytechnic Institute and State University, Department of Sociology, 1969.

Bateson, Patrick, David Barker, Timothy Clutton-Brock, Debal Deb, Bruno D'Udine, Robert A. Foley, Peter Gluckman, et al. "Developmental Plasticity and Human Health." *Nature* 430 (2004): 419–21.

Becker, Anne E. *Body, Self, and Society: The View from Fiji*. Philadelphia, PA: University of Pennsylvania Press, 1995.

Becker, G., A. Butler, and R. Nachtigall. "Resemblance Talk: A Challenge for Parents Whose Children Were Conceived with Donor Gametes in the US." *Social Science and Medicine* 61 (2005): 1300–9.

Belkin, Douglas. "Blue Eyes Are Increasingly Rare in the Americas." *New York Times*, October 18, 2006.

Bilu, Yoram. "From Milah (Circumcision) to Milah (Word): Male Identity and Rituals of Childhood in the Jewish Ultraorthodox Community." *Ethos* 31, no. 2 (2003): 172–203.

Biodynamic Farming and Gardening Association. "About Biodynamics." 2012. Retrieved from www.biodynamics.com/biodynamics.

Blackwell, B, S.S. Bloomfield, and C.R. Buncher. "Demonstration to Medical Students of Placebo Responses and Non-Drug Factors." *Lancet* 1, no. 7763 (1972): 1279–82.

Blair, Thea. "From Bullying to Belonging: How Peer Massage Relieves Social Stress." 2012. Retrieved from http://www.waldorftoday.com/2012/10/from-bullying-to-belonging-how-peer-massage-relieves-social-stress/.

Blakeslee, Sandra. "Cells that Read Minds." *New York Times*, January 10, 2006.

Boas, Franz. "Changes in Bodily Form of Descendants of Immigrants." *American Anthropologist (N.S.)* 14, no. 3 (1912): 530–62

Bogin, Barry. "The Tall and the Short of It." *Discover* 19, no. 2 (1998): 40–4.

Bourgois, Philippe. "The Power of Violence in War and Peace: Post-Cold War Lessons from El Salvador." *Ethnography* 2, no.1 (2001): 5–34.

Bruner, Emiliano, and Giorgio Manzi. "Variability in Facial Size and Shape among North and East African Human Populations." *Italian Journal of Zoology* 71, no. 1 (2004): 51–6.

Burton, John W. *Culture and the Human Body*. Long Grove, IL: Waveland Press, 2001.

California Newsreel. *Finding Hope for the Future by Reclaiming the Past*. San Francisco, CA: California Newsreel, 2008a.

California Newsreel. *Unnatural Causes: In Sickness and in Wealth*. San Francisco, CA: California Newsreel with Vital Pictures, 2008b.

California Newsreel. *Unnatural Causes: When the Bough Breaks*. San Francisco, CA: California Newsreel with Vital Pictures, 2008c.

Cannon, Walter B. "'Voodoo' Death." *American Anthropologist* 44, no. 2 (1942): 169–81.

Carroll, Lewis. *Through the Looking-Glass*. London: Macmillan, 1871.

Case, Anne, and Angus Deaton. "Mortality and Morbidity in the 21st Century." *Brookings Papers on Economic Activity* (2017): 397–443. Spring. Retrieved from https://www.brookings.edu/wp-content/uploads/2017/08/casetextsp17bpea.pdf.

Cassidy, C. "The Good Body: When Bigger Is Better." *Medical Anthropology* 13 (1991): 181–213.

Cohen, Mark Nathan. *Health and the Rise of Civilization*. New Haven, CT: Yale University Press, 1989.

Cormier, Loretta A. *The Ten-Thousand Year Fever: Rethinking Human and Wild-Primate Malarias*. Walnut Creek, CA: Left Coast Press, Inc., 2011.

Cousins, Norman. *Anatomy of an Illness as Perceived by the Patient: Reflections on Healing and Regeneration*. New York: Bantam Books, 1981.

Dailey, Kate, and Abby Ellin. "America's War on the Overweight." *Newsweek*, 2009.

Darwin, Charles R. *On the Origin of Species by Means of Natural Selection, or the Preservation of Favoured Races in the Struggle for Life*. 1st ed. London: John Murray, 1859.

David, Richard J., and James W. Collins. "Differing Birth Weight among Infants of U.S.-Born Blacks, African-Born Blacks, and U.S.-Born Whites." *New England Journal of Medicine* 337, no. 17 (1997): 1209–14.

Devlin, Hannah. "Tracing the Tangled Tracks of Humankind's Evolutionary Journey." *The Guardian* (US edition), February 12, 2018. Retrieved from https://www.theguardian.com/news/2018/feb/12/tracing-the-tangled-tracks-of-humankinds-evolutionary-journey.

Diamond, Jared. "Race without Color." *Discover* 15, no. 11 (1994): 82–9.

Diamond, Jared. *Guns, Germs, and Steel: The Fates of Human Societies*. New York: WW Norton and Company, 1997.

DiAngelo, Robin. "White People Are Still Raised to be Racially Illiterate. If We Don't Recognize the System, Our Inaction Will Uphold It." Retrieved from https://www.nbcnews.com/think/opinion/white-people-are-still-raised-be-racially-illiterate-if-we-ncna906646, September 16, 2018.

Dick, Lyle. "'Pibloktoq' (Arctic Hysteria): A Construction of European-Inuit Relations?" *Arctic Anthropology* 32, no. 2 (1995): 1–42.

Doleac, Jennifer L., and Luke C. D. Stein. 2010. "The Visible Hand: Race and Online Market Outcomes." ssrn.com/abstract=1615149.

Dolgin, Elie. "The Myopia Boom." *Nature* 519, no. 7543 (2015): 276–8

Douglas, Mary. *Purity and Danger: An Analysis of Concepts of Pollution and Taboo*. London: Routledge and Keegan Paul, 1966.

Dressler, William W., and James R. Bindon. "The Health Consequences of Cultural Consonance: Cultural Dimensions of Lifestyle, Social Support and Blood Pressure in an African American Community." *American Anthropologist* 102, no. 2 (2000): 244–60.

Dretzin, Rachel. "Interview—Clifford Nass." Boston: WGBH Educational Foundation. 2010. Retrieved from www.Pbs.Org/Wgbh/Pages/Frontline/Digitalnation/Interviews/Nass.Html.

Eakin, Emily. "The Excrement Experiment." *The New Yorker*, December 1, 2014. Retrieved from https://www.newyorker.com/magazine/2014/12/01/excrement-experiment.

Eaton, S. B., S. B. Eaton III, and M. J. Konner. "Paleolithic Nutrition Revisited: A Twelve-Year Retrospective on its Nature and Implications." *European Journal of Clinical Nutrition* 51 (1997): 207–16.

Economist. "Eyeing up the Collaboration." *Economist* 381, no. 8502 (2006): 90.

Economist. "I Am Just a Poor Boy Though My Story's Seldom Told: Neuroscience and Social Deprivation." *Economist* 391, no. 8625 (2009): 82.

Economist. "Mens Sana in Corpore Sano." *Economist* 396, no. 8689 (2010a): 75–6.

Economist. "Stiletto Stiffness." *Economist* 396, no. 8691 (2010b): 84.

Ehrlich, Paul R. *Human Natures*. New York: Penguin Books, 2006.

Emba, Christine. "White Americans, Welcome to the Club of Being Asked, 'Where Are You Really From?'" *Washington Post*, February 2, 2018. Retrieved from https://www.washingtonpost.com/opinions/white-americans-where-are-you-really-from/2018/02/02/dc324fb2-0843-11e8-b48c-b07fea957bd5_story.html?utm_term=.b54ac88fe7ac.

Engels. "Manchester in 1844 (Excerpted from the Condition of the Working Class in England in 1844)." In *Essential Works of Socialism*, edited by Irving Howe, 58–71. Clinton, MA: Colonial Press, 1970 [1845].

Environmental Health Coalition. "The Science of Precaution: Barrio Logan Residents Use Research and Land Use Planning to Prevent Harm." *Race, Poverty & the Environment* 11, no. 2 (Winter 2004/2005): 53–5.

Eppig, Christopher, Corey L. Fincher, and Randy Thornhill. "Parasite Prevalence and the Worldwide Distribution of Cognitive Ability." *Proceedings of the Royal Society B* (2010). Retrieved from doi:10.1098/rspb.2010.0973

Falls, Susan. *White Gold: Stories of Breast Milk Sharing*. Lincoln, NE: University of Nebraska Press, 2017.

Farmer, Paul. *Pathologies of Power: Health, Human Rights, and the New War on the Poor*. Los Angeles, CA: University of California Press, 2005.

Fluehr-Lobban, Carolyn. *Race and Racism: An Introduction*. Lanham, MD: AltaMira, 2006.

Fuentes, Agustin. *Evolution of Human Behavior*. New York: Oxford University Press, 2009.

Fuentes, Agustin. *The Creative Spark: How Imagination Made Humans Exceptional*. New York: Dutton. 2017.

Galanti, Geri-Ann. *Caring for Patients from Different Cultures: Case Studies from American Hospitals*. 4th ed. Philadelphia, PA: University of Pennsylvania Press, 2008.

Galtung, Johan. "Violence, Peace, and Peace Research." *Journal of Peace Research* 6, no. 3 (1969): 167–91.

Gammeltoft, Tine M., and Ayo Wahlberg. "Selective Reproductive Technologies." *Annual Review of Anthropology* 43 (2014): 201–16.

Garcia, Richard. "The Misuse of Race in Medical Diagnosis." *Chronicle of Higher Education* 49, no. 35 (2003): B15.

Genetic Science Learning Center. "Nutrition & the Epigenome." *Learn.Genetics*, July 15, 2013. Retrieved from https://learn.genetics.utah.edu/content/epigenetics/nutrition/.

Gersten, Omer. "The Path Traveled and the Path Ahead for the Allostatic Framework: A Rejoinder on the Framework's Importance and the Need for Further Work Related to Theory, Data, and Measurement." *Social Science & Medicine* 66, no. 3 (2008): 531–5.

Gill, G. W. "Does Race Exist? A Proponent's Perspective." *Nova* 2000. Retrieved from www.pbs.org/wgbh/nova/first/gill.html.

Gilman, Sander L. *Making the Body Beautiful: A Cultural History of Aesthetic Surgery*. Princeton, NJ: Princeton University Press, 1999.

Goldschmidt, Walter. *The Bridge to Humanity: How Affect Hunger Trumps the Selfish Gene*. New York: Oxford University Press, 2006.

Goodman, Alan H., and Thomas L. Leatherman, eds. *Building a New Biocultural Synthesis: Political Economic Perspectives on Human Biology*. Ann Arbor: University of Michigan Press, 1998.

Goodyear, Dana. "Grub: Must We Eat Insects?" *New Yorker* August 15 and 22, no. 20B (2011): 38–47.

Gorman, Rachael Moeller. "Cooking Up Bigger Brains." *Scientific American*, 2008. Retrieved from https://www.scientificamerican.com/article/cooking-up-bigger-brains/.

Gould, Stephen Jay. "The Geometer of Race." *Discover* 15, no. 11 (1994): 65–9.

Gravlee, Clarence, H. Russell Bernard, and William R. Leonard. "Boas's Changes in Bodily Form: The Immigrant Study, Cranial Plasticity, and Boas's Physical Anthropology." *American Anthropologist* 105, no. 2 (2003): 326–32.

Gross, Daniel R. *Discovering Anthropology*. Mountain View, CA: Mayfield, 1992.

Gura, Trisha. "Nature's First Functional Food." *Science* 345, no. 6198 (2014): 747–9.

Hansen, Helena, and Julie Netherland, 2016. "Is the Prescription Opioid Epidemic a White Problem?" *American Journal of Public Health* 106, no. 12 (2016): 2127–2129.

Hansen, Margaret M., Reo Jones, and Kirsten Tocchini. "Shinrin-Yoku (Forest Bathing) and Nature Therapy: A State-of-the-Art Review." *International Journal of Environmental Research and Public Health* 14, no. 8 (2017): 851.

Harron, Mary. "A Hero for Our Times." *Independent*, January 29, 1995.

Henig, Robin Marantz. "How Science is Helping Us Understand Gender." NationalGeographic.com, 2017. Retrieved from https://www.nationalgeographic.com/magazine/2017/01/how-science-helps-us-understand-gender-identity/.

Herdt, Gilbert. *Guardians of the Flutes: Volume 1. Idioms of Masculinity*. Chicago, IL: University of Chicago Press, 1994 [1981].

Jabr, Ferris. "For the Good of the Gut: Can Parasitic Worms Treat Autoimmune Diseases?" *Scientific American*, December 1, 2010. Retrieved from https://www.scientificamerican.com/article/helminthic-therapy-mucus/.

Joralemon, Donald. *Exploring Medical Anthropology*. 2nd ed. San Francisco, CA: Pearson Allyn and Bacon, 2006.

Katz, J. N. *The Invention of Hetereosexuality*. Chicago, IL: University of Chicago Press, 2007 [1995].

Kidder, Tracy. *Mountains Beyond Mountains: The Quest of Dr. Paul Farmer, a Man Who Would Cure the World*. New York: Random House, 2003.

Kluger, Jeffrey. "The Science of Appetite." *Time*, May 31, 2007.

Kohn, David. "When Gut Bacteria Change Brain Function." *The Atlantic*, June 24, 2015. Retrieved from https://www.theatlantic.com/health/archive/2015/06/gut-bacteria-on-the-brain/395918/.

Kristof, Nicholas D. "Raising the World's IQ." *New York Times*, November 4, 2008, A43.

Kuhn, Thomas S. *The Structure of Scientific Revolutions*. Chicago, IL: University of Chicago Press, 1996 [1962].

Kumar, Pooja. "Indian Surrogacy: Blessing, Business or Both?" *Neem Magazine* 1, no. 4 (2008). Retrieved from www.neemmagazine.com/culture63.

Kuzawa Christopher W., and Elizabeth Sweet. "Epigenetics and the Embodiment of Race: Developmental Origins of US Racial Disparities in Cardiovascular Health." *American Journal of Human Biology* 21, no. 1 (2009): 2–15.

LaFleur, William R. *Liquid Life: Abortion and Buddhism in Japan*. Princeton, NJ: Princeton University Press, 1994.

Langley, Michelle. "Humanity's Story Has no End of Surprising Twists." *Sapiens.org*, March 15, 2018. Retrieved from https://www.sapiens.org/evolution/human-evolution-australia-asia/.

Langness, L L. *The Study of Culture*. 3rd, revised ed. Novato, CA: Chandler and Sharp, 2005.

Larsen, Clark Spencer. "The Agricultural Revolution as Environmental Catastrophe: Implications for Health and Lifestyle in the Holocene." *Quaternary International* 150, no. 1 (2006): 12–20.

Leach, Edmund R. "Time and False Noses." In *Reader in Comparative Religion: An Anthropological Approach*, edited by William A. Lessa and Evon Z. Vogt, 226–9. San Francisco, CA: Harper and Row, 1979 [1961].

Lee, Richard B. "Eating Christmas in the Kalahari." *Natural History* 78 no. 10 (1969):14ff. Retrieved from http://www.naturalhistorymag.com/htmlsite/master.html?http://www.naturalhistorymag.com/htmlsite/1103/1103_feature.html.

Lee, Richard B. "What Hunters Do for a Living, or, How to Make Out on Scarce Resources." In *Nutritional Anthropology*, edited by Alan H. Goodman, Darna L. Dufour and Gretel H. Pelto, 35–46. Mountain View, CA: Mayfield, 2000 [1968].

LeMoine, Genevieve M., Susan A. Kaplan, and Christyann M. Darwent. "Living on the Edge: Inughuit Women and Geography of Contact." *Arctic* 69, Suppl. 1 (2016): 1–12.

Lenkeit, Roberta Edwards. *Introducing Cultural Anthropology*. 4th ed. San Francisco, CA: McGraw Hill, 2009.

Lewontin, Richard. "The Apportionment of Human Diversity." *Evolutionary Biology* 6 (1972): 391–8.

MacDorman, M. F., and T. J Mathews. *Recent Trends in Infant Mortality in the United States. NCHS Data Brief, No. 9.* Hyattsville, MD: National Center for Health Statistics, 2008.

MacKenzie, Debora. "Cystic Fibrosis Gene Protects Against Tuberculosis." *New Scientist*. September 7, 2006. Retrieved from https://www.newscientist.com/article/dn10013-cystic-fibrosis-gene-protects-against-tuberculosis/.

Mallios, Seth. *Born a Slave, Died a Pioneer: Nathan Harrison and the Historical Archaeology of Legend.* New York: Berghahn Press, 2019.

Martinez, Lourdes S., Sharon Hughes, Eric R. Walsh-Buhi, and Ming-Hsiang Tsou. "'Okay, We Get It. You Vape': An Analysis of Geocoded Content, Context, and Sentiment Regarding E-Cigarettes on Twitter." *Journal of Health Communication* 23, no. 6 (2018): 550–62.

Maxted, Ian. 2000. "Cider and Eighteenth-Century Evidence-Based Healthcare." *Devon's Heritage*. Devon Library Services. Retrieved from www.devon.gov.uk/localstudies/100676/1.html.

Mays, V. M., S. D. Cochran, and N. W. Barnes. "Race, Race-Based Discrimination, and Health Outcomes among African Americans." *Annual Review of Psychology* 58 (2007): 201–25.

McElroy, Ann, and Patricia K. Townsend. *Medical Anthropology in Ecological Perspective*. 4th ed. Boulder, CO: Westview Press, 2004.

McKenna, James J., and Thomas McDade. "Why Babies Should Never Sleep Alone: A Review of the Co-Sleeping Controversy in Relation to SIDS, Bedsharing and Breast Feeding." *Paediatric Respiratory Reviews* 6 (2005): 134–52.

Mead, Margaret. *Male and Female*. New York: HarperCollins, 2001 [1949].

Mead, Margaret. *Sex and Temperament in Three Primitive Societies*. New York: HarperCollins, 2001 [1935].

Meigs, Anna S. "Blood Kin and Food Kin." In *Conformity and Conflict*, edited by James P. Spradley and David W. McCurdy, 117–24. Boston, MA: Little, Brown and Company, 1987.

Meigs, Anna S. "Food as a Cultural Construction." In *Food and Culture: A Reader*, edited by C. Counihan and P. Van Esterik, 95–106. New Brunswick, NJ: Rutgers University Press, 1997.

Melby, Melissa K., and Michelle Lampl. "Menopause, A Biocultural Perspective." *Annual Review of Anthropology* 40 (2011): 53–70.

Mennella, Julie A., Coren P. Jagnow, and Gary K. Beauchamp. "Prenatal and Postnatal Flavor Learning by Human Infants." *Pediatrics* 107, no. 6 (2001): e88.

Milo, Ron, and Rob Phillips. "How Big Are Genomes?" In: *Cell Biology by the Numbers*. Garland Science, 2016. Retrieved from http://book.bionumbers.org/how-big-are-genomes/.

Miner, Horace. "Body Ritual among the Nacirema." *American Anthropologist* 58, no. 3 (1956): 503–7.

Moerman, Daniel. *Meaning, Medicine, and the 'Placebo Effect.'* Cambridge: University of Cambridge Press, 2002.

Moerman, Daniel. "Society for the Anthropology of Consciousness Distinguished Lecture: Consciousness, 'Symbolic Healing,' and the Meaning Response." *Anthropology of Consciousness* 23, no. 2 (2012): 192–210.

Mooallem, Jon. "Neanderthals Were People, Too." *New York Times*, 2017. Retrieved from https://www.nytimes.com/2017/01/11/magazine/neanderthals-were-people-too.html.

Morrell, Kayla, and Melissa K. Melby. "Celiac Disease: The Evolutionary Paradox." *International Journal of Celiac Disease* 5, no. 3 (2017): 86–94.

Mosley, Michael. 2015. "The Extraordinary Case of the Guevedoces." *BBC News Magazine*. Retrieved from https://www.bbc.com/news/magazine-34290981.

Mullin, S., and M. Duwell. "Despite Much Progress, Significant Disparities Persist in New York City: New Report Details Disproportionate Burden of Illness and Premature Death among Poor, as Well as Black and Hispanic New Yorkers." Press Release (2004). New York City Department of Health and Mental Hygiene Office of Communications. Retrieved from www.nyc.gov/html/doh/html/press_archive04/pr085-0715.shtml.

National Center for Emerging and Zoonotic Infectious Diseases. "One Health > One Health Basics > History." Web page (2016). Centers for Disease Control and Prevention. Retrieved from https://www.cdc.gov/onehealth/basics/history/index.html.

National Institute of Diabetes and Digestive and Kidney Diseases. *The Pima Indians: Pathfinders for Health.* Bethesda, MD: National Institute of Diabetes and Digestive and Kidney Diseases, National Institutes of Health, 2002.

Navdanya. 2009. "Navdanya: Two Decades of Service to the Earth and Small Farmers." Retrieved from www.Navdanya.Org/Attachments/Navdanya.Pdf.

Nisbett, Richard E., and Takahiko Masuda. "Culture and Point of View." *Proceedings of the National Academy of Sciences* 100, no. 19 (2003): 11163–70.

Nunn, Charles L., David R. Samson, and Andrew D. Krystal. "Shining Evolutionary Light on Human Sleep and Sleep Disorders." *Evolution, Medicine, and Public Health* 2016, no. 1 (2016): 227–43.

Odling-Smee, F. John, Kevin N. Laland, and Marcus W. Feldman. *Niche Construction: The Neglected Process in Evolution*. Princeton, NJ: Princeton University Press, 2003.

O'Keefe, John (Producer), and Time Guthrie (Director). "Mother Kuskokwim." *Creighton Backpack Journalism*, 2014. Retrieved from https://vimeo.com/109125103.

Olff, Miranda, Jessie L. Frijling, Laura D.Kubzansky, Bekh Bradley, Mark A.Ellenbogen, Christopher Cardoso, Jennifer A.Bartz, Jason R.Yee, and Mirjam van Zuiden. "The Role of Oxytocin in Social Bonding, Stress Regulation and Mental Health: An Update on the Moderating Effects of Context and Interindividual Differences." *Psychoneuroendocrinology* 38, no. 9 (2013): 1883–94.

Palmquist, Aunchalee. "5 Milk Sharing Myths Busted." *HuffPost*, December 6, 2017. Retreived from https://www.huffingtonpost.com/aunchalee-palmquist/5-milk-sharing-myths-bust_b_9282510.html.

Park, Alice. "Can Sugar Substitutes Make You Fat?" *Time*, 2008. Retrieved from www.Time.Com/Time/Health/Article/0,8599,1711763,00.Html.

Pembrey, Marcus E, Lars Olov Bygren, Gunnar Kaati, Sören Edvinsson, Kate Northstone, Michael Sjöström, Jean Golding, and The ALSPAC Study Team. "Sex-Specific, Male-Line Transgenerational Responses in Humans." *European Journal of Human Genetics* 14, no. 2 (2006): 159–66.

Perry, Richard J. *Five Key Concepts in Anthropological Thinking*. Upper Saddle River, NJ: Pearson Education, 2003.

Pollan, Michael. *The Omnivore's Dilemma: A Natural History of Four Meals*. New York: Penguin Books, 2006.

Popenoe, Rebecca. *Feeding Desire: Fatness, Beauty, and Sexuality among a Saharan People*. London: Routledge, 2004.

Public Broadcasting System, PBS. *The Evolutionary Arms Race*. Boston, MA: WGBH Educational Foundation and Clear Blue Sky Productions, 2001.

Raymond, J. "GPS Addict? It May Be Eroding Your Brain." *Today Health*, NBC News 2010. Retrieved from today.msnbc.msn.com/id/40138522/ns/health-mental_health/#.UClgvaOs_5E.

Relethford, J. H. "Human Skin Color Diversity is Highest in Sub-Saharan African Populations." *Human Biology* 72, no. 5 (2000): 773–80.

Rich, Alisa L, Laura M. Phipps, Sweta Tiwari, Hemanth Rudraraju, Philip O. Dokpesi. "The Increasing Prevalence in Intersex Variation from Toxicological Dysregulation in Fetal Reproductive Tissue Differentiation and Development by Endocrine-Disrupting Chemicals." *Environmental Health Insights* 10 (2016): 163–71.

Richards, Graham. "Loss of Innocence in the Torres Straits." *The Psychologist* 23, no. 12 (2010): 982–3. Retrieved from https://thepsychologist.bps.org.uk/volume-23/edition-12/loss-innocence-torres-straits.

Ridley, Matt. *Genome: The Autobiography of a Species in 23 Chapters*. New York: Harper Perennial, 1999.

Ridley, Matt. *The Agile Gene*. New York: Perennial, 2003.

Riis, Jacob. *How the Other Half Lives: Studies among the Tenements of New York*. Whitefish, MT: Kessinger, 2004 [1890].

Robbins, Jim. "The Ecology of Disease." *New York Times*, July 14, 2012. Retrieved from https://www.nytimes.com/2012/07/15/sunday-review/the-ecology-of-disease.html?pagewanted=all&_r=0.

Roose, Kevin, and Sheera Frenkel. "Mark Zuckerberg's Reckoning: 'This is a Major Trust Issue'." *New York Times* (online edition), March 21, 2018. Retrieved from https://www.nytimes.com/2018/03/21/technology/mark-zuckerberg-q-and-a.html.

Roscoe, Will. *The Zuni Man-Woman*. Albuquerque, NM: University of New Mexico Press, 1992.

Rose, Todd. *The End of Average: How We Succeed in a World that Values Sameness*. New York: Harper One, 2015.

Roulette, Casey J., Hayley Mann, Brian M. Kemp, Mark Remiker, Jennifer W. Roulette, Barry S. Hewlett, Mirdad Kazanji, Sébastien Breurec, Didier Monchy, Roger J. Sullivan, and Edward H.

Hagen. "Tobacco Use vs. Helminths in Congo Basin Hunter-Gatherers: Self-Medication in Humans?" *Evolution and Human Behavior* 35, no. 5 (2014): 397–407.

Roulette, Casey J., Mirdad Kazanji, Sébastien Breurec, and Edward H. Hagen. "High Prevalence of Cannabis Use among Aka Foragers of the Congo Basin and its Possible Relationship to Helminthiasis: Cannabis Use among Aka Foragers of the Congo Basin." *American Journal of Human Biology* 28, no. 1 (2016): 5–15.

Rubel, Arthur J., Carl W. O'Nell, and Rolando Collado-Ardón. *Susto: A Folk Illness*. Los Angeles, CA: University of California Press, 1984.

Rudrappa, Sharmila. "India Outlawed Commercial Surrogacy – Clinics Are Finding Loopholes." *The Conversation*, October 23, 2017. Retrieved from http://theconversation.com/india-outlawed-commercial-surrogacy-clinics-are-finding-loopholes-81784.

Sample, Ian. "Fifth of Neanderthals' Genetic Code Lives on in Modern Humans." *The Guardian* (US edition), January 29, 2014. Retrieved from https://www.theguardian.com/science/2014/jan/29/fifth-neanderthals-genetic-code-lives-on-humans.

Sanders, Clinton R. "Marks of Mischief. Becoming and Being Tattooed." *Journal of Contemporary Ethnography* 16, no. 4 (1988): 395–432.

Schell, Lawrence M. "Transitioning from Traditional: Pollution, Diet, and the Development of Children." *Collegium Antropologicum* 36, no. 4 (2012): 1129–34.

Schell, Lawrence M., and Melinda Denham. "Environmental Pollution in Urban Environments and Human Biology" *Annual Review of Anthropology* 32, no. 1 (2003): 111–34.

Schell, Lawrence M., Julia Ravenscroft, Maxine Cole, Agnes Jacobs, Joan Newman, and Akwesasne Task Force on the Environment. "Health Disparities and Toxicant Exposure of Akwesasne Mohawk Young Adults: A Partnership Approach to Research." *Environmental Health Perspectives* 113, no. 12 (2005): 1826–32.

Schmidt, D. "The Family as the Unit of Medical Care." *Journal of Family Practice* 7, no. 2 (1978): 303–13.

Schor, Juliet B. *The Overworked American: The Unexpected Decline of Leisure*. New York: Basic Books, 1993.

Schueller, Gretel H. "How Good Gut Bacteria Could Transform Your Health." *Eating Well*, July/August 2014. Retrieved from http://www.eatingwell.com/article/283417/how-good-gut-bacteria-could-transform-your-health/.

Sender, Ron, Shai Fuchs, and Ron Milo. "Revised Estimates for the Number of Human and Bacteria Cells in the Body." *PLoS Biology* 14, no. 8 (2016): e1002533.

Sharp, Lesley A. *Strange Harvest Organ Transplants, Denatured Bodies, and the Transformed Self.* Los Angeles, CA: University of California Press, 2006.

Sheikh, Knvul. "How Gut Bacteria Tell Their Hosts What to Eat." *Scientific American*, April 25, 2017. Retrieved from https://www.scientificamerican.com/article/how-gut-bacteria-tell-their-hosts-what-to-eat/.

Shenk, David. *The Genius in All of Us: New Insights into Genetics, Talent, and IQ*. New York: Anchor Books, 2010.

Simmons, Ann M. "Where Fat is a Mark of Beauty." *Los Angeles Times*, September 30, 1998.

Smith, Yolanda. "Old Friends Hypothesis." *News Medical*, August 23, 2018. Retrieved from https://www.news-medical.net/health/Old-Friends-Hypothesis.aspx.

Snowdon, W., and A. M. Thow. "Trade Policy and Obesity Prevention: Challenges and Innovation in the Pacific Islands." *Obesity Reviews* 14, Suppl. 2 (2013): 150–8.

Sobo, Elisa J. *One Blood: The Jamaican Body*. Albany, NY: State University of New York Press, 1993.

Sobo, Elisa J., and Martha Loustaunau. *The Cultural Context of Health, Illness and Medicine*. 2nd ed. Santa Barbara, CA: Praeger, 2010.

Speier, Amy. *Fertility Holidays: IVF Tourism and the Reproduction of Whiteness*. New York: NYU Press, 2016.

Stothard, Ellen R., Andrew W. McHill, Christopher M. Depner, Monique K. LeBourgeois, John Axelsson, and Kenneth P. Wright. "Circadian Entrainment to the Natural Light-Dark Cycle across Seasons and the Weekend." *Current Biology* 27 (2017):508–13.

Steiner, Rudolf. *The Philosophy of Freedom: The Basis for a Modern World Conception*. Trans. Michael Wilson. Forest Row, UK: Rudolf Steiner Press, 2008 [1894].

Steiner, Rudolf. *Study of Man (General Education Course): Fourteen Lectures Given in Stuttgart between 21 August and 5 September 1919*. Translated by Daphne Harwood and Helen Fox. Forest Row, UK: Rudolf Steiner Press, 2007.

Swarns, Rachel L. “DNA Tests Offer Immigrants Hope or Despair.” *New York Times*, April 10, 2007, A1.

Theophanous, C., B. S. Modjtahedi, M. Batech, D. S. Marlin, T. Q. Luong, and D. S. Fong. “Myopia Prevalence and Risk Factors in Children.” *Clinical Ophthalmology* 12 (2018): 1581–7.

Thomas, Anthony. “Fat.” In *Frontline*, Season 16, Episode 17, 1998.

Tohono O’odham Community Action. 2010. Retrieved from www.tocaonline.org.

Tomasello, Michael. “Human Culture in Evolutionary Perspective.” In *Advances in Culture and Psychology*, edited by M. Gelfand, 5–51. New York: Oxford University Press, 2011.

Trevathan, Wenda R. “Evolutionary Obstetrics.” In *Evolutionary Medicine*, edited by Wenda R. Trevathan, E.O. Smith, and James J. McKenna, 183–207. New York: Oxford University Press, 1999.

Turner, Victor W. “Betwixt and Between: The Liminal Period in ‘Rites De Passage.’” In *Reader in Comparative Religion: An Anthropological Approach*, edited by William A. Lessa and Evon Z. Vogt, 234–43. San Francisco, CA: Harper and Row, 1979 [1964].

Twenge, Jean. “Teenage Depression and Suicide Are Way Up — And so is Smartphone Use.” *The Washington Post*, November 19, 2017. Retrieved from https://www.washingtonpost.com/national/health-science/teenage-depression-and-suicide-are-way-up--and-so-is-smartphone-use/2017/11/17/624641ea-ca13-11e7-8321-481fd63f174d_story.html?noredirect=on&utm_term=.37890639d15c.

Ulrich, Roger S. “View through a Window May Influence Recovery from Surgery.” *Science* 224, no. 4647 (1984): 420–1.

Ulrich, Roger S. “Evidence-Based Healthcare Architecture.” *Lancet* 368, Supplement 1 (2006): S38–9.

Vangay, Pajau, Abigail J. Johnson, Tonya L. Ward, Gabriel A. Al-Ghalith, Robin R. Shields-Cutler, Benjamin M. Hillmann, Sarah K. Lucas, Lalit K. Beura, Emily A. Thompson, Lisa M. Till, Rodolfo Batres, Bwei Paw, Shannon L. Pergament, Pimpanitta Saenyakul, Mary Xiong, Austin D. Kim, Grant Kim, David Masopust, Eric C. Martens, Chaisiri Angkurawaranon, Rose McGready, Purna C. Kashyap, Kathleen A. Culhane-Pera, and Dan Knights. “US Immigration Westernizes the Human Gut Microbiome, *Cell* 175, no. 4 (2018): 962–72.e10.

Van Gennep, Arnold. *The Rites of Passage*. Translated by Monika B. Vizedom and Gabrielle L. Caffee. Chicago, IL: University of Chicago Press, 1960 [1908].

Waber, R. L., B. Shiv, Z. Carmon, and D. Ariely. “Commercial Features of Placebo and Therapeutic Efficacy.” *Journal of the American Medical Association* 299, no. 9 (2008): 1016–17.

Waddington, C. H. “The Epigenotype.” *Endeavour* 1 (1942): 18–20.

Wade, Nicholas. “Human Culture, an Evolutionary Force.” *New York Times*, March 2, 2010, D1.

Wallace, Susan. “Inuit Health: Selected Findings from the 2012 Aboriginal Peoples Survey.” *Statistics Canada*, Catalogue no. 89 653 X — No. 003 (August 2014). Retrieved from https://www150.statcan.gc.ca/n1/en/pub/89-653-x/89-653-x2014003-eng.pdf?st=UrHGZb_I.

Watkins, A. “Contemporary Context of Complementary and Alternative Medicine: Integrated Mind-Body Medicine.” In *Fundamentals of Complementary and Alternative Medicine*, edited by M. Micozzi, 49–63. New York: Churchill Livingstone, 1996.

Weller, Susan C., Trenton K. Ruebush II, and Robert E. Klein. “An Epidemiological Description of a Folk Illness: A Study of Empacho in Guatemala.” In *Anthropological Approaches to the Study of Ethnomedicine*, edited by Mark Nichter, 19–31. New York: Gordon and Breach, 1992.

Whitelaw, Emma. “Epigenetics: Sins of the Fathers, and Their Fathers.” *European Journal of Human Genetics* 14, no. 2 (2006): 131–2.

Whiting, John W. M., and Irvin L. Child. *Child Training and Personality: A Cross-Cultural Study*. New Haven, CT: Yale University Press, 1953.

Wilder, Billy. “Some Like It Hot.” United Artists, 1959.

Wilkinson, Richard, and Kate Pickett. *The Spirit Level: Why Greater Equality Makes Societies Stronger*. New York: Bloomsbury Press, 2009.

Williams, Jonathan, and Eric Taylor. “The Evolution of Hyperactivity, Impulsivity and Cognitive Diversity.” *Journal of the Royal Society Interface* 3 (2006): 399–413.

Willyard, Cassandra. “An Epigenetics Gold Rush: New Controls for Gene Expression.” *Nature* 542, no. 43 (2017):406–8.

Wong, Kate. “Case for (Very) Early Cooking Heats Up.” *Scientific American*, 2013. Retrieved from https://www.scientificamerican.com/article/case-for-very-early-cooking-heats-up/.

Wrangham, Richard. *Catching Fire: How Cooking Made Us Human*. New York: Basic Books, 2009.

Wyatt, S. B., D. R. Williams, R. Calvin, F. C. Henderson, E. R. Walker, and K. Winters. "Racism and Cardiovascular Disease in African Americans." *American Journal of the Medical Sciences* 325, no. 6 (2003): 315–31.

Yuki, Masaki, William W. Maddux, and Takahiko Masuda. "Are the Windows to the Soul the Same in the East and West? Cultural Differences in Using the Eyes and Mouth as Cues to Recognize Emotions in Japan and the United States." *Journal of Experimental Social Psychology* 43, no. 2 (2007): 303–11.

Zink, Katherine D., and Daniel E. Lieberman. "Impact of Meat and Lower Palaeolithic Food Processing Techniques on Chewing in Humans." *Nature* 531 (2016): 500–3.

INDEX

Page numbers in **bold** denote tables, those in *italic* denote figures.